Springer Series in Materials Science

Volume 276

The Springer Series in Materials Science covers the complete spectrum of materials physics, including fundamental principles, physical properties, materials theory and design. Recognizing the increasing importance of materials science in future device technologies, the book titles in this series reflect the state-of-the-art in understanding and controlling the structure and properties of all important classes of materials.

More information about this series at http://www.springer.com/series/856

Ping-Heng Tan
Editor

Raman Spectroscopy of Two-Dimensional Materials

Springer

Editor
Ping-Heng Tan
State Key Laboratory of Superlattices
and Microstructures
Institute of Semiconductors, Chinese
Academy of Sciences
Beijing, China

Center of Materials Science
and Opto-Electronics Engineering & CAS Center
of Excellence in Topological Quantum Computation
University of Chinese Academy of Sciences
Beijing, China

ISSN 0933-033X ISSN 2196-2812 (electronic)
Springer Series in Materials Science
ISBN 978-981-13-1827-6 ISBN 978-981-13-1828-3 (eBook)
https://doi.org/10.1007/978-981-13-1828-3

Library of Congress Control Number: 2018959852

This Springer imprint is published by the registered company Springer Nature Singapore Pte Ltd.
The registered company address is: 152 Beach Road, #21-01/04 Gateway East, Singapore 189721, Singapore

Contents

Contributors

Xuhong An School of Physics, Southeast University, Nanjing, China

Luiz Gustavo Cançado Departamento de Física, Universidade Federal de Minas Gerais, Belo Horizonte, MG, Brazil

Hyeonsik Cheong Department of Physics, Sogang University, Seoul, Korea

Sandhya Chintalapati Centre for Advanced 2D Materials, National University of Singapore, Singapore, Singapore

Chunxiao Cong School of Information Science and Technology, Fudan University, Shanghai, China

M. S. Dresselhaus (deceased)

H. Guo College of Sciences, Liaoning Shihua University, Fushun, China

S. Huang Electrical Engineering Department, Pennsylvania State University, University Park, PA, USA

Ado Jorio Departamento de Física, Universidade Federal de Minas Gerais, Belo Horizonte, MG, Brazil

Jae-Ung Lee Department of Physics, Sogang University, Seoul, Korea

Miao-Ling Lin State Key Laboratory of Superlattices and Microstructures, Institute of Semiconductors, Chinese Academy of Sciences, Beijing, China

College of Materials Science and Opto-Electronic Technology, University of Chinese Academy of Science, Beijing, China

C. H. Lui Department of Physics and Astronomy, University of California, Riverside, CA, USA

Xin Luo Department of Applied Physics, The Hong Kong Polytechnic University, Hong Kong, People's Republic of China

Xin Lu Division of Physics and Applied Physics, School of Physical and Mathematical Sciences, Nanyang Technological University, Singapore, Singapore

Current Address: Department of Physics, Emory University, Atlanta, GA, USA

Zhenhua Ni School of Physics, Southeast University, Nanjing, China

Su Ying Quek Centre for Advanced 2D Materials, National University of Singapore, Singapore, Singapore

Department of Physics, National University of Singapore, Singapore, Singapore

R. Saito Department of Physics, Tohoku University, Sendai, Japan

Zexiang Shen Division of Physics and Applied Physics, School of Physical and Mathematical Sciences, Nanyang Technological University, Singapore, Singapore

Ping-Heng Tan State Key Laboratory of Superlattices and Microstructures, Institute of Semiconductors, Chinese Academy of Sciences, Beijing, China

Center of Materials Science and Opto-Electronics Engineering & CAS Center of Excellence in Topological Quantum Computation, University of Chinese Academy of Sciences, Beijing, China

Qing-Hai Tan State Key Laboratory of Superlattices and Microstructures, Institute of Semiconductors, Chinese Academy of Sciences, Beijing, China

College of Materials Science and Opto-Electronic Technology, CAS Center of Excellence in Topological Quantum Computation, University of Chinese Academy of Sciences, Beijing, China

Y. Tatsumi Department of Physics, Tohoku University, Sendai, Japan

Lianming Tong Center for Nanochemistry, College of Chemistry and Molecular Engineering, Peking University, Beijing, China

Yanlong Wang Division of Physics and Applied Physics, School of Physical and Mathematical Sciences, Nanyang Technological University, Singapore, Singapore

Key Laboratory of Chemical Lasers, Dalian Institute of Chemical Physics, Chinese Academy of Sciences, Dalian, China

Jiang-Bin Wu State Key Laboratory of Superlattices and Microstructures, Institute of Semiconductors, Chinese Academy of Sciences, Beijing, China

College of Materials Science and Opto-Electronic Technology, University of Chinese Academy of Science, Beijing, China

Juanxia Wu Center for Nanochemistry, College of Chemistry and Molecular Engineering, Peking University, Beijing, China

Qihua Xiong Division of Physics and Applied Physics, School of Physical and Mathematical Sciences, Nanyang Technological University, Singapore, Singapore

MajuLab, CNRS-UNS-NUS-NTU International Joint Research Unit, Singapore, Singapore

T. Yang Shenyang National Laboratory for Materials Science, Institute of Metal Research, Chinese Academy of Sciences, Shenyang, China

Ting Yu Division of Physics and Applied Physics, School of Physical and Mathematical Sciences, Nanyang Technological University, Singapore, Singapore

Jin Zhang Center for Nanochemistry, College of Chemistry and Molecular Engineering, Peking University, Beijing, China

Jun Zhang State Key Laboratory of Superlattices and Microstructures, Institute of Semiconductors, Chinese Academy of Sciences, Beijing, China

College of Materials Science and Opto-Electronic Technology, CAS Center of Excellence in Topological Quantum Computation, University of Chinese Academy of Sciences, Beijing, China

Shishu Zhang Center for Nanochemistry, College of Chemistry and Molecular Engineering, Peking University, Beijing, China

L. Zhou School of Chemistry and Chemical Engineering, Shanghai Jiao Tong University, Shanghai, China

Department of Electrical Engineering and Computer Science, Massachusetts Institute of Technology, Cambridge, MA, USA

The original version of this book was revised. A correction to this book is available at https://doi.org/10.1007/978-981-13-1828-3_12

Chapter 1
Raman Spectroscopy of Monolayer and Multilayer Graphenes

Jiang-Bin Wu, Miao-Ling Lin, and Ping-Heng Tan

Abstract The discovery of monolayer graphene in 2004 has triggered a great effort to investigate the fundamental properties and applications of all two-dimensional materials (2DMs). Monolayer graphene (1LG) can be stacked layer by layer in a particular way (AB, ABC and twist) to form multilayer graphene (MLG), whose properties vary according to the stacking. Raman spectroscopy is a useful tool to reveal the chemical and physical properties of graphene materials. In this chapter, we review the systematic development of the Raman spectroscopy of pristine 1LG and MLG. The essential Raman scattering processes of the entire first and second order modes in intrinsic 1LG are addressed in detail. We further introduce the concept of double resonance Raman scattering in graphene. Moreover, a series of works on the shear (C), layer-breathing (LB) and 2D modes of MLGs with different stacking orders are discussed. Finally, various types of resonance Raman spectroscopy of 1LG and MLG are also presented. The Raman spectroscopy of graphene materials can serve as a typical example in studying the Raman spectroscopy of other 2DMs and introducing the fundamental physical concepts for 2DMs.

The original version of this chapter was revised. A correction to this chapter is available at https://doi.org/10.1007/978-981-13-1828-3_12

J.-B. Wu · M.-L. Lin
State Key Laboratory of Superlattices and Microstructures, Institute of Semiconductors, Chinese Academy of Sciences, Beijing, China

College of Materials Science and Opto-Electronic Technology, University of Chinese Academy of Science, Beijing, China

P.-H. Tan (✉)
State Key Laboratory of Superlattices and Microstructures, Institute of Semiconductors, Chinese Academy of Sciences, Beijing, China

Center of Materials Science and Opto-Electronics Engineering & CAS Center of Excellence in Topological Quantum Computation, University of Chinese Academy of Sciences, Beijing, China
e-mail: phtan@semi.ac.cn

P.-H. Tan (ed.), *Raman Spectroscopy of Two-Dimensional Materials*, Springer Series in Materials Science 276, https://doi.org/10.1007/978-981-13-1828-3_1

1.1 Introduction

There are many members in the two-dimensional (2d) material family. Monolayer graphene (1LG) is a truly 2d system; the atoms in one plane are connected via strong covalent bonds, forming sp^2 carbon hexagonal networks [1]. 1LG can be stacked layer by layer in the Bernal (AB) or rhombohedral (ABC) manner via van der Waals (vdW) coupling to form multilayer graphene (MLG). AB-stacked and ABC-stacked MLG are denoted AB-MLG and ABC-MLG, respectively. Each N-layer graphene (NLG) exhibits distinct electronic band structure dependent on the layer number N, interlayer coupling and stacking order [2]. When the physical properties related to the interlayer coupling and stacking order are discussed in this chapter for NLG, N is assumed to be no less than the smallest layer number of the corresponding MLG; thus, $N>1$ for AB-stacked NLG (AB-NLG), and $N>2$ for ABC-stacked NLG (ABC-NLG). High-quality NLG flakes can be produced by several methods, such as micromechanical exfoliation (ME), chemical vapour deposition (CVD) and epitaxial growth from the SiC surface [3–5]. Graphene materials exhibit many remarkable properties, making them the ideal materials for next-generation nanoscale devices [6–9]. On the other hand, 1LG and MLG are usually the typical essential building blocks for vdW heterostructures (vdWHs), which can be formed by vertically stacking various 2d materials (2DMs) via vdW coupling but without any constraints of lattice matching and fabrication compatibility, [7, 10] offering significant opportunities for designing functionalities [7, 8]. The graphene materials in vdWHs are usually employed as electrodes to form various high-performance devices, such as field-effect tunnelling transistors, logic transistors, photovoltaics and memory devices [7, 8, 10]. All the facts stated above imply that graphene materials can serve as a typical prototype to introduce the physical properties and fundamental concepts of 2DMs.

Raman spectroscopy has historically played an important role in nondestructively characterizing the lattice structure and electronic, optical and phonon properties of graphene materials with high resolution [9, 11, 12]. NLG exhibits N-dependent and stacking-orientation dependent Raman features, providing information on the phonon properties and on their unique band structures [13, 14]. In addition, quantum interference effects can serve as the main factor to modify the Raman intensity of NLG [15–17]. Therefore, it is easy to determine the physical properties of graphene materials and the performance mechanism in graphene-based devices using Raman spectroscopy [9].

In this chapter, we review the systematic development of Raman spectroscopy of 1LG and MLG. We further introduce the concept of double resonance Raman scattering in graphene in detail. A series of studies about the shear (C), layer-breathing (LB) and 2D modes of MLGs with different stacking orders are discussed. The shear modes are usually referred to as the C modes in MLGs because it provides a direct measurement of the interlayer $Coupling$. Various types of resonance Raman spectroscopy of 1LG and MLG are also presented. A more systematical review on the developments in the Raman spectroscopy of graphene-based materials from both fundamental research and practical (i.e., device applications) perspectives can be found in our recent review paper [9].

1.2 Raman Spectroscopy of Monolayer Graphene

1.2.1 Band Structure of Monolayer and Multilayer Graphenes

There are two inequivalent carbon atoms, A and B, in a unit cell of 1LG, each forming a triangular in-plane network. The distance between these two atoms, also called the length of the carbon-carbon covalent bond, is 0.142 nm, leading to a 0.246 nm lattice constant, as shown in Fig. 1.1a. MLG is a layered structure obtained from 1LG by stacking in the out-of-plane direction (*c*-axis) with an interlayer distance of 0.334 nm. AB stacking is the most common and stable stacking in MLG, in which the two adjacent graphene layers are vertically stacked with 60° rotation. Thus, the vacant centres of the hexagons on one layer are occupied by the carbon atoms at hexagonal corner sites on the adjacent layers from the top view, as shown in Fig. 1.1b. The two next nearest layers are in superposition from the top view in AB-stacked *N*LG. Moreover, the unit cell of graphite can also be shown as Fig. 1.1b, whose lattice constants are $a = 0.246$ nm and $c = 0.67$ nm. The reciprocal space of 1LG is a plane, as shown in Fig. 1.1c. The high symmetry points within the Brillouin zone (BZ) of 1LG are also shown in Fig. 1.1c, with the Γ point at the zone centre, M points in the middle of the hexagonal sides, and K and K′ points at the corners of the hexagons. The K and K′ points are inequivalent.

Graphene is a zero-gap semi-metal with a linear electronic band structure, which causes its unique optical properties from the terahertz to the visible range [19]. Furthermore, any laser excitation from the visible to the ultraviolet can induce a quantum interference effect in the Raman scattering process, making the Raman spectrum of graphene remarkable [15]. As N increases, *N*LG exhibits significant N-dependent electronic band structures due to the interlayer vdW coupling and different symmetries [2]. Micromechanical exfoliation was the first method used to produce *N*LG flakes with high quality. The exfoliated graphene flakes are usually deposited onto a Si/SiO_2 substrate with a specific thickness (90 or 300 nm), as shown in Fig.1.1d, which makes the graphene visible and enhances Raman signals of graphene due to the optical interference effect in the air/*N*LG/SiO_2/Si multilayered structure [17, 20, 21]. Figure 1.1e–h shows the linear band structure of 1LG and the corresponding band structures of AB-2LG, -3LG and -4LG. The difference in the band structures in *N*LG can be reflected in their peculiar Raman features [9, 13]. Thus, Raman spectroscopy can be used to probe the band structure of 2LG [14] and to identify 1–4LG [22].

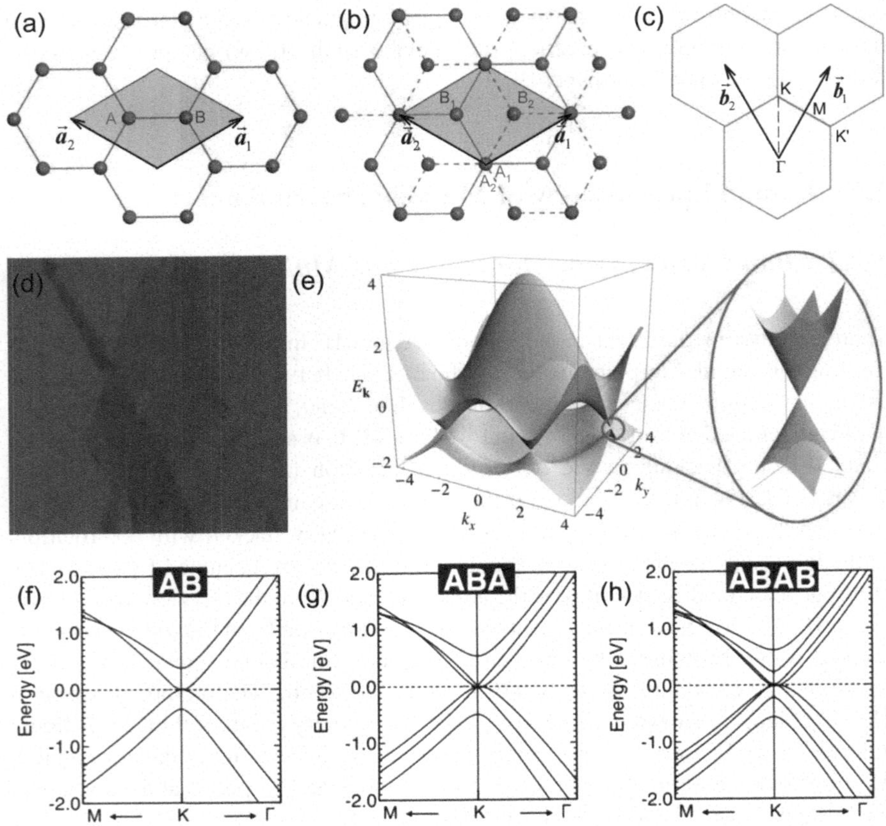

Fig. 1.1 (**a**) A top view of the unit cell of a 1LG lattice. (**b**) A top view of AB-stacked 2LG. (**c**) The unit cell in reciprocal space of 1LG. The primitive vectors and high symmetry points are shown. (**d**) Optical image of exfoliated graphene flakes on a Si/SiO_2 substrate. (**e**) Band structure of 1LG. (Reproduced with permission from Ref. [18]). Band structure of AB-stacked 2LG (**f**), 3LG (**g**) and 4LG (**h**) in the vicinity of K. (Reproduced with permission from Ref. [2])

1.2.2 Phonon Dispersion and the Raman Spectrum of Graphene

The phonon modes (lattice vibrations) are usually presented by the irreducible representation based on the symmetry group of the crystals, which are essential to the interpretation of Raman spectra. The point group symmetries of 1LG and graphite are both D_{6h}. As two and four atoms are in the unit cells of 1LG and graphite, six and twelve phonon modes are expected, respectively. Thus, the lattice vibrations of 1LG and graphite at Γ can be expressed as $\Gamma_{1LG} = A_{2u} + B_{2g} + E_{1u} + E_{2g}$ and $\Gamma_{bulk} = 2(A_{2u} + B_{2g} + E_{1u} + E_{2g})$ [25]. One A_{2u} mode and one doubly degenerate E_{1u} mode are the three acoustic modes in 1LG and bulk. As for the optical

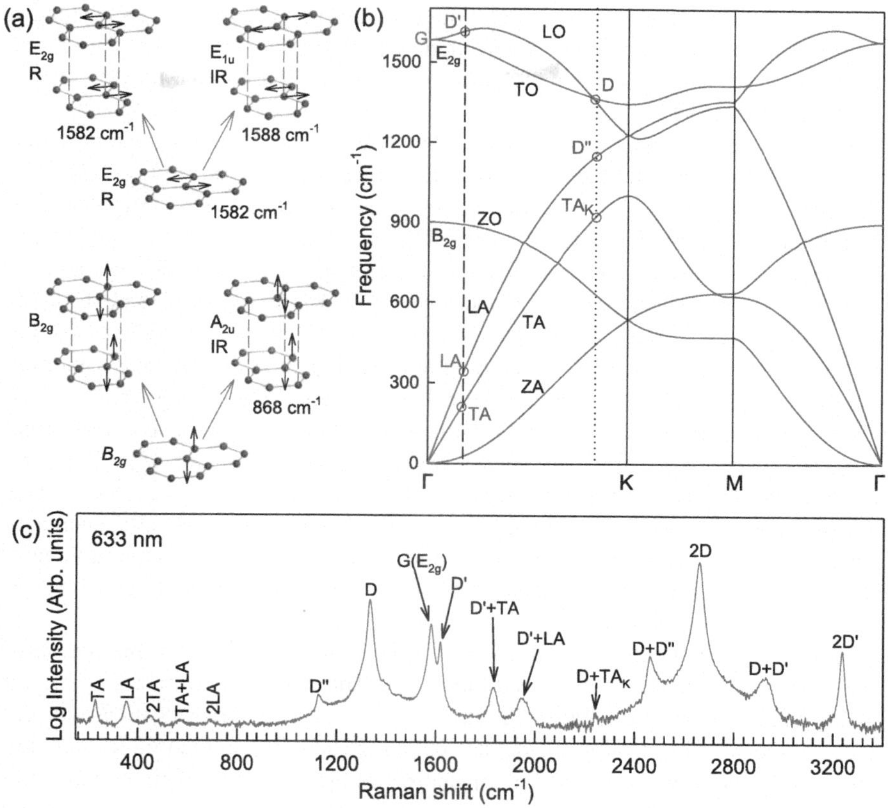

Fig. 1.2 (**a**) Phonon-displacement pattern at the Γ-point for graphene and graphite. Long arrows show how each phonon mode in graphene gives rise to two phonon modes of graphite. (**b**) The solid curves represent the dispersion of phonon modes in graphene calculated by density functional perturbation theory. (Reproduced with permission from Ref. [23]). The attribution of each branch is labelled. (**c**) 633 nm excited Raman spectrum under excitation at 633 nm of a graphite whisker, which can be considered as the assembly of individual 1LG. (Reproduced with permission from Ref. [24])

phonon, 1LG contains the doubly degenerate in-plane mode, E_{2g}, and one out-of-plane mode, B_{2g}, as shown in Fig. 1.2a. In graphite, several Davydov doublets are present because the two graphene layers in the unit cell of graphite are inequivalent, [11, 12] *e.g.*, the E_{2g} mode in 1LG generates an infrared-active mode, E_{1u}, and a Raman-active (R) mode, E_{2g}, in graphite, and the B_{2g} mode in 1LG divides into an infrared-active (IR) mode, A_{2u}, and an inactive mode, B_{2g}, in graphite, as shown in Fig. 1.2a. The corresponding frequencies and atomic displacements are also depicted. In graphite, the E_{2g} mode at $\sim$1582 cm^{-1} is the characteristic peak of graphene-related materials, named the G mode. The remaining E_{2g} and one B_{2g} mode are the C and LB modes, corresponding to the in-plane and out-of-plane interlayer vibration of the rigid atom layers, respectively [26, 27]. Due to the weak interlayer coupling, the peak position of the C mode (Pos(C)) is very low, located at 43.5 cm^{-1} [26, 28]. The LB modes are silent and can not be observed in the Raman spectrum. Pos(LB) can be estimated to be ~125.3 cm^{-1} based on the experimental Pos(LB) in twisted MLG [27].

The phonon dispersions of 1LG calculated by density functional perturbation theory (DFPT) within the local density approximation are shown in Fig. 1.2b [23]. There are three acoustic branches and three optical branches in 1LG inherited from the six vibrational modes at the Γ point. The in-plane longitudinal acoustic (LA) and transverse acoustic (TA) branches have linear dispersion and higher frequencies than the out-of-plane acoustic (ZA) mode around Γ. The strong in-plane C-C bonds give large slopes of the LA (21.36 km/s) and TA (13.79 km/s) branches, leading to the high in-plane sound velocity of graphene and thus the ultra-high thermal conductivity [29]. The three optical branches: in-plane longitudinal optical (LO), in-plane transverse optical (TO) and out-of-plane optical (ZO) branches, correspond to the irreducible representations (Fig. 1.2a) E_{2g} (LO and TO) and B_{2g} (ZO) at Γ, respectively. The LO and TO branches split in frequency when the wave vector is away from Γ, and the frequency of the TO branch monotonically decreases along Γ-K. The phonon dispersions of solid materials are usually experimentally detected via inelastic X-ray, neutron scattering and high-resolution electron energy-loss spectroscopy [30, 31].

Figure 1.2c shows the Raman spectra of the graphite whisker in the region of 150–3400 cm^{-1} under 633 nm excitation [24]. The spiral structure and cone tip of the graphite whisker suggest the existence of a twist angle between adjacent layers in the whisker, leading to weaker interlayer coupling compared to graphite. Thus, the graphite whisker can be considered as the assembly of individual 1LG units. Therefore, graphite whiskers can be considered as a prototype to show the abundant Raman modes that should be observed in 1LG [9]. Indeed, many weak first- and second-order Raman modes have been revealed in the Raman spectrum, as indicated in Fig. 1.2c. In principle, Raman spectroscopy can usually be used to probe the phonon modes in the BZ centre. Indeed, the G peak corresponds to the E_{2g} mode at Γ, located at ~1582 cm^{-1}. The other Raman peaks observed in graphite whiskers or 1LG should correspond to phonons away from Γ. The observation of these phonons can be attributed to double resonance (DR) or triple resonance (TR) Raman processes in 1LG, as depicted in Fig. 1.3.

1.2.3 Double and Triple Resonance Raman Processes in Graphene

A resonance Raman process occurs when the excitation energy is chosen to match or nearly match an optical transition bandgap of the crystal, and the Raman intensity can be enhanced by 2 to 6 orders of magnitude. Because the two linear electronic bands of intrinsic 1LG cross at the Fermi energy, an incoming photon with energy

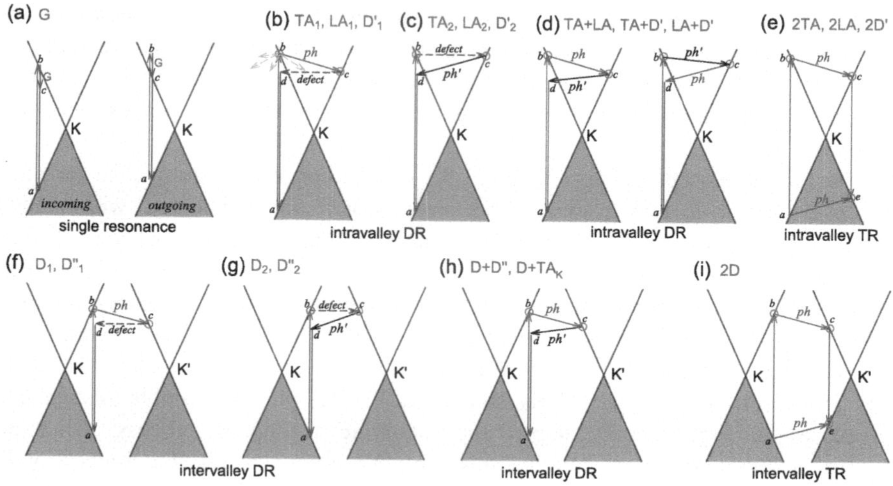

Fig. 1.3 (**a**) Incoming and outgoing single resonance processes of the G mode in 1LG. (b-i) Resonance processes of dispersive Raman modes in 1LG, including an electron-hole pair excited by an incident laser photon, inelastic scattering of the electron and hole by phonon emission/absorption, and the elastic scattering of the electron mediated by the defect, and recombination of the electron-hole pair. (b-d) The intravalley double resonance processes. (**e**) An intravalley triple resonance process. (f-h) The inner intervalley double resonance processes. (**i**) An inner intervalley triple resonance process. The Raman modes in graphite whiskers activated by each resonance process are labelled accordingly in each panel. (Reproduced with permission from Ref. [9])

ε_L can always excite a resonant transition from the state in the valence band to another state in the conduction band to generate an electron-hole pair. The excited electron can be scattered by the E_{2g} phonons to recombine with a hole, satisfying the incoming single resonance process. The outgoing single resonance process can also be satisfied for the G mode of 1LG. Both resonance processes are shown in Fig. 1.3a. The excited electron can also be scattered by phonons of an arbitrary wave vector, as shown in Fig. 1.3b by the dashed arrows. The scattering probability, however, will be particularly high if the phonon scatters the electron from one real electronic state b into another real state c. Such one-phonon first-order Raman scattering is forbidden by selection rules. However, the above scattered electron in state c can be scattered back to state d with the same wave vector as state b by crystal defects, which then leads to emission of a photon through recombination with a hole in the initial state. The above resonance process is one-phonon DR Raman scattering, in which the two events consist of one inelastic scattering event by emitting a phonon and one elastic scattering event by crystal defects. The two scattering processes can also start from an elastic scattering event followed by an inelastic scattering event, shown in Fig. 1.3c. If the elastic scattering of defects is displaced by another inelastic phonon scattering, the process becomes two-phonon DR Raman scattering, as illustrated in Fig. 1.3d. Further, if the valence and conduction bands are almost mirror bands, a triple resonance (TR) Raman process can occur. For example, for the special case of 1LG, as shown in Fig. 1.3e, the electron-hole generation is a resonance process, and both electron and hole scattering will be resonant. Finally, the electron-hole recombination will also be resonant. Therefore, for the TR Raman process, all steps in the normal double resonance process become resonant [9].

The resonance processes in Fig. 1.3b–e are intravalley double resonance processes because two resonant electronic states are connected within the same Dirac cones at the K point (or the K′ point) of 1LG. Instead, if the resonance process connects two associated resonant electronic states within the two inequivalent Dirac cones at the K and K′ points of 1LG, the corresponding double resonance mechanism is called an intervalley process. By analogy to the intravalley DR and TR processes in Fig. 1.3b–e, the intervalley DR and TR processes also occur, [9] as shown in Fig. 1.3f–i. The double resonance processes in Fig. 1.3 only depict the incoming resonance process. After a careful analysis, there are two inequivalent DR processes for the fundamental and combination Raman modes mediated by the scattering of electrons, and there is only one double resonant process for the overtone mediated by the scattering of electrons [32]. It should be noted that the inelastic scattering of the hole by phonon emission/absorption and elastic scattering of the hole mediated by the defect can also be involved in the DR Raman process [33].

In principle, many different initial electronic states around the Dirac point and phonons with different symmetries and wave vectors can satisfy the DR conditions. However, considering (1) the existence of singularities in the density of phonon states that satisfy the DR condition, (2) the angular dependence of the electron-phonon scattering matrix elements, and (3) destructive interference effects when the Raman transition probability is calculated, only a few specific DR processes contribute to the observed Raman modes. Because there are two inequivalent DR Raman processes for both intravalley and intervalley DR Raman processes, each ε_L can select two phonon wave vectors for the fundamental and combination modes and one phonon wave vector for the overtones near the Γ and K points, respectively. Therefore, both the fundamental modes and the combination modes activated by the DR Raman process can be fitted by two Lorentzian peaks, while the overtone only exhibits a single Lorentzian peak [9, 24, 34].

Because the phonon energy in 1LG is much smaller than ε_L, for each ε_L, the phonons with almost equal wave vectors at different branches of the phonon dispersion curves can be probed by the DR Raman process. These are linked by the vertical dashed and dotted lines close to the Γ and K points, respectively, for the intravalley and intervalley DR Raman processes in Fig. 1.2b. Moreover, for the intervalley process, just the inner processes are shown in Fig. 1.3. The outer process can also exist to satisfy the DR conditions. The inner processes correspond to the phonon along the K-Γ line, while the outer processes correspond to the phonon along the K-M line. The most intense contribution to both the D and 2D intensity is due to phonons along the high-symmetry direction K-Γ line (inner phonons), according to the numerical calculation [33].

The observed Raman spectra in Fig. 1.2b can be well understood by the DR and TR Raman processes, as addressed above. The D (D″) peak is the phonon at the TO (LA) branch around the Brillouin zone corner K activated by a defect in the DR Raman process [32, 35]. The 2D peak is the overtone of the D peak, in which momentum conservation is satisfied by two phonons with opposite wave vectors. Thus, no defects are required for their activation, and these peaks are always present in the pristine 1LG. The DR Raman process can also involve two phonons from different branches, making the combination modes observable, such as D+D″ and D+TA_K [9]. The D+D″ mode is assigned as a combination mode of a D mode from the TO phonon branch and a D″ mode belonging to the LA branch near the K point, seen at ~1,100 cm^{-1} [33, 36, 37]. The D+TA_K mode is assigned as a combination mode of a D phonon and a phonon belonging to the TA branch near the K point [33]. All the DR-related modes mentioned above occur between two different valleys (K and K′). These intervalley DR processes involve the phonons near the K point as linked by the dotted line in Fig.1.2b. The DR Raman process can also occur within the same valley, that is, connecting two points belonging to the same Dirac cone around K (or K′). Such an intravalley DR process associated with the LO (D′) phonon near Γ and a defect gives the so-called D′ peak, whose overtone is 2D′ [34, 38]. The LA and TA phonons near Γ can also be activated by the defect to form LA and TA peaks in the Raman spectra, respectively, whose overtones are 2TA and 2TA. The typical intravalley combination modes are D′+TA and D′+LA. The LA and TA peaks can also be combined to form the LA+TA peak. All the above intravalley DR Raman modes (i.e., LA, TA, D′+TA, D′+LA and LA+TA) associated with LA and TA phonons have been observed in the Raman spectra of a graphite whisker [24]. These intravalley DR processes involve the phonons near Γ, as linked by the dashed line in Fig. 1.2b.

For the DR or TR Raman processes, as depicted in Fig. 1.3, the wave vector of the involved phonons is determined by ε_L. Thus, the peak position of the corresponding Raman modes is dependent on ε_L. The dispersion for the 2D mode is almost linear with ε_L, with a slope of ~100 cm^{-1}/eV, due to the linear band structure and the almost linear dispersion of the TO branch near K point. By neglecting the phonon energy, the dispersion slope of the frequency of the combination mode of phonon mode 1 (ph_1) and phonon mode 2 (ph_2) can be estimated as $(v_q^{ph_1} + v_q^{ph_2})/v_f$, where $v_q^{ph_1}$ and $v_q^{ph_2}$ are the phonon group velocity of ph_1 and ph_2 and $v_f = 10^6$ m/s is the Fermi velocity of 1LG. Thus, by detecting the DR modes with multi-wavelength Raman spectroscopy, the dispersion of the LO, TO, LA and TA branches can be observed [32–34]. Furthermore, the physical properties related to these branches, such as the Kohn anomaly [40], can be obtained. The doublet asymmetric profile in fundamental and combination peaks can be confirmed by the D mode of disordered 1LG, as shown in Fig. 1.4c. Obviously, the D peak can be fitted by two peaks, the D_1 and D_2 peaks. The D_1 peak is from the process plotted in Fig. 1.3f, and the D_2 peak is from the process plotted in Fig. 1.3g. The overtone modes, such as 2D and 2D′, give a symmetric peak because just one phonon is involved. In addition,

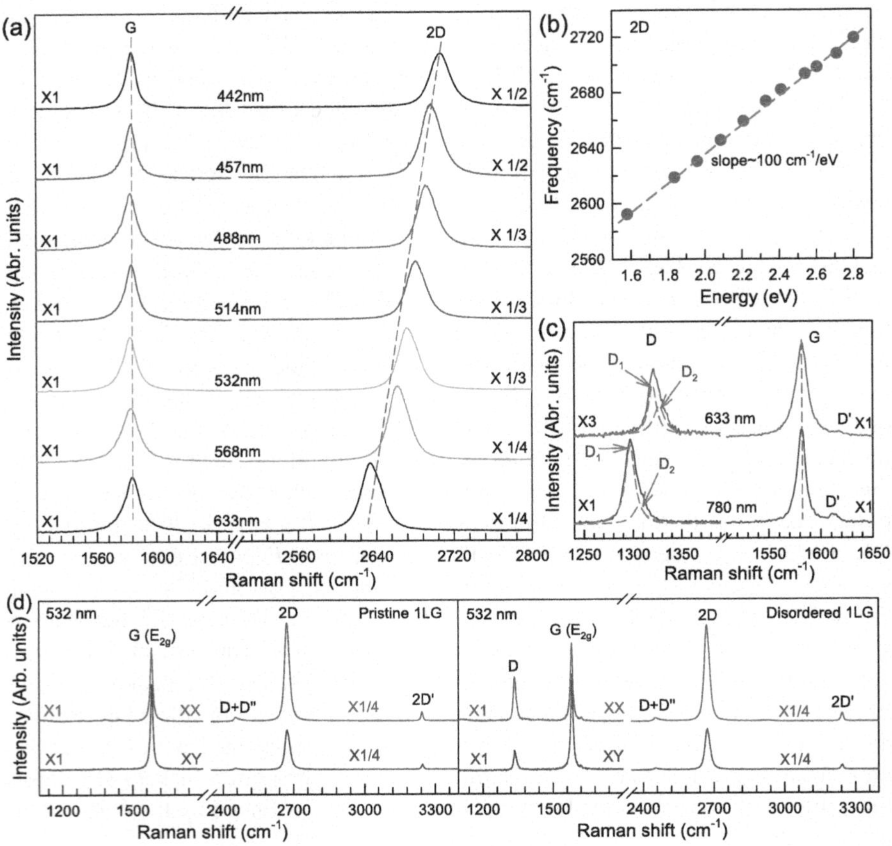

Fig. 1.4 (**a**) Raman spectra of 1LG measured at various excitations in the G and 2D spectral region. (**b**) 2D-peak position as a function of excitation energy, and the slope is $100\,cm^{-1}/eV$. (**c**) Comparison of Raman spectra with D, G and D bands under 633 and 780 nm excitations. The two components (D_1 and D_2) of Lorentzian fitting for the D band in graphene are shown. (**d**) Polarized Raman spectra of the pristine and disordered 1LG under 532 nm excitation. (Reproduced with permission from Ref. [9])

the intensities of the DR peaks are dependent on ε_L because of the ε_L-dependent photon absorption and electron-phonon coupling in the Raman scattering process [33]. Furthermore, the DR peaks of pristine and disordered 1LG show polarized behaviour, as shown in Fig. 1.2d, due to the selective optical absorption and emission mediated by electron-phonon coupling in the DR Raman process [33, 39].

Apart from graphene, the DR Raman process is also present in other graphene-based materials, such as graphite, carbon nanotubes and multilayer graphenes, although they do not have the linear band structures. Thus, phonons of graphene-based materials away from Γ can also be probed by Raman spectroscopy via the double resonant Raman process [32, 34, 35].

1.3 Raman Spectroscopy of Multilayer Graphene

1.3.1 Group Theory of AB-Stacked Multilayer Graphene

One of the prerequisites of the DR Raman process is the permission of the selection rules, including electron-photon and electron-phonon couplings. These selection rules are determined by group theory. The selection rule of electron-photon coupling is determined by the symmetries of the π electron and photon, and the selection rule of electron-phonon coupling is determined by the symmetries of the π electron and phonon. The symmetries of the electron and phonon can be obtained from space group analysis. Table 1.1 presents the space groups and wave-vector point groups for 1LG, AB-NLG, and graphite at the high symmetry points [11]. Because the phonon along the Γ-K axis is the main contribution of the DR process [33], the symmetry information of the Γ-K (T) axis is also shown.

The irreducible representations for the lattice vibrations (Γ_{vib}) of 1LG, even N layer graphene (ENLG) and odd N layer graphene (ONLG) at Γ and along the Γ-K axis in the BZ are found in Table 1.2. Both Bethe and Mulliken notations are adopted here for the phonon modes. A more complete group theory analysis at each high symmetry point inside the BZ and their N-dependent behaviour can be found in Ref. [41].

Momentum conservation makes phonons at Γ be observable in the first order Raman process. The phonon modes at Γ for 1LG, AB-2LG and AB-3LG are summarized with Mulliken notation:

$$\Gamma_{1LG} = A_{2u} + B_{2g} + E_{1u} + E_{2g}$$
$$\Gamma_{AB-2LG} = 2(A_{1g} + E_g + A_{2u} + E_u)$$
$$\Gamma_{AB-3LG} = 2A'_1 + 4A''_2 + 4E' + 2E''$$

Compared to 1LG, the G band in AB-2LG belongs to the E_g irreducible representation, which is doubly degenerate and represents the atomic motion of

Table 1.1 The space groups and wavevector point groups for NLG and graphite at all related points in the Brillouin zone

	Space group	Γ	K(K′)	M	T(Γ-K)
1LG	$P6/mmm$	D_{6h}	D_{3h}	D_{2h}	C_{2v}
NLG even	$P\bar{3}m1$	D_{3d}	D_3	C_{2h}	C_2
NLG odd	$P\bar{6}m2$	D_{3h}	C_{3h}	C_{2v}	C_{1h}
Graphite	$P6_3/mmc$	D_{6h}	D_{3h}	D_{2h}	C_{2v}

Table 1.2 The irreducible representations for the lattice vibrations of 1LG, ENLG and ONLG at Γ and along the T(Γ-K) axis. Both Bethe and Mulliken notations are given

	1LG	ENLG	ONLG
Γ	$\Gamma_2^- + \Gamma_5^- + \Gamma_4^+ + \Gamma_6^+$	$N(\Gamma_1^+ + \Gamma_3^+ + \Gamma_2^- + \Gamma_3^-)$	$(N-1)\Gamma_1^+ + (N+1)\Gamma_2^- + (N+1)\Gamma_3^+ + (N-1)\Gamma_3^-$
	$A_{2u}+B_{2g}+E_{1u}+E_{2g}$	$N(A_{1g}+E_g+A_{2u}+E_u)$	$(N-1)A'_1+(N+1)A''_2+(N+1)E'+(N-1)E''$
T(Γ-K)	$2T_1+T_2+2T_3+T_4$	$3N(T_1+T_2)$	$(3N+1)T^+ + (3N-1)T^-$
	$2A_1+A_2+2B_1+B_2$	$3N(A+B)$	$(3N+1)A'+(3N-1)A''$

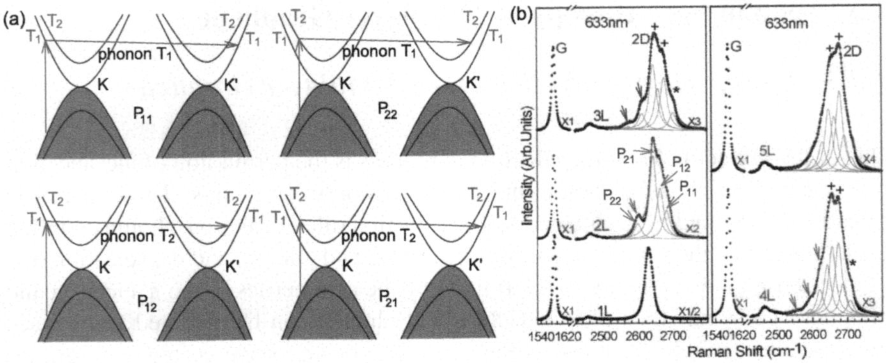

Fig. 1.5 (**a**) The four allowed DR processes in AB-2LG, denoted as P_{11}, P_{22}, P_{12} and P_{21}. (**b**) Raman spectra of the G and 2D peaks for 1LG and AB-stacked 2-5LGs excited by a 633 nm laser. The 2D peaks are fitted by Lorentzian line shapes (gray peaks). The arrows, crosses and stars on the 2D peak indicate its weak shoulders at the low energy side, strongest components and weak shoulders at the high energy side, respectively. (Reproduced with permission from Ref. [22])

nearest neighbour carbon atoms moving against each other within the plane and in phase between the two layers, denoted as the symmetric G mode [11]. The E_u representation, which is not Raman active in AB-2LG because of the centre of inversion symmetry in the D_{3d} point group, can become Raman active if this inversion symmetry operation is broken by the presence of twisting stacking [42, 43], doping [22, 44, 45] and strain [46]. The E_u mode is also represented by the atomic motion of the nearest neighbour carbon atoms moving against each other within the plane but out-of-phase between the two layers, denoted as the antisymmetric G band [11]. There is also a low-frequency E_g mode (31 cm^{-1}), corresponding to the interlayer rigid shear vibration, the so-called C mode [26]. The other two A_{1g} irreducible representations are phonon modes at ~90 cm^{-1} (LB mode) and ~867 cm^{-1} [9]. For AB-3LG, the corresponding G and C bands should be assigned to E′ and E″, and the out-of-plane vibrational modes assigned to A_1' and A_2''.

1.3.2 Raman Spectra of AB-Stacked Multilayer Graphene

For intrinsic multilayer graphene, the characteristic G peak exhibits similar features located at ~1582 cm^{-1}. However, the DR-related Raman peaks exhibit distinguishing properties from 1LG due to the different band structures in MLG. In the DR Raman process, the symmetry of electrons and phonons in MLG should be taken into account. Taking AB-2LG as an example, the number of allowed DR processes will be different from those in 1LG since both electronic and phonon branches are doubled. The transitions of $T_1 \rightleftharpoons T_2$ along the T axis in the BZ are allowed for AB-2LG, as shown in Fig. 1.5a [41]. The TO phonons for AB-2LG have T_1 and

T_2 symmetries. For the electron scattering by a T_1 phonon, the allowed process should occur in the K and K′ electronic bands with the same symmetry, $T_1 \rightarrow T_1$ or $T_2 \rightarrow T_2$. However, the T_2 phonon connects conduction bands of different symmetries, i.e., $T_1 \rightleftharpoons T_2$. Thus, there are four possible DR processes, as shown in Fig. 1.5a.

Figure 1.5b shows Raman spectra of 1LG and AB-stacked 2-5LGs excited by a 633 nm laser. All the G peaks can be fitted by a single Lorentzian peak at ~1582 cm^{-1}. 1LG has a single 2D peak at 2629.7 cm^{-1}. The 2D band of AB-2LG can be fitted by four Lorentzian peaks, corresponding to the four DR processes in Fig. 1.5a [13, 41]. For AB-3LG, the 2D bands are expected to consist of 15 Lorentzian peaks based on the symmetry analysis, and those of AB-4LG and AB-5LG are more complex [41]. However, the number of observed components is much less than the theoretical one because of the possible degeneracy and overlapping of these peak components. Six and eight components are sufficient to fit the 2D bands of AB-3LG and AB-4LG, respectively, in which the peak widths of all the components are kept as a fixed constant of 24 cm^{-1} [22]. As shown in Fig. 1.5b by arrows, crosses and stars, five and six 2D components can be clearly identified in the Raman spectra of AB-3LG and AB-4LG excited by the 633 nm laser, respectively. The 2D band of AB-4LG under 633 nm excitation shows more distinct spectral features than that under 532 nm excitation. AB-3LG and AB-4LG can be distinguished under 633 nm excitation by verifying the peak positions and number of shoulders at the lower energy side and the intensity ratio of the strongest two components of their 2D bands, as indicated by arrows and crosses in Fig. 1.5b. Thus, Raman spectroscopy can be used to clearly identify 1LG and AB-stacked 2-4LG under 633 nm excitation [22].

Now, we can present the detailed insights into the first order Raman scattering process in AB-NLG and graphite. In graphite, there are now two Raman-active E_{2g} modes, and each is doubly degenerate. The high-energy E_{2g} mode due to the in-plane vibrations is widely studied as the so-called G peak in all graphite systems [12, 13]. The low-frequency E_{2g} mode is the C mode, which results from the relative motion of atoms in the adjacent planes. There is no C mode in 1LG because there is only one layer. The C mode is observed at 43.5 cm^{-1} in bulk graphite, whose frequency is determined by the interlayer coupling force constant [28]. Figure 1.6a plots the C and G modes of AB-stacked 2–8LG and bulk graphite, while Fig. 1.6b shows Pos(C) and Pos(G) of AB-NLG as a function of 1/N. The G peaks exhibits a Lorentzian lineshape, while the C mode displays an asymmetrical profile, which can be well fitted by the Breit-Wigner-Fano (simplified as Fano) lineshape, as shown in Fig. 1.6c for AB-3LG and bulk graphite. The unusual lineshape originates from the quantum interference between a Raman-active phonon and the continuum electronic transitions, as shown in Fig. 1.6d.

In contrast to the constant Pos(G) of ~1582 cm^{-1} in AB-NLG, Pos(C) decreases monotonously with decreasing N. The N-dependent Pos(C) in AB-NLG can be well explained by the linear chain model (LCM), [26] in which each graphene layer is considered as one ball for the rigid interlayer shear vibration and only the nearest-neighbour interlayer interactions are taken into account. There are N-1 pairs of

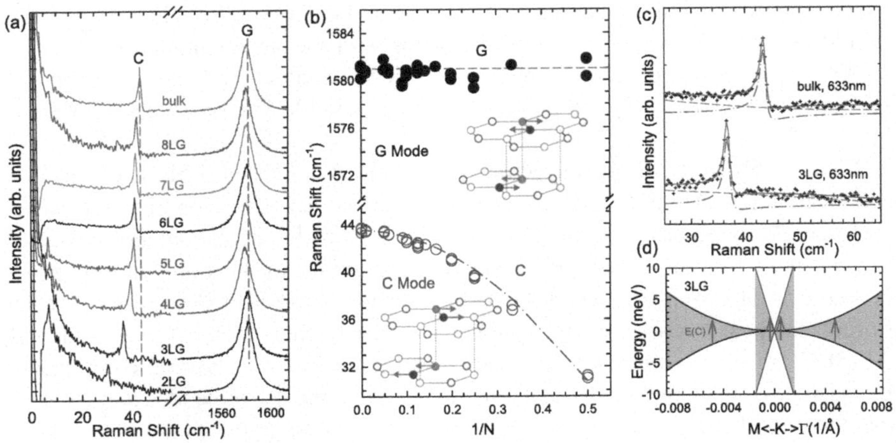

Fig. 1.6 (**a**) Raman spectra of the C and G peaks for AB-stacked 2-8LG and bulk graphite. The intensity of all the C modes is increased by 10 times. (**b**) Pos(G) and Pos(C) of AB-*N*LG as a function of 1/*N*. The insets show the atomic displacements of the C and G modes. (**c**) The C peak of AB-3LG and bulk graphite fitted by the Fano lineshape. The solid lines, dash lines and dash-dot lines are the fitted curve, background and Fano component, respectively. (**d**) Schematic band structure of 3LG near the K point. The transitions denoted by arrows have the same energy as the C mode. (Reproduced with permission from Ref. [26])

degenerate C modes, denoted as C_{NN-i}, i=1,2,. . . ,N-1. Here, the C mode with the highest frequency is assumed as C_{N1} ($i = N$-1), and the lowest one is C_{NN-1} ($i = 1$). The interlayer shear coupling can be described by an interlayer force constant per unit area, $\alpha^{\parallel}$. The frequencies ω (in cm^{-1}) and the displacement patterns of N-1 shear modes can be calculated by solving the corresponding $N \times N$ (tridiagonal) dynamical matrix as follows [42]:

$$(\omega_i)^2 \mathbf{u}_i = \frac{1}{2\pi^2 c^2 \mu} \mathbf{D} \mathbf{u}_i \tag{1.1}$$

where $\mu = 7.6 \times 10^{-27}$ kgÅ^{-2} is the monolayer mass per unit area, $\mathbf{u}_i$ is the phonon eigenvector of mode i with frequency ω_i, $c = 3.0 \times 10^{10}$ cm/s is the speed of light and **D** the shear part of the force constant matrix. Thus, the frequency of the C_{NN-i} mode can be given by [26]

$$\begin{aligned} \omega(C_{NN-i}) &= \frac{1}{\pi c}\sqrt{\alpha^{\parallel}/\mu} sin(i\pi/2N) \\ &= \sqrt{2}\omega(C_{21}) sin(i\pi/2N) \\ &= \omega(C_{bulk}) sin(i\pi/2N), \end{aligned} \tag{1.2}$$

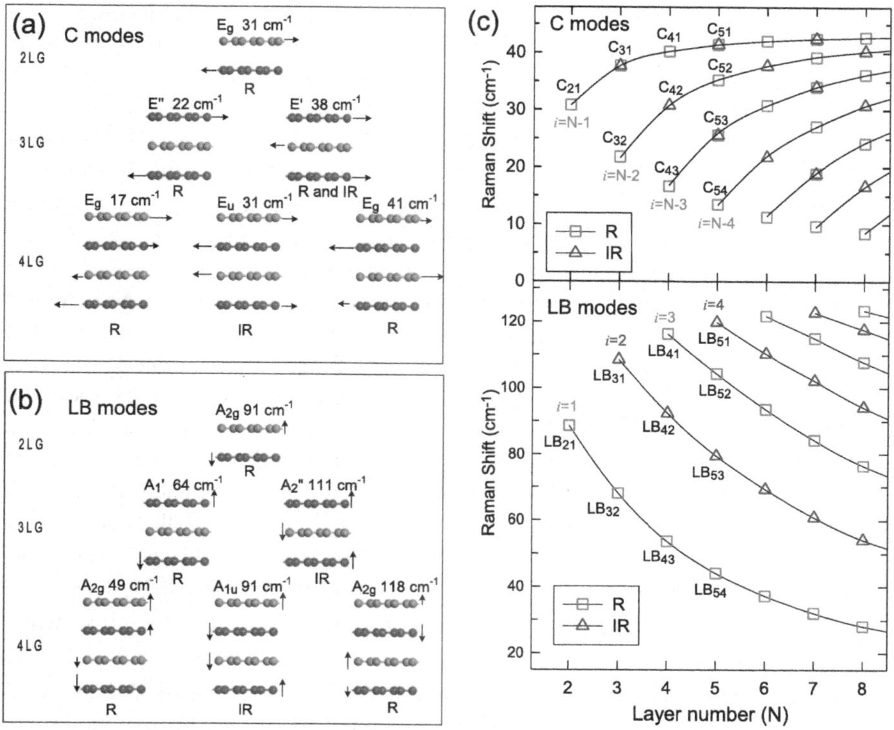

Fig. 1.7 Symmetry, frequency, Raman activity and normal mode displacement for each C mode (**a**) and LB mode (**b**) of AB-stacked (2–4)LG. (**c**) Pos(C) and Pos(LB) of AB-*N*LG as a function of *N* calculated by LCM and 2LCM, respectively. The rectangles and triangles indicate Raman (R) and infrared (IR) active modes, respectively

where i=1,2,$\cdots$,N-1. The relation of $\omega(C_{bulk}) = \sqrt{2}\omega(C_{21})$ is in line with the experimental results, Pos(C) ~31 cm^{-1} for AB-2LG and Pos(C) ~43.5 cm^{-1} for bulk graphite. Based on the experimental values of $\omega(C_{bulk})$ or $\omega(C_{21})$, we can obtain the only unknown parameter, the interlayer shear coupling strength $\alpha^{\parallel}$ ~12.8×10^{18} N/m^3.

Figure 1.7a depicts the frequencies and the corresponding displacement patterns of N-1 C modes in AB-NLG (N = 2,3,4). The C modes can be Raman-active (R), infrared-active (IR) or both, depending on the symmetry and N. As stated above, ENLG and ONLG belong to D_{3d} and D_{3h}, respectively [41, 47]. Thus, N-1 C modes in AB-stacked ENLG and ONLG are represented by $\frac{N}{2}E_g + \frac{N-2}{2}E_u$ and $\frac{N-1}{2}E'' + \frac{N-1}{2}E'$, respectively. E_g and E'' are Raman active, E_u is IR active and E' is both R and IR active. The detailed information of the assignments can be found in Ref. [47]. However, only the highest-frequency C modes were observed in the Raman spectra of AB-NLG (Fig. 1.6a). Other Raman active modes were not detected due to the weak electron-phonon coupling (EPC) and the unsuitable polarization configuration [26].

In AB-NLG, there is another interlayer mode, the LB mode, due to the relative motions of the adjacent layers perpendicular to the basic plane. Similar to the C modes, with a given N, there are N-1 LB modes, denotes as LB_{NN-i}, where $i = 1,2,\cdots,N$-1, in which the LB_{N1} represents the LB modes at the highest frequency. Unlike the C modes, the LB modes in AB- and ABC-stacked NLG cannot be observed at room temperature directly due to the weak EPC [26]. However, in twisted-NLG (tNLG), the LB modes can be detected under the resonance condition [27, 48, 49]. Replacing $\alpha^{\parallel}$ with the interlayer layer-breathing force constant $\alpha^{\perp}$ in Eq. 1.2, the corresponding Pos(LB) can be obtained. However, Wu *et al.* found that LCM underestimated the frequencies of the LB_{NN-i} ($i = 1,2,\ldots$,N-2) modes in tNLG, implying that the second nearest layer-breathing force constant ($\beta^{\perp}$) should be taken into account to reproduce the LB modes. Considering the first and second nearest interlayer coupling and then diagonalizing the corresponding $N \times N$ (tridiagonal) dynamical matrix addressed above (Eq. 1.2), we can obtain the frequencies and the displacement patterns for all the LB modes. This improved LCM is denoted as 2LCM [27]. The interlayer coupling in the perpendicular direction is determined, $\alpha^{\perp} \sim 106\times10^{18}$ N/m^3 and $\beta^{\perp} \sim 9.3\times10^{18}$ N/m^3 [27, 50]. Then N-1 LB modes are explicit, as shown in Fig. 1.7b, considering AB-(2–4)LG as examples. N-1 LB modes in AB-stacked ENLG and ONLG are represented by $\frac{N}{2}A_{2g} + \frac{N-2}{2}A_{1u}$ and $\frac{N-1}{2}A_2'' + \frac{N-1}{2}A'$, respectively, in which A_{2g} and A_1' are Raman active, while A_{1u} and A_2'' are IR active.

Figure 1.7c plots Pos(C) and Pos(LB) of AB-NLG as functions of N reproduced by LCM and 2LCM, respectively. The Raman-active and infrared-active modes are also identified. Overall, LCM and 2LCM can predict the N-dependent Pos(C) and Pos(LB) of AB-NLG, respectively. Furthermore, because Pos(C_{21}) and Pos(LB_{21}) are linked with the corresponding interlayer force constant between two adjacent layers, the good agreement between LCM/2LCM and experimental data for AB-NLG indicate a constant interlayer force constant from 2LG to bulk graphite. Moreover, 2LCM may only be suitable for one atom thickness 2D materials, such as graphene and BN; however, the LCM discussed above can be potentially applied to all 2d materials, such as MoS_2, WSe_2, black phosphorus (BP) and even 2d vdWHs [51–53]. Because the positions of the C and LB modes in 2DMs are strongly dependent on the layer number, the C and LB modes can be used for thickness identification of 2DMs [52–54].

1.3.3 Raman Spectra of ABC-Stacked Multilayer Graphene

In addition to N, the stacking order also has a great influence on the electrical and optical properties of NLG. Apart from the AB stacking, ABC (rhombohedral) stacking is also common in NLG. ABC-NLG comprises ~15% of the area in exfoliated samples, [55] which is in good agreement with a previous X-ray diffraction study of bulk graphite [56]. The Raman spectra of AB- and ABC-NLG

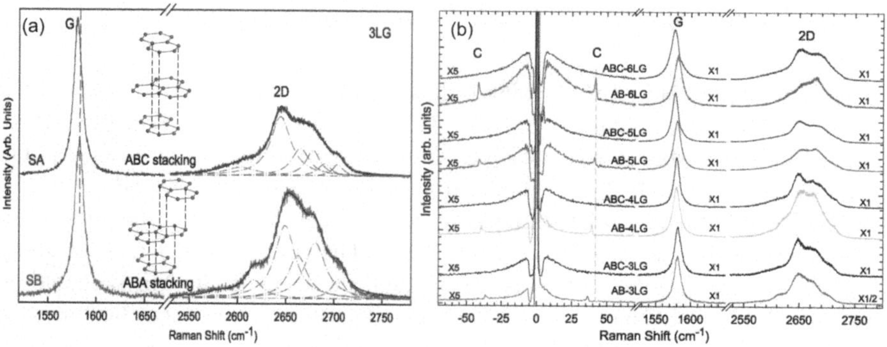

Fig. 1.8 (**a**) Raman spectra of 3LG measured at the ABA and ABC zones in the G and 2D mode regions. Schematic diagrams of ABA and ABC structures are shown. (**b**) Raman spectra of AB- and ABC-stacked 3–6LG in the C, G and 2D peak spectral regions at room temperature. The C modes can be observed in AB-*N*LG, but not in ABC-*N*LG. (Reproduced with permission from Ref. [47])

are quite different from each other. Pos(G) of ABC-*N*LG is 1–5 cm^{-1} lower than that of AB-*N*LG, depending on *N* [47]. The 2D mode profile of ABC-*N*LG shows distinct characteristics from those of AB-*N*LG, [47, 57–59] as shown in Fig. 1.8a for the case of 3LG, which is ascribed to the different electronic structures between them. As discussed above, six sub-peaks can be used to closely fit the 2D peak of AB-3LG, while the 2D mode of AB-3LG exhibits a more asymmetric feature. Further characterization of the 2D mode profile in ABC- and AB-stacked 4–6LG can be found in Refs. [47, 58, 59]. The differences in the 2D mode profiles between AB- and ABC-stacked *N*LG depend on the excitation energy (E_{ex}) and become more complicated with increasing *N*.

In the low frequency region, the C_{N1} modes can only be observed in AB-*N*LG at room temperature but not in ABC-*N*LG, [47] as shown in Fig. 1.8b. The absence of the C_{N1} mode in ABC-*N*LG was attributed to the Raman inactivity induced by the intrinsic symmetry and the weak EPC. Lui *et al.* observed the C_{31} mode at $\sim$33 cm^{-1} in AB-3LG and the C_{32} peak at $\sim$19 cm^{-1} in ABC-3LG suspended on quartz substrate at a high temperature of approximately 800 K by laser heating [58]. However, all the C_{NN-i} modes in ABC-*N*LG cannot be observed all the time on SiO_2/Si substrate, even when the laser power is as high as 10 mW, because a high substrate background emerges, [47] which is quite different from that on a quartz substrate [55]. The results showed that the C mode can also be utilized to identify AB and ABC stacking orders in *N*LG.

1.3.4 Raman Spectra of Twisted Multilayer Graphene

In addition to AB and ABC stacking orders, *n*LGs, *m*LGs, *p*LGs, ... (AB-*n*LG, AB-*m*LG, AB-*p*LG, ... if $n>1, m>1, p>1, \ldots, N=n+m+p+\ldots$) can be stacked together with twist angles (θ_t) at each twist interface to form twisted *N*LG (t*N*LG, or t$(n+m+p+\ldots)$LG). For example, t$(m+n)$LG is assembled by *m*LG ($m\geq1$) and *n*LG ($n\geq1$) flakes. Twisted graphene layers occur naturally at the surface of crystalline graphite, so that t*N*LG can be generated naturally by accidentally folding graphene layers onto themselves or other graphene flakes in the ME preparation process [42, 43]. t*N*LG has been widely observed to occur in non-controlled manner in graphene flakes grown by CVD [48, 49, 60–63]. The simplest t*N*LGs, twisted bilayer graphenes (t2LGs), can even exhibit novel physical properties due to the periodically modulated interaction between the two Dirac electron gases with a large Moiré supercell, as shown in Fig. 1.9a. They have a Dirac-like linear dispersion with

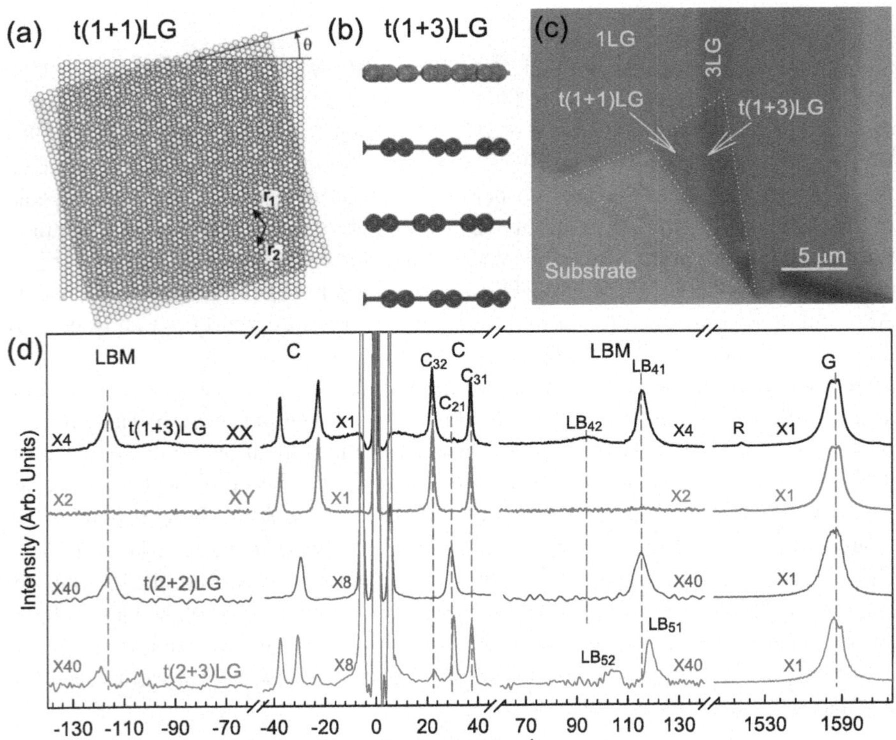

Fig. 1.9 (**a**) Moiré pattern of t(1+1)LG. (**b**) Schematic diagram of t(1+3)LG. (**c**) Optical image of a flake comprising t(1+1)LG and t(1+3)LG. (**d**) Stokes/anti-Stokes Raman spectra in the C and LB spectral range, and Stokes Raman spectra in the G peak region for t(1+3)LG, t(2+2)LG and t(2+3)LG. Polarized Raman spectra of t(1+3)LG are also shown. (Reproduced with permission from Ref. [42] and Ref. [27])

a Fermi velocity lower than that in 1LG [64]. By changing θ_t, it is possible to tune the optical absorption [65]. θ_t can be expressed in terms of the twist vector, [66, 67] (p, q), defined as the coordinates with respect to the basis vectors of 1LG. The positions of the so-called R and R′ Raman bands have been used to probe θ_t in tNLG [48, 49, 68, 69].

In addition to θ_t, the stacking sequence is also important for modulating the physical properties in tNLG. For example, t(2+3)LG has significantly different properties compared to AB-5LG, t(1+4)LG, or t(1+1+3)LG, etc., even though all these have the same $N=5$. For a given total N, the choice of m, n, etc. (with $m+n+\ldots=N$) and relative angles of each interface between mLG, nLG, … leads to a family of systems with different optical and electronic properties [27, 42, 68, 69]. The band structure of t(m+n)LG can be considered as the overlap between those of mLG and nLG. Figure 1.9b shows the diagram of t(1+3)LG, and Fig. 1.9c presents an optical image of t(1+1)LG and t(1+3)LG formed by the ME method.

From the perspective of the Raman spectroscopy of tNLG, several features are usually observed. The G peaks are usually significantly enhanced, and two further sub-peaks (G^+ and G^-) can be observed with the proper laser excitation if mLG ($m>1$) is one of the constituents [42,43]. The enhancement of the G mode originates from the match between the excitation laser and energy gap of the van Hove singularities (VHSs) in the joint density of state (JDOS) of all the optically allowed transitions ($JDOS_{OAT}$) [42]. In the low-frequency region, tNLG also exhibits some distinguishing characteristics from AB-NLG. Because the twisted interface would obstruct the interlayer shear coupling due to the periodicity mismatch between the two twisted layers, the C modes locate in the AB-stacked constituent [42]. However, the twisted interface would not affect the interlayer breathing coupling; thus, the LB modes are from the collective motions of all the stacked layers [27]. Therefore, Pos(C) of tNLG is mainly determined by the layer number of each AB-stacked constituent, while its Pos(LB) is mainly determined by N, as shown in Fig. 1.9d. Most of the C modes and one or two LB modes can be observed under the resonant condition in t($m+n$)LG, because the twist-induced lower symmetry makes them Raman active [27, 42]. The N-dependent Pos(C) of AB-NLG can be predicted by LCM [26, 70]. However, in tNLG, a softening factor is introduced for the C mode at the twisted interface for shear coupling, $\alpha_t^{\parallel}/\alpha_0^{\parallel}$ ~0.2, and another softening factor is considered at the AB-stacked interface next to the twist interface, $\alpha_{0t}^{\parallel}/\alpha_0^{\parallel}$ ~0.9, producing the so-called twist LCM (tLCM) [42]. The 2LCM used in AB-NLG can be also adopted for tNLG [27]. tLCM and 2LCM can accurately reproduce the experimental Pos(C) and Pos(LB) in tNLG, respectively [27, 42].

1.4 Resonance Raman Spectroscopy of Monolayer and Multilayer Graphene

Raman scattering is a process related to the generation of electron-hole pairs by photon excitation, electron-phonon interaction, electron-hole recombination and photon emission. Thus, in addition to the selection rule and Raman tensor, the intensities and profiles of Raman modes are mainly determined by the EPC, electron-photon interaction and electronic band structure of solid materials. The intensity of the Raman mode in 1LG can be calculated by second order perturbation theory [33]. As discussed above, most of the Raman modes in graphene are involved in resonance Raman scattering because of the unique band structure [9]. Even the Raman intensity of the G mode, I(G), in 1LG is dominated by the quantum interference effect [15]. In AB-NLG, the quantum interference between the C mode and a continuum of electronic transitions near the K point plays an important role in the Fano lineshape of the C modes, [26] as shown in Fig. 1.6c, d. For tNLG, significant enhancements of first order modes (C, LB and G modes) are always observed when the laser excitation energy matches that of the VHSs in $JDOS_{OAT}$ [42]. In the following subsections, we will discuss the three resonant Raman effects in 1LG and MLG.

1.4.1 Quantum Interference of the G and 2D Modes

In the DR process in Fig. 1.2, only the one-dimensional wave vectors are considered, which means that the wave vectors of the involved phonons or defects are always reversed (***q***, -***q***). A theorical work found that the integration in the two-dimensional BZ does not alter the DR phonon wave vectors determined from the one-dimensional integration [71]. This is because, when the wave vectors of involved phonons or defect are not reversed, the contribution of two wave vectors would be cancelled by interference. In particular, when the two wave vectors are in the same direction, the contribution could be almost totally destructive interference. Thus, it is reasonable to describe the DR process by the one-dimensional model.

The quantum interference of the G mode was first predicted by theory, [16] and then proven by an electrostatic doping experiment [15]. Under electrostatic doping, I(G) increases dramatically. This unexpected phenomenon presents evidence of a new understanding of resonance Raman scattering in graphene. Figure 1.10a shows I(G) as a function of $2|E_F|$, where E_F is the Fermi level of graphene tuned by the electrostatic doping. The observation cannot be interpreted well by the general assumption that the pathway of G mode Raman scattering is only on resonance (I in Fig. 1.10a). To further understand the phenomenon, pathways are extended that can be close to resonance (II in Fig. 1.10a) and off resonance. In this condition, all pathways interfering with each other have different quantum mechanical amplitudes, including phase and magnitude. For any specific one-

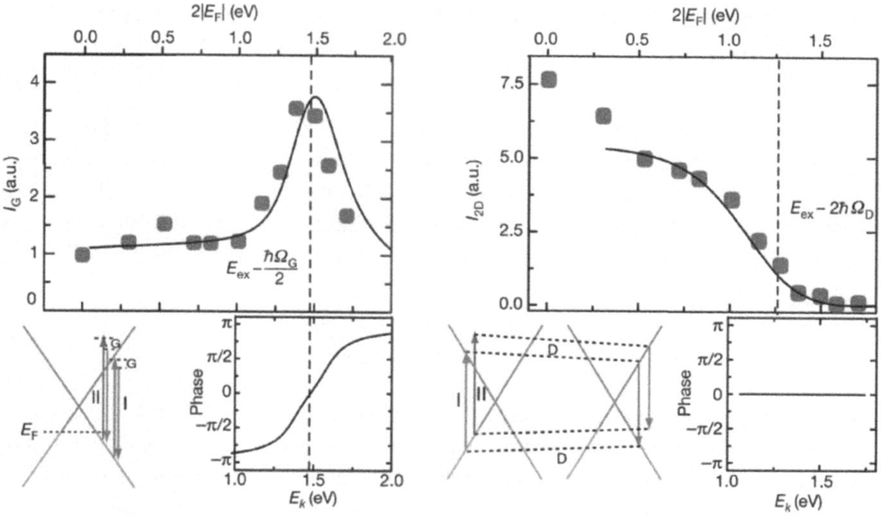

Fig. 1.10 Upper panel: Doping dependence of I(G) (left) and I(2D) (right). The black solid line is the theoretical prediction. Lower left panel: schematic of two representative pathways (I and II) of the G mode through two different intermediate excited states and the quantum phase of Raman pathway for the G mode as a function of intermediate excitation energy. Lower right panel: schematic of two representative pathways (I and II) of the 2D mode through two different intermediate excited states and the quantum phase of Raman pathway for the 2D mode as a function of intermediate excitation energy. (Reproduced with permission from Ref. [15])

phonon and one-photon state, I(G) is described by

$$I(G) = |\sum_k C_k R_k|^2$$
$$R_k = \frac{1}{(E_{ex} - E_k - i\gamma)(E_{ex} - \hbar\omega_G - E_k - i\gamma)} \qquad (1.3)$$

where C_k and R_k are the matrix element and resonance factor, respectively. Its phase $\omega = \arg(R_k)$ is plotted as a function of transition energy (E_k) in Fig. 1.10a. With E_k above and below the energy of E_{ex}-$\hbar\omega_G/2$, the pathways have a phase difference of approximately π. With a small enough $|E_F|$, all pathways are allowed and interfere destructively. With a large enough $|E_F|$, all pathways are blocked. The two situations lead to a weak overall Raman signal. When $2|E_F| = E_{ex}$-$\hbar\omega_G/2$, I(G) reaches a maximum value because just half of the transitions with the same phase sign (positive or negative) are allowed, and destructive interference would not occur. In addition to the G mode, I(2D) is also determined by the quantum interference effect. The different pathways have the same phase, presented in Fig. 1.10b, resulting in the monotonic decrease of I(2D).

1.4.2 Fano Resonance of the C and G Mode in Multilayer Graphene

The C modes of the Raman spectrum in AB-*N*LG exhibit a Fano lineshape, [26] as shown in Fig. 1.6c. Historically, the Fano lineshape was observed in the Raman spectra of several systems, such as degenerate silicon systems, [72] metallic carbon nanotubes, [73], nanowires, [74] topological insulators [75] and 2d semiconductors [76]. The Fano lineshape of a Raman mode can be expressed by [26]

$$I(\omega) = I_0 \frac{[1 + 2(\omega - \omega_0)/(q\Gamma)]^2}{[1 + 4(\omega - \omega_0)^2/\Gamma^2]} \tag{1.4}$$

where I_0, ω_0, Γ and $1/|q|$ are the intensity, uncoupled mode frequency, broadening parameters and coupling coefficient, respectively. The peak maximum is at $\omega_{max} = \omega_0 + \Gamma/2q$, while the full-width at half-maximum (FWHM) is $\Gamma(q^2+1)/|q^2 - 1|$. The Fano lineshape exhibits high asymmetry dependent on the coupling coefficient. It will recover to a Lorentzian lineshape when $1/|q|$ approaches to 0. Because Pos(G) remains constant and the C peak at 77 K has the same coupling coefficient, $1/|q|$, as that at room temperature, [26] the laser-induced electron-hole plasma and the multiphonon resonance can be ruled out as the reason for the resulting Fano lineshape. In general, the Fano lineshape of the C mode in AB-*N*LG can be attributed to the quantum interference between the C mode and a continuum of electronic transitions near the K point [26]. As in the case of AB-3LG in Fig. 1.6d, the C phonon can efficiently couple with the active electronic transitions with energies close to the C mode energy (E(C), ~5 meV). The charge transferred from the substrate and the air molecules can raise the Fermi level (especially for E_F >E(C)/2). This can largely reduce the coupling coefficient of the Fano resonance, particularly for few-layer graphenes, which has been confirmed by the *N*-dependent $1/|q|$ [26].

The G band of 1LG should exhibit an asymmetric Fano-like lineshape near the charge neutrality point, whereas the line shape is symmetric when the graphene flake is doped by electrons or holes [77]. However, the G band of AB-2LG does not exhibit any Fano resonance, regardless of doping. The Fano resonance observed in the G band lineshape of 1LG originates from the interferences between the phonon and excitonic many-body spectra, whereas the absence of a Fano resonance in the G band lineshape of AB-2LG can be elucidated by the unexpected excitonic interaction in AB-2LG [77, 78]. However, compared with the imperceptible asymmetric C modes with $1/|q|$ ~0.3, [26] the lineshape of the G mode in neutral 1LG just shows a very slightly asymmetric behaviour due to the much smaller $1/|q|$ of ~0.07 [77] The Fano resonance of the G mode can also be observed in the infrared spectra of phonons in *N*LG because of the strong coupling between phonons and interband electronic transitions [79].

1.4.3 Resonance Raman Spectroscopy of Twisted Multilayer Graphene

The band structure of tNLG can be modified by several twist-related parameters, such as the layer number of the constitutes and the twist angles, showing novel and interesting optical properties [42]. For example, the intensity of the first order modes can be enhanced strongly. As shown in Fig. 1.11a, the Raman signals of t(1+1)LG would be enhanced when E_{ex} matches its θ_t-dependent VHS energies of JDOS$_{OAT}$ [60, 61]. The relationship between the VHS energy of JDOS$_{OAT}$ and θ_t can be estimated by the formula [61] $E_{VHSs} \approx 4\pi\theta_t\hbar v_f/\sqrt{3}a$, where a is the lattice constant of graphene (2.46 Å), $\hbar$ is the reduced Planck's constant, and v_f is the Fermi velocity of 1LG (10^6 m/s). JDOS$_{OAT}$ of NLG can be calculated by the following equation [42]:

$$JDOS_{OAT}(E) \propto \sum_{ij}\sum_{k} |M_{ij}(\mathbf{k})|^2 \delta\left(E_{ij}(\mathbf{k}) - E\right), \tag{1.5}$$

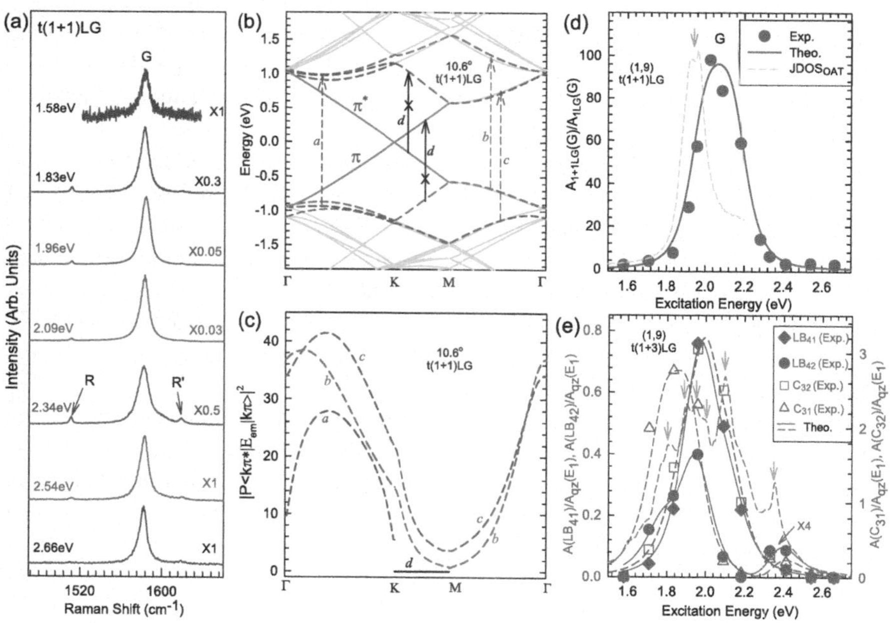

Fig. 1.11 (**a**) G mode of t(1+1)LG for seven E_{ex}. The R and R′ modes are indicated by arrows. (**b**) The band structure of (1,9) t(1+1)LG. The optically allowed transitions are marked by dashed arrows, and the forbidden transitions are indicated by solid arrows with crosses. (**c**) Squared optical matrix elements of the corresponding band pairs in (**b**). (**d**) A(G) of t(1+1)LG and (**e**) A(C_{31}), A(C_{32}), A(LB_{41}) and A(LB_{42}) of t(1+3)LG as a function of E_{ex}. The dashed lines indicated with arrows in (**d**) and (**e**) are JDOS$_{OAT}$ in t(1+1)LG and t(1+3)LG along Γ-K-M-Γ, respectively. (Reproduced with permission from Ref. [42])

where $M_{ij}(\mathbf{k})$ is the optical matrix element between the ith conduction and jth valence bands, and $E_{ij}(\mathbf{k})$ gives the transition energy of an $i \rightarrow j$ band pair at the wavevector $\mathbf{k}$. The band structure of t(1+1)LG with 10.6° ($(p,q)=(1,9)$) is shown in Fig. 1.11b, and the optically allowed transitions are marked by dashed arrows. The squared optical matrix elements of the corresponding transitions are shown in Fig. 1.11c. JDOS$_{OAT}$ for t(1+1)LG is shown in Fig. 1.11d by dashed lines indicated with arrows, showing only one distinct VHS in JDOS$_{OAT}$ labelled by arrows. The major contribution to the Raman intensity comes from the term of Eq. 1.5 when E_{ex} matches the VHS energies of JDOS$_{OAT}$. Therefore, the intensity of Raman modes in t(m+n)LG as a function of E_{ex} can be evaluated by second order perturbation theory [33, 68],

$$A \propto \left| \sum_j \frac{M_j}{(E_{\mathrm{ex}} - E_{\mathrm{VHS}}(j) - i\gamma)\left(E_{\mathrm{ex}} - E_{ph} - E_{\mathrm{VHS}}(j) - i\gamma\right)} \right|^2, \quad (1.6)$$

where M_j are constants treated as fitting parameters that encompass the product of the electron-phonon and electron-phonon interaction matrix elements for the jth VHS in JDOS$_{OAT}$, E_{ph} is the phonon energy, and γ gives the energy uncertainty related to the lifetime of the excited state. Here, we adopt $\gamma = 0.15$ eV to reflect the effect of summation along all the high symmetry path. The E_{ex}-dependent G mode area intensity (A(G)) of t(1+1)LG is shown in Fig. 1.11d. The calculated results are in agreement with the experimental data.

In t(1+1)LGs, the VHS energy of the electronic density of states (DOS) is similar to that of JDOS, due to the simple band structure [42, 61]. Therefore, the resonant excitation energy can be estimated from the VHS energy of either DOS or JDOS. However, in t(m+n)LG ($m \geq 1$, $n > 1$), the resonance process is more complicated because of its more complex band structure and EPC. More VHS energies exist in JDOS$_{OAT}$. Taking the (1,9) t(1+3)LG as an example, the E_{ex}-dependent A(C), A(LB) and A(G) fitted by the VHS energies are in good agreement with the experimental data, [42] as shown in Fig. 1.11e.

Moreover, the DR process in t(1+1)LG is much more complicated, due to the folding of the band structure. Similar to 1LG, the 2D peak of most t(1+1)LGs exhibits a single Lorentzian peak. The FWHM, frequency and area of the 2D peak in t(1+1)LG depend on θ_t [61, 80]. Therefore, the profile of the 2D peak is not suitable for identifying the N of tNLG.

1.5 Conclusion

Graphene and MLG have extraordinary electronic, optoelectronic and mechanical properties as well as appealing applications in devices. In this chapter, we have reviewed the basic Raman spectroscopy of pristine 1LG and MLG, including the origin of the G mode; all the dispersive Raman modes in intrinsic 1LG; novel

Raman modes and new phenomena found in AB-, ABC- and twisted-MLG; how the layer number, stacking sequences and twist angle affect the Raman spectra of the C, LB and 2D modes; and the peculiar resonance Raman effects in pristine 1LG and MLG. The related Raman concepts reviewed for 1LG and MLG, such as the double resonance Raman spectroscopy, low-frequency interlayer Raman modes and the linear chain model, may be applicable to other 2DMs to understand the interlayer coupling in 2DMs. This chapter may be helpful to gain a deeper understanding of the Raman scattering process in various 2DMs.

Acknowledgements We acknowledge support from the National Key Research and Development Program of China (Grant No. 2016YFA0301204), the National Natural Science Foundation of China (Grant No. 11474277, 11874350 and 11434010), and the Beijing Municipal Science and Technology Commission.

References

1. K.S. Novoselov, A.K. Geim, S.V. Morozov, D. Jiang, Y. Zhang, S.V. Dubonos, I.V. Grigorieva, A.A. Firsov, Science **306**(5696), 666 (2004)
2. S. Latil, L. Henrard, Phys. Rev. Lett. **97**(3), 036803 (2006)
3. K.S. Novoselov, D. Jiang, F. Schedin, T.J. Booth, V.V. Khotkevich, S.V. Morozov, A.K. Geim, P. Natl. Acad. Sci. USA **102**, 10451 (2005)
4. X. Li, W. Cai, J. An, S. Kim, J. Nah, D. Yang, R. Piner, A. Velamakanni, I. Jung, E. Tutuc, et al., Science **324**(5932), 1312 (2009)
5. A. Reina, X. Jia, J. Ho, D. Nezich, H. Son, V. Bulovic, M.S. Dresselhaus, J. Kong, Nano Lett. **9**(1), 30 (2008)
6. A.K. Geim, K.S. Novoselov, Nat. Mater. **6**(3), 183 (2007)
7. A.K. Geim, I.V. Grigorieva, Nature **499**(7459), 419 (2013)
8. K.S. Novoselov, A. Mishchenko, A. Carvalho, A.H.C. Neto, Science **353**, 6298 (2016)
9. J.B. Wu, M.L. Lin, X. Cong, H.N. Liu, P.H. Tan, Chem. Soc. Rev. **47**, 1822 (2018)
10. H. Li, J.B. Wu, F. Ran, M.L. Lin, X.L. Liu, Y. Zhao, X. Lu, Q. Xiong, J. Zhang, W. Huang, H. Zhang, P.H. Tan, ACS Nano **11**, 11714 (2017)
11. L. Malard, M. Pimenta, G. Dresselhaus, M. Dresselhaus, Phys. Rep. **473**(5), 51 (2009)
12. A.C. Ferrari, D.M. Basko, Nat. Nanotechnol. **8**(4), 235 (2013)
13. A. Ferrari, J. Meyer, V. Scardaci, C. Casiraghi, M. Lazzeri, F. Mauri, S. Piscanec, D. Jiang, K. Novoselov, S. Roth, et al., Phys. Rev. Lett. **97**(18), 187401 (2006)
14. L. Malard, J. Nilsson, D. Elias, J. Brant, F. Plentz, E. Alves, A.C. Neto, M. Pimenta, Phys. Rev. B **76**(20), 201401 (2007)
15. C.F. Chen, C.H. Park, B.W. Boudouris, J. Horng, B. Geng, C. Girit, A. Zettl, M.F. Crommie, R.A. Segalman, S.G. Louie, et al., Nature **471**(7340), 617 (2011)
16. D. Basko, New J. Phys. **11**(9), 095011 (2009)
17. D. Yoon, H. Moon, Y.W. Son, J.S. Choi, B.H. Park, Y.H. Cha, Y.D. Kim, H. Cheong, Phys. Rev. B **80**(12), 125422 (2009)
18. A.C. Neto, F. Guinea, N.M. Peres, K.S. Novoselov, A.K. Geim, Rev. Mod. Phys. **81**(1), 109 (2009)
19. J.M. Dawlaty, S. Shivaraman, J. Strait, P. George, M. Chandrashekhar, F. Rana, M.G. Spencer, D. Veksler, Y. Chen, Appl. Phys. Lett. **93**(13), 131905 (2008)
20. P. Blake, E. Hill, A.C. Neto, K. Novoselov, D. Jiang, R. Yang, T. Booth, A. Geim, Appl. Phys. Lett. **91**(6), 063124 (2007)

21. X.L. Li, X.F. Qiao, W.P. Han, Y. Lu, Q.H. Tan, X.L. Liu, P.H. Tan, Nanoscale **7**(17), 8135 (2015)
22. W. Zhao, P. Tan, J. Zhang, J. Liu, Phys. Rev. B **82**(24), 245423 (2010)
23. M. Lazzeri, C. Attaccalite, L. Wirtz, F. Mauri, Phys. Rev. B **78**(8), 081406 (2008)
24. P. Tan, C. Hu, J. Dong, W. Shen, B. Zhang, Phys. Rev. B **64**(21), 214301 (2001)
25. S. Reich, C. Thomsen, Philos. Transact. A Math Phys. Eng. Sci. **362**(1824), 2271 (2004)
26. P. Tan, W. Han, W. Zhao, Z. Wu, K. Chang, H. Wang, Y. Wang, N. Bonini, N. Marzari, N. Pugno, et al., Nat. Mater. **11**(4), 294 (2012)
27. J.B. Wu, Z.X. Hu, X. Zhang, W.P. Han, Y. Lu, W. Shi, X.F. Qiao, M. Ijias, S. Milana, W. Ji, et al., ACS Nano **9**(7), 7440 (2015)
28. P.H. Tan, J.B. Wu, W.P. Han, W.J. Zhao, X. Zhang, H. Wang, Y.F. Wang, Phys. Rev. B **89**(23), 235404 (2014)
29. D. L. Nika, A. A. Balandin, J. Phys. Condes. Matter **24**, 233203 (2012)
30. M. Mohr, J. Maultzsch, E. Dobardžić, S. Reich, I. Milošević, M. Damnjanović, A. Bosak, M. Krisch, C. Thomsen, Phys. Rev. B **76**(3), 035439 (2007)
31. S. Siebentritt, R. Pues, K.H. Rieder, A.M. Shikin, Phys. Rev. B **55**(12), 7927 (1997)
32. R. Saito, A. Jorio, A. Souza Filho, G. Dresselhaus, M. Dresselhaus, M. Pimenta, Phys. Rev. Lett. **88**(2), 027401 (2001)
33. P. Venezuela, M. Lazzeri, F. Mauri, Phys. Rev. B **84**(3), 035433 (2011)
34. P. Tan, L. An, L. Liu, Z. Guo, R. Czerw, D.L. Carroll, P.M. Ajayan, N. Zhang, H. Guo, Phys. Rev. B **66**(24), 245410 (2002)
35. C. Thomsen, S. Reich, Phys. Rev. Lett. **85**(24), 5214 (2000)
36. P. Tan, Y. Deng, Q. Zhao, Phys. Rev. B **58**(9), 5435 (1998)
37. P. May, M. Lazzeri, P. Venezuela, F. Herziger, G. Callsen, J.S. Reparaz, A. Hoffmann, F. Mauri, J. Maultzsch, Phys. Rev. B **87**(7), 075402 (2013)
38. D. Basko, S. Piscanec, A. Ferrari, Phys. Rev. B **80**(16), 165413 (2009)
39. D. Yoon, H. Moon, Y.W. Son, G. Samsonidze, B.H. Park, J.B. Kim, Y. Lee, H. Cheong, Nano Lett. **8**(12), 4270 (2008)
40. S. Piscanec, M. Lazzeri, F. Mauri, A. Ferrari, J. Robertson, Phys. Rev. Lett. **93**(18), 185503 (2004)
41. L. Malard, M. Guimaraes, D. Mafra, A. Jorio, Phys. Rev. B **79**(12), 125426 (2009)
42. J.B. Wu, X. Zhang, M. Ijäs, W.P. Han, X.F. Qiao, X.L. Li, D.S. Jiang, A.C. Ferrari, P.H. Tan, Nat. Commun. **5**, 5309 (2014)
43. C. Cong, T. Yu, Nat. Commun. **5**, 4709 (2014)
44. L. Malard, D. Elias, E. Alves, M. Pimenta, Phys. Rev. Lett. **101**(25), 257401 (2008)
45. W. Zhao, P.H. Tan, J. Liu, A.C. Ferrari, J. Am. Chem. Soc. **133**(15), 5941 (2011)
46. J. Zabel, R.R. Nair, A. Ott, T. Georgiou, A.K. Geim, K.S. Novoselov, C. Casiraghi, Nano Lett. **12**(2), 617 (2012)
47. X. Zhang, W.P. Han, X.F. Qiao, Q.H. Tan, Y.F. Wang, J. Zhang, P.H. Tan, Carbon **99**, 118 (2016)
48. J.B. Wu, H. Wang, X.L. Li, H. Peng, P.H. Tan, Carbon **110**, 225 (2016)
49. M.L. Lin, T. Chen, W. Lu, Q.H. Tan, P. Zhao, H.T. Wang, Y. Xu, P.H. Tan, J. Raman Spectrosc. **49**, 46 (2018)
50. H. Wang, M. Feng, X. Zhang, P.H. Tan, Y. Wang, J. Phys. Chem. C **119**(12), 6906 (2015)
51. X. Zhang, X.F. Qiao, W. Shi, J.B. Wu, D.S. Jiang, P.H. Tan, Chem. Soc. Rev. **44**(9), 2757 (2015)
52. X.L. Li, W.P. Han, J.B. Wu, X.F. Qiao, J. Zhang, P.H. Tan, Adv. Funct. Mater. **27**(19), 1604468 (2017)
53. L. Liang, J. Zhang, B.G. Sumpter, Q. Tan, P.H. Tan, V. Meunier, ACS Nano **11**(12), 11777 (2017)
54. X.F. Qiao, X.L. Li, X. Zhang, W. Shi, J.B. Wu, T. Chen, P.H. Tan, Appl. Phys. Lett. **106**(22), 223102 (2015)
55. C.H. Lui, Z. Ye, C. Keiser, E.B. Barros, R. He, Appl. Phys. Lett. **106**(4), 041904 (2015)
56. H. Wilhelm, B. Croset, G. Medjahdi, Carbon **45**(12), 2356 (2007)

57. C. Cong, T. Yu, K. Sato, J. Shang, R. Saito, G.F. Dresselhaus, M.S. Dresselhaus, ACS Nano **5**(11), 8760 (2011)
58. C.H. Lui, Z. Li, Z. Chen, P.V. Klimov, L.E. Brus, T.F. Heinz, Nano Lett. **11**(1), 164 (2010)
59. T. Nguyen, J. Lee, D. Yoon, H. Cheong, Sci. Rep. **4**, 4630 (2014)
60. R.W. Havener, H. Zhuang, L. Brown, R.G. Hennig, J. Park, Nano Lett. **12**(6), 3162 (2012)
61. K. Kim, S. Coh, L.Z. Tan, W. Regan, J.M. Yuk, E. Chatterjee, M. Crommie, M.L. Cohen, S.G. Louie, A. Zettl, Phys. Rev. Lett. **108**(24), 246103 (2012)
62. J. Campos-Delgado, L.G. Cançado, C.A. Achete, A. Jorio, J.P. Raskin, Nano Res. **6**(4), 269 (2013)
63. A. Jorio, L.G. Cançado, Solid State Commun. **175**, 3 (2013)
64. Z. Ni, Y. Wang, T. Yu, Y. You, Z. Shen, Phys. Rev. B **77**(23), 235403 (2008)
65. P. Moon, M. Koshino, Phys. Rev. B **87**(20), 205404 (2013)
66. G. Trambly de Laissardière, D. Mayou, L. Magaud, Nano Lett. **10**(3), 804 (2010)
67. K. Sato, R. Saito, C. Cong, T. Yu, M.S. Dresselhaus, Phys. Rev. B **86**(12), 125414 (2012)
68. V. Carozo, C. Almeida, B. Fragneaud, P. Bedê, M. Moutinho, J. Ribeiro-Soares, N. Andrade, A. Souza Filho, M. Matos, B. Wang, et al., Phys. Rev. B **88**(8), 085401 (2013)
69. V. Carozo, C.M. Almeida, E.H. Ferreira, L.G. Cançado, C.A. Achete, A. Jorio, Nano Lett. **11**(11), 4527 (2011)
70. X. Zhang, W. Han, J. Wu, S. Milana, Y. Lu, Q. Li, A. Ferrari, P. Tan, Phys. Rev. B **87**(11), 115413 (2013)
71. J. Maultzsch, S. Reich, C. Thomsen, Phys. Rev. B **70**(15), 155403 (2004)
72. F. Cerdeira, T. Fjeldly, M. Cardona, Phys. Rev. B **8**(10), 4734 (1973)
73. H. Lin, L.A. Wray, Y. Xia, S. Xu, S. Jia, R.J. Cava, A. Bansil, M.Z. Hasan, Nat. Mater. **9**(7), 546 (2010)
74. R. Gupta, Q. Xiong, C. Adu, U. Kim, P. Eklund, Nano Lett. **3**(5), 627 (2003)
75. J. Zhang, Z. Peng, A. Soni, Y. Zhao, Y. Xiong, B. Peng, J. Wang, M.S. Dresselhaus, Q. Xiong, Nano Lett. **11**(6), 2407 (2011)
76. Q.H. Tan, Y.J. Sun, X.L. Liu, Y. Zhao, Q. Xiong, P.H. Tan, J. Zhang, 2D Mater. **4**(3), 031007 (2017)
77. D. Yoon, D. Jeong, H.J. Lee, R. Saito, Y.W. Son, H.C. Lee, H. Cheong, Carbon **61**, 373 (2013)
78. E.H. Hasdeo, A.R. Nugraha, M.S. Dresselhaus, R. Saito, Phys. Rev. B **90**(24), 245140 (2014)
79. T.T. Tang, Y. Zhang, C.H. Park, B. Geng, C. Girit, Z. Hao, M.C. Martin, A. Zettl, M.F. Crommie, S.G. Louie, et al., Nat. Nanotechnol. **5**(1), 32 (2010)
80. R. He, T.F. Chung, C. Delaney, C. Keiser, L.A. Jauregui, P.M. Shand, C. Chancey, Y. Wang, J. Bao, Y.P. Chen, Nano Lett. **13**(8), 3594 (2013)

Chapter 2
Raman Spectroscopy of Isotropic Two-Dimensional Materials Beyond Graphene

Xin Lu, Qing-Hai Tan, Qihua Xiong, and Jun Zhang

In this chapter, we will focus on the isotropic (or rather less anisotropic) two-dimensional layered materials, including the layered transition metal dichalcogenides, the topologically insulating Bi_2X_3 (X = Se, Te) etc.

The layered transition metal dichalcogenides (TMDs), such as MoS_2 and WSe_2, have been attracting lots of interests in the past decade, because of their unique properties and potential applications [1–5]. Monolayer TMDs are excellent candidates for the study of valley pseudospin [5]. Valley polarization [6–8] and valley Hall effect [9] have both been observed and reported. Later, electrical control of valley polarization and valley Hall effect has been realized as well [10, 11], which increases the possibility of potential applications in valleytronics. The observation of valley polarization on spectroscopy, to some extent, benefits from the strong photoluminescence in monolayer. The bulk TMD crystals are indirect-gap semiconductors. As thickness decreases, there is an indirect-to-direct gap transition.

X. Lu
Division of Physics and Applied Physics, School of Physical and Mathematical Sciences, Nanyang Technological University, Singapore, Singapore

Current Address: Department of Physics, Emory University, Atlanta, GA, USA

Q.-H. Tan · J. Zhang (✉)
State Key Laboratory of Superlattices and Microstructures, Institute of Semiconductors, Chinese Academy of Sciences, Beijing, China

College of Materials Science and Opto-Electronic Technology, CAS Center of Excellence in Topological Quantum Computation, University of Chinese Academy of Sciences, Beijing, China
e-mail: zhangjwill@semi.ac.cn

Q. Xiong (✉)
Division of Physics and Applied Physics, School of Physical and Mathematical Sciences, Nanyang Technological University, Singapore, Singapore

MajuLab, CNRS-UNS-NUS-NTU International Joint Research Unit, Singapore, Singapore
e-mail: qihua@ntu.edu.sg

P.-H. Tan (ed.), *Raman Spectroscopy of Two-Dimensional Materials*, Springer Series in Materials Science 276, https://doi.org/10.1007/978-981-13-1828-3_2

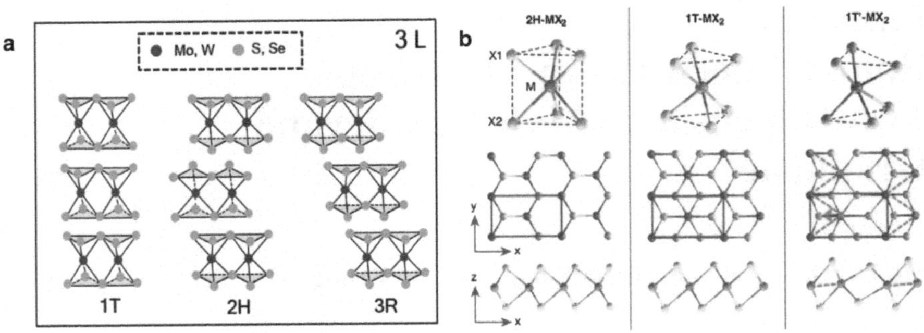

Fig. 2.1 Structural polytypism in MX_2 (M = Mo, W; X = S, Se, Te). (**a**) Trilayer (3 L) MX_2 (M = Mo, W; X = S, Se) with 1T, 2H, and 3R stacking configurations. (Reproduced with permission from Ref. [14], © Tsinghua University Press and Springer-Verlag Berlin Heidelberg 2016). (**b**) Comparison of 2H, 1T, and 1T' structures of monolayer $MoTe_2$. (Reproduced with permission from Ref. [15], © American Association for the Advancement of Science 2014)

Monolayer TMDs possess a direct gap, which gives an emerging photoluminescence [12, 13].

Among all the TMDs, the semiconducting group VI ones (MX_2, M = Mo, W; X = S, Se, Te) are widely studied, especially the 2H phase MoS_2 [6–13, 16, 17]. Besides the stable 2H stacking, 1T and 3R phases are also common in the S- and Se-based MX_2 (M = Mo, W; X = S, Se) [1, 18]. We note that the monolayer counterparts of 2H and 3R phases are the same, also referred to as 1H. The spatial inversion symmetry is broken in the 1H stacking, but preserved in the 1T structure, as shown in Fig. 2.1a. The difference between 2H and 3R starts from the bilayer (2 L). Besides the layer shift, which also occurs in the 3R phase, the *N*th layer of the 2H stacking is rotated by 180° with respect to the (*N−1*)th layer. The 3R stacking is noncentrosymmetric from monolayer (1L) to the bulk, giving rise to the valley polarization regardless of the thickness [19]. In contrast, the 2H stacking with an even number of layers possesses an inversion symmetry point, while that with an odd number of layers is noncentrosymmetric. In Te-based TMDs (MTe_2, M = Mo, W), 1T' phase is also thermodynamically stable, as the large size of the telluride atom distorts the structure [15, 20]. The Raman spectra of $MoTe_2$ have shown dependence on excitation lasers [21–23], which will be discussed in Chap. 5. On the other hand, the phonon properties in WTe_2 are more complicated than other compounds [24, 25]. Here we will focus on the S- and Se-based TMDs.

Each layer of 2H-MX_2 (M = Mo, W; X = S, Se) consists of X–M–X tri-atomic planes, with the plane of Mo/W atoms sandwiched in two hexagonal planes of chalcogenide atoms. The primitive unit cell of bulk 2H-MX_2 has two layers, *i.e.*, 6 inequivalent atoms. Thus, there are 18 Brillouin zone center (Γ) phonons (3 * number of atoms), with 15 optical phonons and 3 acoustic phonons. The irreducible representation of the zone center phonons in bulk 2H-MX_2 (M = Mo, W; X = S, Se) is $\Gamma = A_{1g} + 2A_{2u} + B_{1u} + 2B_{2g} + E_{1g} + 2E_{1u} + E_{2u} + 2E_{2g}$. As the thickness decreases from bulk to few- and mono-layer, we have to consider the difference in

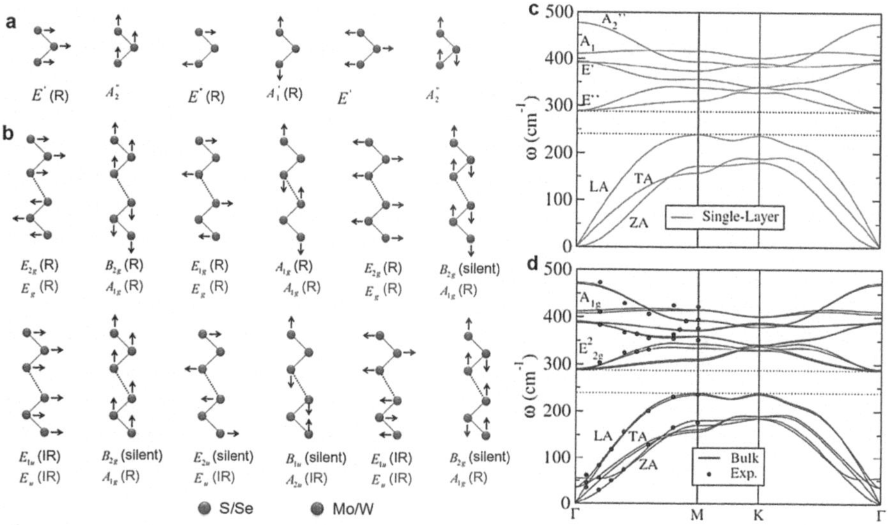

Fig. 2.2 (**a** and **b**) Normal displacements of vibrational modes for 1 L (**a**) and 2 L & bulk (**b**) 2H-MX_2 (M = Mo, W; X = S, Se). (Reproduced with permission from Ref. [26], © American Chemical Society 2013). (**c** and **d**) Phonon dispersion curves of single-layer (**c**) and bulk MoS_2 (**d**). Points in (**d**) are experimental data extracted from Ref. [27]. (Reproduced with permission from Ref. [28], © American Physical Society 2011)

symmetry between films with even and odd number of layers (even N and odd N). The irreducible representations in odd N and even N are $\Gamma = \frac{3N-1}{2}\left(A_1' + E''\right) + \frac{3N+1}{2}\left(A_2'' + E'\right)$ ($N = 1, 3, 5\ldots$) and $\Gamma = \frac{3N}{2}\left(A_{1g} + A_{2u} + E_g + E_u\right)$ ($N = 2, 4, 6\ldots$), respectively. Here E denotes the doubly degenerate in-plane modes, while A and B represent the out-of-plane vibration modes. Among these modes, only A_{1g}, E_{1g}, and E_{2g} are Raman-active. Note that the A_{1g} mode for bulk 2H-MX_2 (~408 cm^{-1} for MoS_2) is A_1' in 1L due to the change of symmetry, though the corresponding mode is still A_{1g} in 2L. Modes that are neither Raman active nor infrared (IR) active are called the silent modes, such as B_{2g} mode in bulk MoS_2. The normal mode displacements for 1L, 2L, and bulk 2H-MX_2 (M = Mo, W; X = S, Se) are shown in Fig. 2.2a, b.

In the following, we will take 2H-MoS_2 as an example to further discuss the phonon modes. The calculated phonon dispersion curves for 1L and bulk 2H-MoS_2 are displayed in Fig. 2.2c, d. There are 3 acoustic branches in 1L: the in-plane longitudinal acoustic (LA), the transverse acoustic (TA), and the out-of-plane acoustic (ZA) modes. The 6 high frequency optical branches are separated from the acoustic branches by a gap of ~50 cm^{-1} for MoS_2 [28]. The optical phonon dispersion curves for bulk are similar to those for 1L MoS_2, as a result of the weak van der Waals interlayer interactions. Meanwhile, we can observe the Davydov splitting of optical modes from Γ point in Fig. 2.2d, *e.g.* the mode at ~384 cm^{-1} (with symmetry of E_{2g}^1 in bulk) and the mode at ~385.5 cm^{-1} (with symmetry of E_{1u} in bulk). Davydov splitting occurs when there is more than one interacting layer from the system [29–31]. The difference within the Davydov pair, E^1_{2g}/E_{1u}, lies in the displacement between the adjacent sulfur atoms. The sulfur atoms of different layers move in opposite direction for E^1_{2g} mode, while sulfur atoms of neighbouring sheets are moving in phase for E_{1u} mode. Due to the weak interlayer coupling, the frequency difference between the two modes is ought to be small, but the out-of-phase displacement for E^1_{2g} mode is supposed to slightly increase the frequency, compared to that for E_{1u} mode. The acoustic branches for bulk, which are related to interlayer interaction, differ from those in 1L. Chap. 9 will discuss the interlayer shear and breathing modes in detail.

Raman spectra of bulk MoS_2 were collected at different excitation wavelengths, shown in Fig. 2.3a. Only two Raman-active modes (E^1_{2g} and A_{1g}) can be seen with 488 nm and 532 nm excitations. Many additional modes, mostly overtones and combinations, appear in the Raman spectra when 594 nm (2.09 eV) and 633 nm (1.96 eV) lasers are used. Resonant Raman spectra of MX_2 (M = Mo, W; X = S, Se, Te) will be further discussed in Chap. 5. Here we focus on the E^1_{2g} and A_{1g} modes.

The Raman tensors for E^1_{2g} and A_{1g} modes in bulk 2H-MX_2 (M = Mo, W; X = S, Se) can be written as:

$$A_{1g}: \begin{pmatrix} a & 0 & 0 \\ 0 & a & 0 \\ 0 & 0 & b \end{pmatrix}, \quad E_{2g}: \begin{pmatrix} b & 0 & 0 \\ 0 & -b & 0 \\ 0 & 0 & 0 \end{pmatrix}, \begin{pmatrix} 0 & b & 0 \\ b & 0 & 0 \\ 0 & 0 & 0 \end{pmatrix}.$$

Correspondingly, A_{1g} mode has A_1' symmetry in odd N, while it is still A_{1g} in even N. E_{2g} mode has E'' symmetry and E_g symmetry in odd N and even N, respectively. Polarization dependent Raman spectra can distinguish the in-plane E mode and the out-of-plane A mode. As shown in Fig. 2.3b, c, A_{1g} mode is suppressed under the cross-polarized configuration; In contrast, E^1_{2g} mode appears under both cross- and parallel-polarized configurations. Reason for the difference is related to the quantity $e_i \cdot \tilde{R} \cdot e_s$, where e_i is the polarization vector of the incident light, e_s is that of the scattered light and $\tilde{R}$ is the Raman tensor. $e_i \cdot \tilde{R} \cdot e_s$ is given by:

$$e_i \cdot \tilde{R} \cdot e_s = (x \;\; y \;\; z) \begin{pmatrix} \alpha_{xx} & \alpha_{xy} & \alpha_{xz} \\ \alpha_{yx} & \alpha_{yy} & \alpha_{yz} \\ \alpha_{zx} & \alpha_{zy} & \alpha_{zz} \end{pmatrix} \begin{pmatrix} x \\ y \\ z \end{pmatrix}$$

For the parallel-polarization $\overline{z}(xx)z$, $y = z = 0$, and x is non-zero; For the cross-polarization $\overline{z}(xy)z$, the incident light is polarized along x direction, while the outgoing scattering light is polarized along y direction. Take the Raman tensor of A_{1g} mode into calculation. It shows a nonzero value when it is $\overline{z}(xx)z$ configuration. However, the calculational result is zero when $\overline{z}(xy)z$ configuration is applied. Therefore, the intensity of A_{1g} mode at ~408 cm^{-1} depends on the polarization of excitation light.

Layer-dependent Raman spectra of MoS_2 is displayed in Fig. 2.3d. As thickness increases, A_{1g} mode shifts to higher frequency (blue shift, stiffening), and E^1_{2g} mode

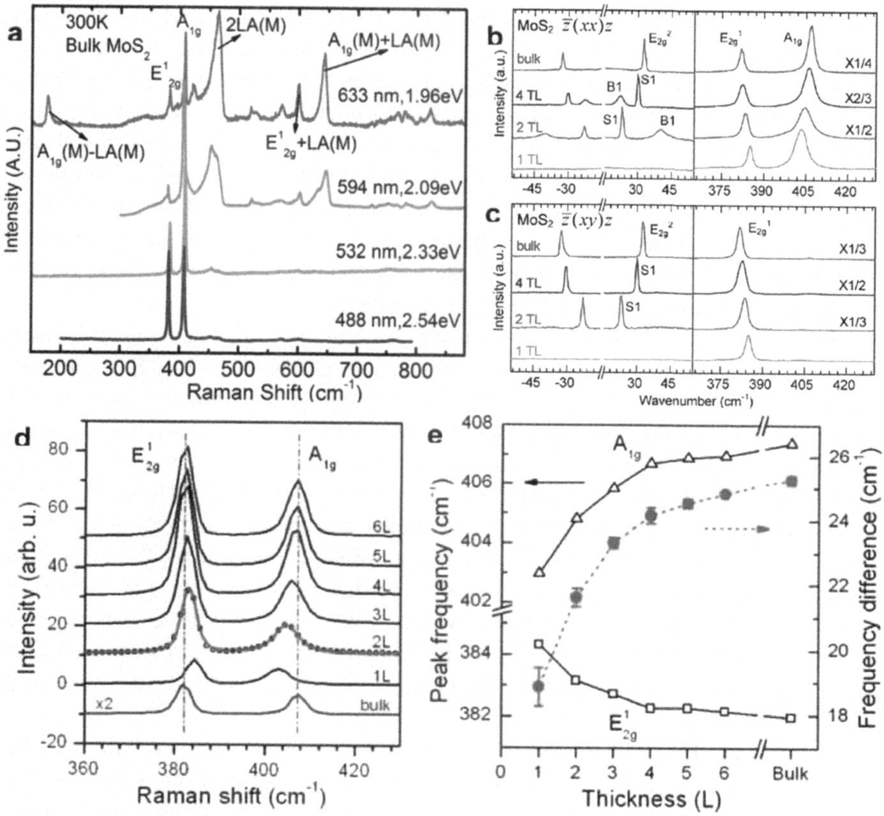

Fig. 2.3 (**a**) Raman spectra of bulk MoS_2 crystal under different excitation wavelengths at room temperature. (Reproduced with permission from Ref. [38], © AIP Publishing LLC 2014). (**b**) Raman spectra of few-layer and bulk MoS_2 films (excitation laser: 514.5 nm). The solid line for the 2 L spectrum is a double Voigt fit through data (circles for 2L, solid lines for the rest). (**c**) Frequencies of E^1_{2g} and A_{1g} Raman modes (left vertical axis) and their difference (right vertical axis) as a function of layer thickness. (Reproduced with permission from Ref. [32], © American Chemical Society 2010)

is shifted towards lower frequency (red shift, softening) [32–35]. The frequency difference is shown in Fig. 2.3e. In general, the difference increases in thicker sample, and it is around 19 cm^{-1} for the mechanically exfoliated monolayer MoS_2 [32]. While MoS_2 is not an exception, similar results have also been found in WS_2, WSe_2, and $MoSe_2$ [18, 36, 37]. The frequency difference between E^1_{2g} and A_{1g} modes provides a novel and rapid way in the identification of layer number in thin film (<6L) TMDs. However, the opposite thickness evolutional trend between E^1_{2g} and A_{1g} modes deserves further investigation and discussion.

The phonon dispersion curves in Fig. 2.2c, d show that the A_{1g} mode is softened away from the Γ point, but the E^1_{2g} mode is dispersionless. Therefore, a change in the frequency of the E^1_{2g} mode cannot be explained by the Richter-Wang-Ley (RWL) theory [39], which was widely used in explaining the size dependence of the asymmetric broadening and frequency shift in nanomaterials [39–42]. In MoS_2, the relative displacements of sulfur atoms from adjacent layers are out-of-phase, as shown in Fig. 2.2b. According to the intuitive harmonic oscillator model, the frequencies of both E^1_{2g} and A_{1g} modes should be blue-shifted when additional layers are added, owing to the larger accumulated restoring force. This effect will be referred to "the thickness effect". The frequency-shift for the A_{1g} mode is consistent with the thickness effect. On the other hand, the opposite frequency trend for the E^1_{2g} mode is not consistent with the thickness effect.

The differences in lattice constant and interatomic distances as thickness decreases might be a possible reason for the anomalous shift of E^1_{2g} mode, but it has been ruled out by Molina-Sanchez et al. [28]. In the Raman community, it was widely believed that the anomalous shift of the E^1_{2g} mode in 2H-MoS_2 is related to a thickness dependence of the dielectric screening, a conclusion based on early calculations on dielectric screening by Molina-Sanchez et al. [28]. However, Lin et al. later showed experimentally that the frequencies of the E^1_{2g} and A_{1g} modes have negligible dependence on the dielectric constant of the environment [43], consistent with Ref. [33], as well as a recent review by Molina-Sanchez et al. [44]. Here we think that it is the effect of larger M-X (M = Mo, W; X = S, Se) force at the surface of the thin film that makes a difference, which contributes to the anomalous red shift of E^1_{2g} mode with increasing thickness. Details on surface effect will be discussed in Chap. 7.

Besides the 2H-stacked MX_2 (M = Mo, W; X = S, Se), it will also be interesting to investigate how stacking order affects the phonon properties. As 1T phase is metastable, most previous studies just focus on the 2H and 3R stacking configurations in MX_2 (M = Mo, W; X = S, Se). It is shown that in MoS_2, $MoSe_2$, and WSe_2, the intalayer vibrational modes do not show observable dependent on stacking orders; while the low frequency interlayer shear modes are very sensitive to the phase difference [18, 45–47]. Similarly, polytypism in the isotropic layered material ReS_2 shows distinguishable spectral features on the low-frequency Raman spectra, but there is not much difference on the intralayer vibrations [48, 49]. Stacking-dependent low-frequency interlayer modes will be further discussed in Chap. 9.

However, $MoTe_2$, different from the S- and Se-based TMDs, demonstrates stacking-dependent intralayer vibrational modes [20, 50], as shown in Fig. 2.4. Take the intralayer vibration A_{1g} mode as an example. It locates at 162–164 cm^{-1} for the 1T' phase, and 170–172 cm^{-1} for the 2H stacking. As a comparison, the high frequency A_{1g} mode has the same energy in 2H- and 3R- stacked MoS_2. It red-shifts by 4 cm^{-1} in the 1T phase [51]. As we have discussed in Fig. 2.1, the monolayer counterparts of 2H and 3R phases are the same, and the difference between the two phases lies in the interlayer stacking and rotation, which hardly affects the intralayer A_{1g} mode. Compare 1T and 2H monolayers; we can see that one of the S atoms has 180° rotation (Fig. 2.1a), which would certainly influence the Mo-S force and shift

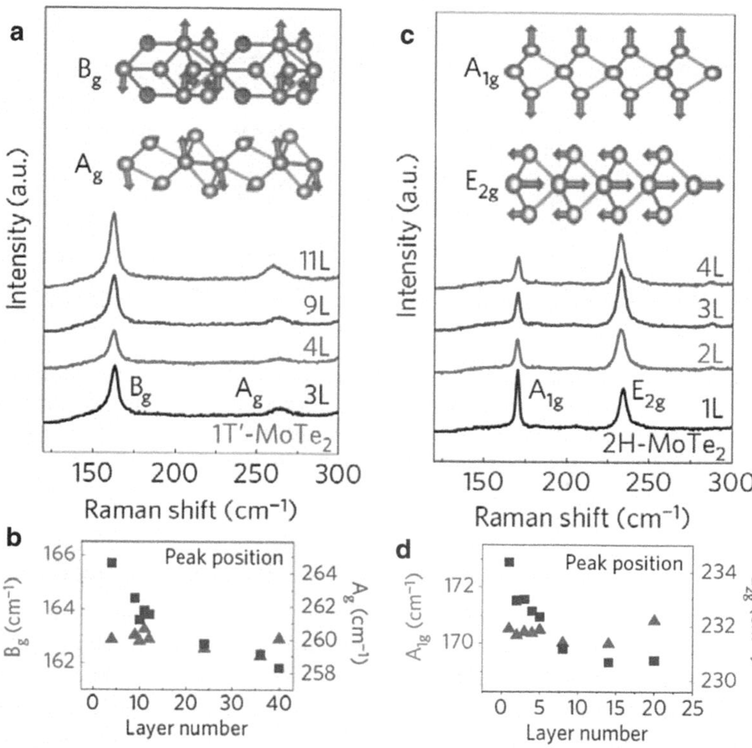

Fig. 2.4 (**a**) Layer-dependent Raman signals for 1T'-$MoTe_2$. The normal displacements of vibrational modes are illustrated in the insets. (**b**) Peak positions in 1T'-$MoTe_2$ with various numbers of layers. (**c**) Layer-dependent Raman signals for 2H-$MoTe_2$. The normal displacements of vibrational modes are illustrated in the insets. (**d**) Peak positions in 2H-$MoTe_2$ with various numbers of layers. (Reproduced with permission from Ref. [20], © Nature Publishing Group 2015)

the energy of A_{1g} mode. Besides having one chalcogenide atom rotated, 1T' $MoTe_2$ possesses a distorted structure. The combined effects render the energy difference between 2H and 1T' phases to reach almost 10 cm^{-1} in A_{1g} mode.

In addition to thickness and stacking, the phonon properties of TMDs can also be influenced by the excitation laser [52, 53], strain [54], doping [55], and temperature [56–59]. Temperature-dependent Raman scattering can be used to estimate the thermal conductivity in MoS_2, thanks to the strong Raman intensities of E^1_{2g} and A_{1g} modes [56–59]. Similarly, Raman spectroscopy has been used to measure the thermal conductivity in graphene [60], while it is not suitable for BN due to the weak Raman signal.

A moderate power of laser is indispensable for the accurate estimation of thermal conductivity, as laser spot can act as a heat source and cause local heating [56–59, 61]. The measurements on monolayer, few-layer, and bulk MoS_2 have been conducted by different groups [56–59], using both suspended and (silicon-, sapphire-, glass-) supported samples. Most groups observed linear softening of

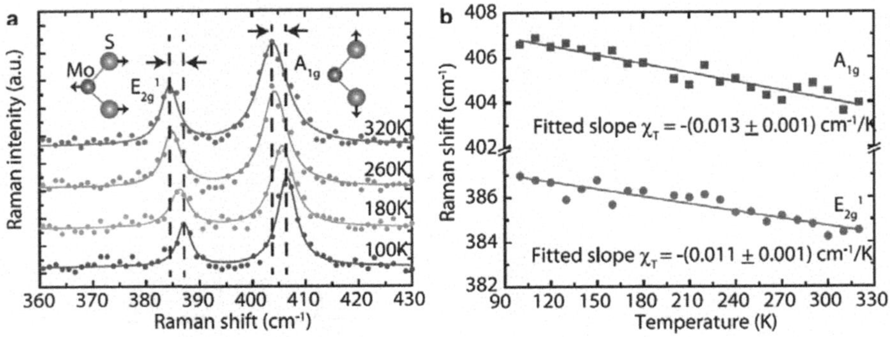

Fig. 2.5 (**a**) Four example Raman spectra of suspended, monolayer MoS_2 collected at 100, 180, 260, and 320 K. Spectra offset vertically for clarity. (**b**) Raman peak frequencies of both A_{1g} (squares) and E_{2g}^{1} (circles) modes as a function of temperature. Fit lines and resulting linear temperature coefficients χ_T are shown. (Reproduced with permission from Ref. [56], © American Chemical Society 2014)

E_{2g}^{1} and A_{1g} modes as temperature increases, and the change of frequencies are within (1.1–1.7) $\times$ 10^{-2} cm^{-1} K^{-1} regardless of thickness and substrate. The temperature-dependent Raman shift is mainly attributed to the thermal expansion (volume contribution) and the anharmonic contributions to the ionic interaction potential in the two-dimensional system. It is worth noting that the expansion of (substrate-supported) MoS_2 lattice due to increased temperature sometimes is hindered by the cooler substrate. Strain would appear and influence the in-plane E_{2g}^{1} mode, while A_{1g} mode is less sensitive to strain [54, 62, 63]. Thus it will be more accurate to use A_{1g} mode to obtain thermal conductivity. Figure 2.5 shows an example of temperature-dependent Raman shift in monolayer MoS_2. In general, nonlinear coefficients are not considered, given that they are only supposed to impact at higher temperatures [57, 60]. However, two groups have observed the nonlinear temperature-dependent Raman shift at low temperature (<100 K) from MoS_2 [59] and WS_2 [64]. Taube et al. believe the nonlinearity is related with the optical phonon decay and lattice potential anharmonicity in the monolayer samples [59]. On the other hand, Yan et al. observed the nonlinearity from power-dependent Raman scattering measurements at room temperature. They consider that the appearance of nonlinear effects results either from the nonlinearity of absorption or higher orders of the temperature-dependent coefficients [56]. Combing the linear temperature (Fig. 2.5) and power-dependent coefficients of the A_{1g} mode, Yan et al. extracted the thermal conductivity $\kappa = (34.5 \pm 4)$ W mK^{-2} from suspended monolayer MoS_2 [56], which is in accordance with theoretical prediction [65].

The new class of quantum matter topological insulators (TIs) have been in the research highlight during the past decade. The bulk phases of TIs are ordinary insulators but their surface states are gapless with nondegenerate spins [66–68]. Many of the TIs are layered materials, including bismuth chalcogenides Bi_2X_3 (X = Se, Te) and 2D HgTe/CdTe quantum wells. Among all, Bi_2Te_3, Bi_2Se_3, and

Sb_2Te_3 are the most widely studied [69]. Here we will focus on Bi_2X_3 (X = Se, Te) in this chapter.

Compared to TMDs, graphene, and phosphorene which are relatively simple in terms of crystal structure, the rhombohedral Bi_2X_3 (X = Se, Te) has five atomic planes within a single layer. Thus, one layer Bi_2X_3 (X = Se, Te) can also be called one quintuple layer (1 QL) [69, 70]. Here we take Bi_2Se_3 as an example. In fact, Bi_2Se_3 has been known for its large thermoelectric effect before the topological surface states have attracted lots of interests [71]. Layered Bi_2Se_3 (with crystal symmetry $R\bar{3}m$) has a rhombohedral crystal structure belonging to the space group D_{3d}^5. The unit cell consists of five atomic layers, Se-Bi-Se-Bi-Se. The hexagonal close-packed Se-Bi-Se-Bi-Se atomic layers periodically arranged along the c axis, as shown in Fig. 2.6a [72]. Similar to other layered materials, the weakly bound van der Waals forces connect different QLs, but the van der Waals interaction in Bi_2Se_3 has a slightly covalent nature.

There are 15 dynamical modes at the center of Brillouin zone, among which 12 are the optical modes and other 3 are the acoustic modes. The irreducible representations of the zone center phonons in bulk Bi_2Se_3 is $\Gamma = 3E_u + 3A_{2u} + 2E_g + 2A_{1g}$. A_{1g} and E_g are Raman-active, while E_u and A_{2u} are Raman-inactive but infrared-active. As thickness reduces from bulk to few- and mono-layer, the irreducible representations in odd and even number of layer are $\Gamma = \frac{5N-1}{2}\left(A_{1g} + E_g\right) + \frac{5N+1}{2}\left(A_{2u} + E_u\right)$ $(N = 1, 3, 5\ldots)$ and $\Gamma = \frac{5N}{2}\left(A_{1g} + A_{2u} + E_g + E_u\right)$ $(N = 2, 4, 6\ldots)$, respectively. Phonon dispersion curves of Bi_2Se_3 along the Γ–Z (where Z corresponds to (0, 0, 3π/c) in reciprocal space) direction of the Brillouin zone is shown in Fig. 2.5b. Only E_g^2 mode exhibits a dispersionless trend among the four Raman-active modes. A_{1g}^1, A_{1g}^2, and E_g^1 all red-shift from Γ to Z. The corresponding vibration patterns of the Raman-active modes are displayed in Fig. 2.6c. According to the outlined normal modes, we remark that the A_{1g}^1 mode at ~72 cm^{-1} is supposed to very sensitive to thickness because it reflects the out-of-plane vibrations of the Se and Bi atoms, and the interlayer van der Waals interactions influence the effective restoring forces acting on these atoms. Detailed discussion will be presented later in this chapter.

Figure 2.7a displays the Stokes and anti-Stokes Raman spectra taken from few-QL and bulk Bi_2Se_3 crystals. A_{1g}^1, A_{1g}^2, and E_g^1 modes red-shift with decreasing thickness, while there is almost no change in the energy of E_g^2 mode. As a further comparison, zoom-in views of A_{1g}^1 and E_g^2 modes are shown in Fig. 2.7b, c, respectively. The slight red-shift in Fig. 2.7c is probably due to systematic error. In contrast, the red-shift is more pronounced in the A_{1g}^1 mode (Fig. 2.7b). This observation is in good agreement with the phonon dispersion curve shown in Fig. 2.6b, and it also suggests that the phonon confinement effect may take a critical role in determining the frequency of phonon modes when thickness decreases.

The phonon confinement effect is a type of quantum size effect that occurs when the phonon wave function is confined into a small volume with one or more dimensions of the same magnitude as its wavelength. In Raman spectroscopy, the direct sign of this effect is an asymmetrically broadened Raman peak which shifts

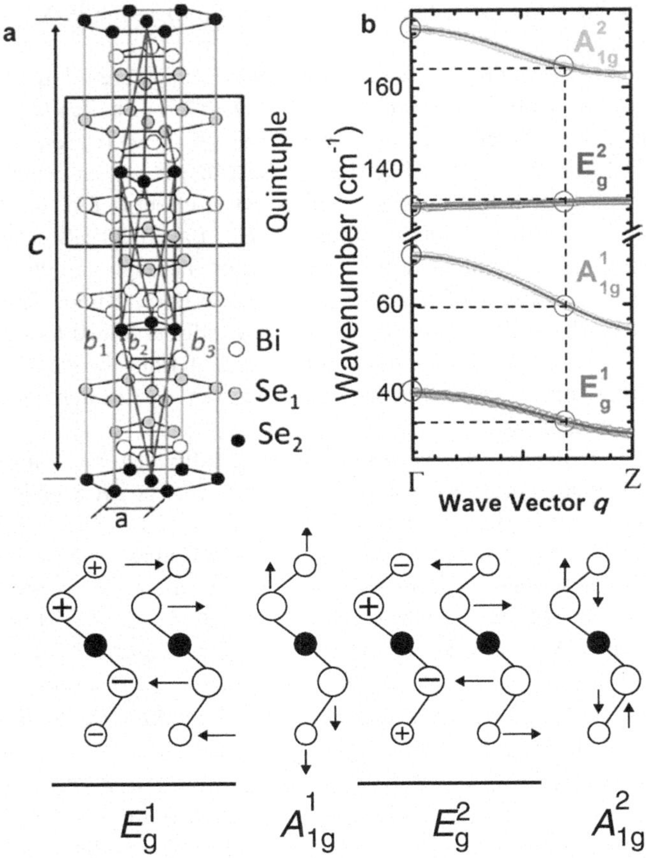

Fig. 2.6 (**a**) Crystalline structure ($R\overline{3}m$) of Bi_2Se_3. The lattice vectors and the quintuple layer of the rhombohedral unit cell are indicated. (**b**) Phonon dispersion curves of Bi_2Se_3 along the Γ−Z direction of the Brillion zone. (**c**) The vibrational normal modes (only Raman active) of a quintuple layer Bi_2Se_3. (Reproduced with permission from Ref. [72], © American Chemical Society 2011)

to lower frequency in nanomaterials. In 1981, Richter et al. proposed a model to explain this effect in microcrystalline silicon (RWL theory) [39]. Later, the RWL theory was applied to explain the nanowire diameter dependence of the asymmetric broadening and frequency shift of the one phonon band in Si nanowires observed at ∼520 cm^{-1} [40–42]. Confinement effect on phonons has also been observed in nanoparticles [73, 74].

The RWL theory [39] can be understood from the Heisenberg uncertainty principle. Take Bi_2Se_3 as an example. Following the relation $\Delta p \cdot \Delta x \geq \hbar/2$ with $\Delta p = \hbar \cdot \Delta k$ and Δx is equal to the crystal size or thickness Δd. The uncertainty of the phonon momentum (Δp) increases with decreased thickness Δd. It is an indication that phonon modes with finite non-zero momentum can be involved in the Raman scattering for small delta x. Based on this principle, we can obtain the

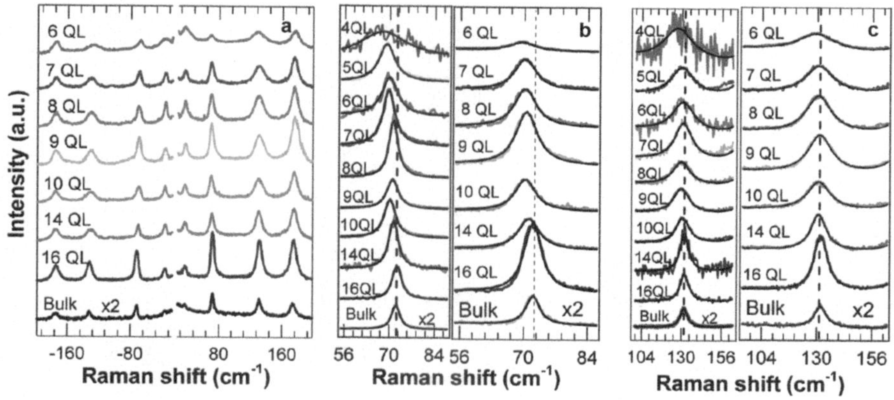

Fig. 2.7 (**a**) Stokes and anti-Stokes Raman shifts of Bi_2Se_3 nanoplatelets for different thicknesses excited by a 632.8 nm laser. (**b** and **c**) Zoom-in view of the A^1_{1g} (**b**) and E^2_g (**c**) modes for various thicknesses of QLs excited by 532 nm (left panel) and 632.8 nm (right panel) lasers. (Reproduced with permission from Ref. [72], © American Chemical Society 2011)

frequency shift Δω once we know the corresponding phonon dispersion $\omega(k)$. This explains why the frequency of E^2_g mode is insensitive to thickness, while the A^1_{1g} mode softens with decreasing thickness. The phonon confinement effect has also been suggested to explain the shifts of A^1_{1g} and A^2_{1g} modes in Bi_2Te_3 [75].

We also note that the A^1_{1g} mode of Bi_2Se_3 exhibits distinct asymmetry towards higher frequency, which is different from that described in the RWL theory. Such asymmetry towards the higher frequency has been well captured by the Breit-Wigner-Fano (BWF) lineshape as investigated by Zhang et al. in wet chemistry synthesized Bi_2Se_3 nanoplates [72]. The BWF Raman lineshape originates from the quantum interference between a discrete phonon state and a continuum of electronic states. The constructive or destructive interferences at different frequencies result in the frequency and lineshape renormalization of a Lorentzian Raman peak:

$$I(\omega) = A \cdot \frac{[q + 2(\omega - \omega_0)/W]^2}{1 + [4(\omega - \omega_0)/W]^2} \tag{2.1}$$

where A, ω_0, W, and q are the peak intensity, frequency, linewidth, and asymmetry parameter, respectively. $|1/q|$ is often used to depict the strength of electron-phonon coupling, and when $|1/q|$ approaches zero, the above equation is reduced to a Lorentzian lineshape.

The asymmetry is more distinct in thinner nanoplates, as shown by the clear contrast between 2 QL and 7 QL (Fig. 2.8). The thickness-dependent asymmetry is shown in Fig. 2.8c, where the electron phonon coupling strength $1/q$ monotonically increases with decreasing thickness [76]. The detailed information on the continuum states that are coupled with the A^1_{1g} mode and the thickness-dependent coupling strength are unclear yet and require further investigations. Similar reports on

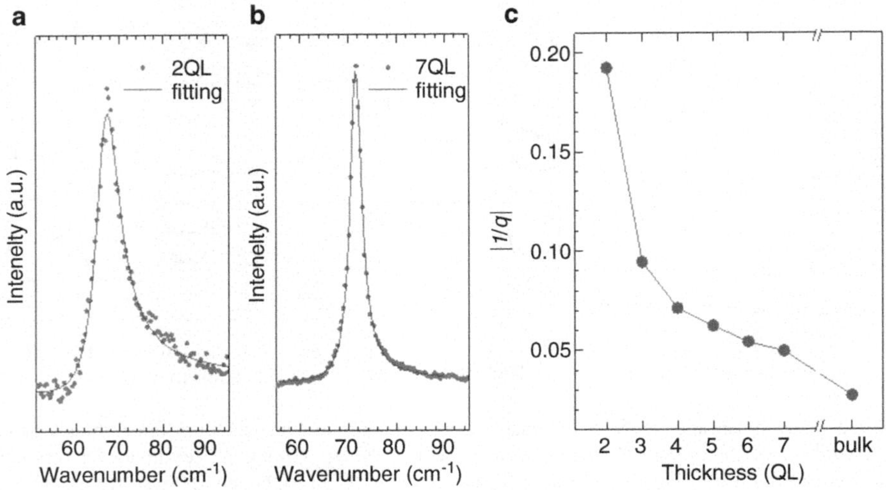

Fig. 2.8 (**a** and **b**) Fano fitting in 2 QL (**a**) and 7 QL (b) Bi_2Se_3. (**c**) The asymmetric parameter |*1/q*| versus crystal thickness. (Reproduced with permission from Ref [14], © Tsinghua University Press and Springer-Verlag Berlin Heidelberg 2016)

the strong electron-phonon interactions in Bi_2Se_3 have also been found through other methods such as infrared reflection and transmission [77, 78], angle-resolved photoemission spectroscopy [79], and electron transport measurements [80].

It will also be interesting to explore the alloy layered materials on Raman spectroscopy. Figure 2.9 shows the Raman spectra of alloy nanoplates $Bi_2Te_{3-x}Se_x$ (different anion) and $Bi_xSb_{2-x}Te_3$ (different cation). Figure 2.8b displays the extracted phonon frequencies versus Se composition for $Bi_2Te_{3-x}Se_x$. In general, the phonon modes should blue-shift with increasing concentration of Se, as Se is lighter than Te. Compared to A^2_{1g} and E^1_g modes which locate at higher frequency, the lower energy peaks A^1_{1g} and E^2_g are less sensitive to Se doping because both modes arises due to the in-phase vibrations of Bi–Te/Se. From the schematic in Fig. 2.6c, it is clear that the central Te/Se atoms are the center of mass of the Raman active vibrations. Consequently, for small doping of Se atoms, which preferentially replaces the central Te atoms, only a slight shifting in the Raman modes is anticipated for $Bi_2Te_{2.9}Se_{0.1}$, $Bi_2Te_{2.84}Se_{0.16}$, and $Bi_2Te_{2.7}Se_{0.3}$ [71]. Likewise, A^2_{1g} mode is the most sensitive to cation (Bi) doping, since A^2_{1g} mode involves the Bi/Sb-Te out-of-phase vibrations along the *c* axis.

There also are plenty of studies on phonon modes in TMD alloy [82–89], with samples including either cation or anion doping such as $Mo_{1-x}W_xSe_2$ and $MoS_{2(1-x)}Se_{2x}$. Figure 2.10 demonstrates an example on monolayer and bilayer $Mo_{1-x}W_xSe_2$ alloy. As the Tungsten (W) composition x increases from 0 to 1, the $MoSe_2$-like E^1_{2g} shifts to lower frequency (E^1_{2g} mode locates at 287 cm^{-1} and ~250 cm^{-1} in monolayer $MoSe_2$ and WSe_2, respectively [90, 91]) and the A_{1g} mode continuously shifts from 240.1 cm^{-1} to 249.3 cm^{-1} in the monolayer

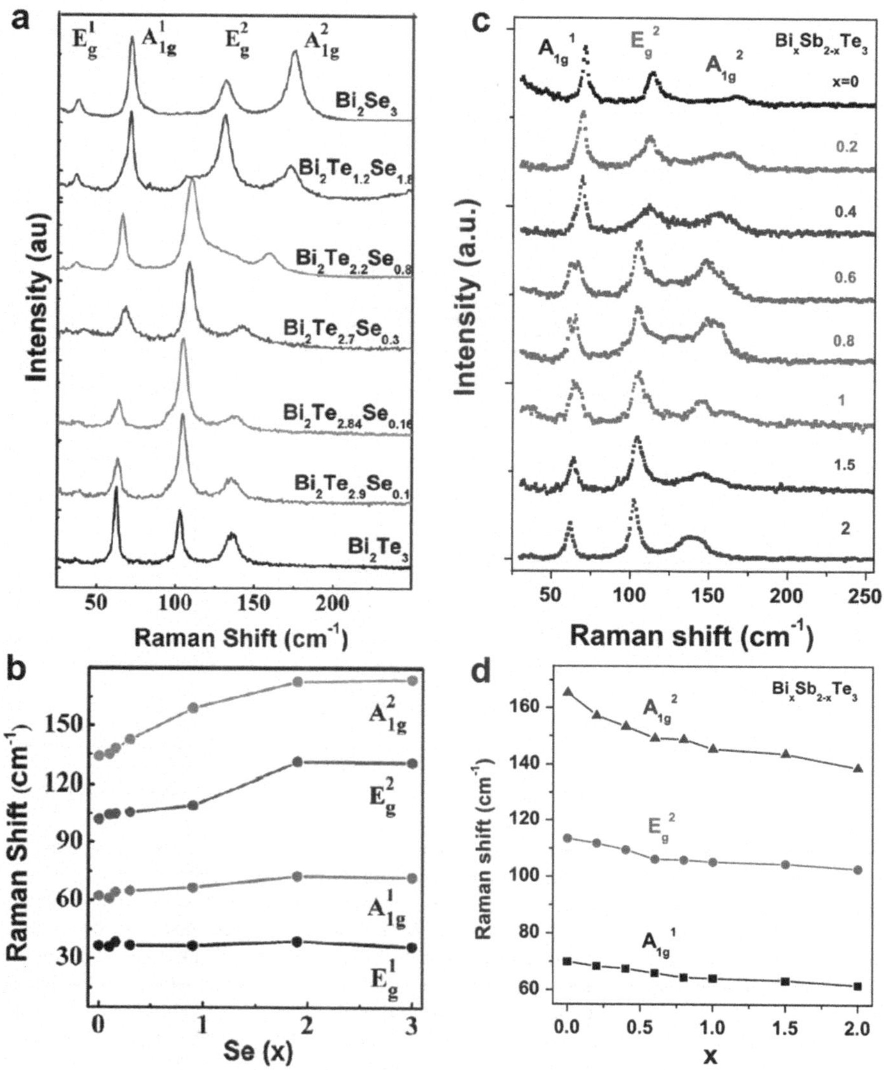

Fig. 2.9 (**a**) Raman spectra of pure and doped $Bi_2Te_{3-x}Se_x$ nanoplates. (**b**) Phonon frequencies versus Se composition. (Reproduced with permission from Ref. [71], © American Chemical Society 2012). (**c**) Raman spectra of $Bi_xSb_{2-x}Te_3$ nanoplatelets with different Bi composition. The nanoplatelets are deposited on SiO_2/Si substrate. (**d**) Phonon frequencies plotted versus the composition parameter x of $Bi_xSb_{2-x}Te_3$. (Reproduced with permission from Ref. [81], © ScienceDirect 2015)

alloy (A_{1g} mode locates at 241 cm^{-1} and ~250 cm^{-1} in monolayer $MoSe_2$ and WSe_2, respectively [90, 91]), shown in Fig. 2.10a, b. In addition, the second order Raman peaks also shift and showed two-mode behaviors as the composition of

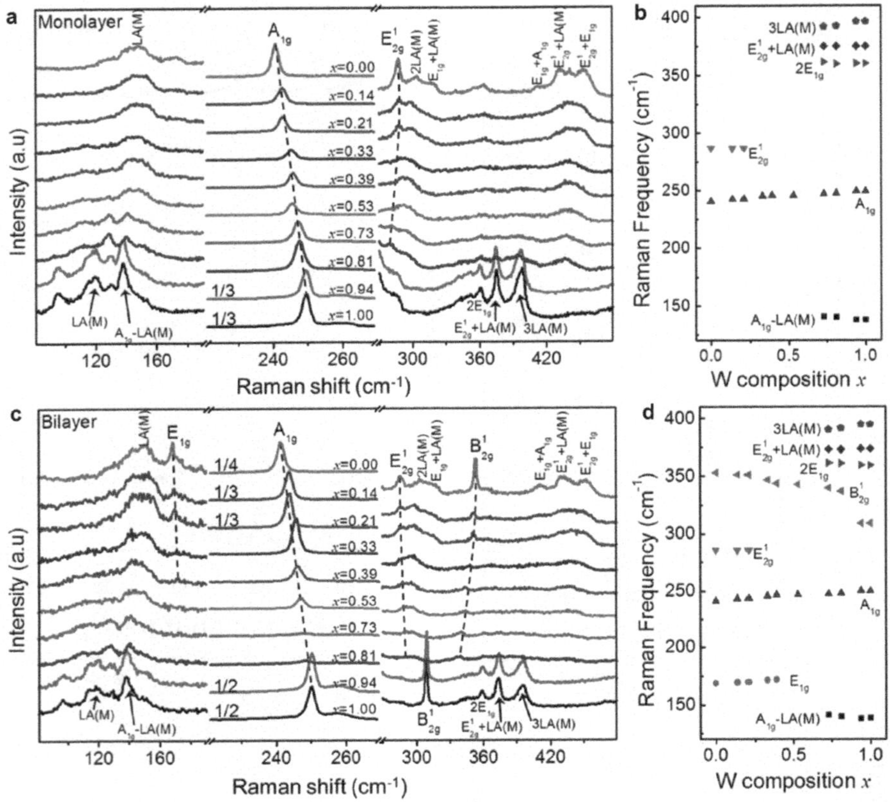

Fig. 2.10 (**a**) Raman spectra of $Mo_{1-x}W_xSe_2$ monolayers with different W composition x in the range 80–480 cm^{-1}. (**b**) Composition-dependent Raman frequencies of $Mo_{1-x}W_xSe_2$ monolayer alloys. (**c**) Raman spectra of $Mo_{1-x}W_xSe_2$ bilayers with different W composition x. (**d**) Composition-dependent Raman frequencies of $Mo_{1-x}W_xSe_2$ bilayer alloys. All the spectra are calibrated with the 520 cm^{-1} Raman peak from the Si substrate. The black dashed lines in (**a**) and (**c**), guides for the eye, show the frequency shift of the first-order Raman modes with different W compositions x. For clarity, the intensity of spectra in the range 220–270 cm^{-1} is magnified by a factor labelled on the corresponding spectra. The excitation laser wavelength was 514.5 nm. (Reproduced with permission from Ref. [82], © American Chemical Society 2015)

W changes; that is, there are two branches of $E^1_{2g}+LA(M)$ mode: one branch is related to WSe_2 (at ~373.6 cm^{-1} in WSe_2) and the other branch is from $MoSe_2$ (at~432.7 cm^{-1} in $MoSe_2$). Similar phenomena can be observed in the bilayer alloy as well (Fig. 2.10c, d). The B^1_{2g} mode, which is optical inactive in bulk and Raman inactive in monolayer (with symmetry of A''_2), is Raman active in bilayer with symmetry of A_{1g} due to lower symmetry in few-layers [91]. When the W composition increases, the $MoSe_2$-like B^1_{2g} mode, which can be observed for x in the range 0–0.81, shifts to lower frequency. While the WSe_2-like B^1_{2g} mode can be observed only for the last two composition alloys ($x = 0.94$ and 1.00). The B_{2g}^{1} mode is at 352 cm^{-1} and 311 cm^{-1} in bilayer $MoSe_2$ and WSe_2, respectively [90, 91].

The modified random element isodisplacement (MREI) model [89]can be used to fit the composition-dependent Raman frequencies of the A_{1g} mode for $Mo_{1-x}W_xSe_2$ monolayers and bilayers. The composition-dependent A_{1g} frequency, analysed using the MREI model, can be expressed as:

$$\omega_{A_{1g}} = \left[\frac{(1-x)\,F_1 + F_2}{m_{Se}}\right]^{1/2} = \left[\frac{(1-x)\,(1-\theta x)\,F_{10} + x\,(1-\theta x)\,F_{20}}{m_{Se}}\right]^{1/2} \tag{2.2}$$

Where m_{Se} is the atomic mass of Se; F_1 and F_2 are the force constants of Mo-Se and W-Se interactions, respectively; F_{10} and F_{20} are limiting values of F_1 and F_2 as x is reduced to 0; and θ is a parameter that is associated with the variation of force constant in the alloys. The fitting result gives a more accurate composition-dependent frequency shift of the A_{1g} mode [82],

$$\omega_{A_{1g}}^{1L} = 240.1\left(1 + 0.057x - 0.023x^2\right) \tag{2.3}$$

$$\omega_{A_{1g}}^{2L} = 241.0\left(1 + 0.052x - 0.018x^2\right) \tag{2.4}$$

Besides the group VI TMDs and Bi_2X_3 (X = Se, Te), there are other isotropic layered materials attracting research focus, such as niobium diselenide ($NbSe_2$) and hexagonal boron nitride (h-BN). $NbSe_2$ is one of the most studied van der Waals layered materials that exhibit both superconductivity (SC) and charge density wave (CDW) at low temperatures [92, 93]. Raman studies on $NbSe_2$ [93–97] mostly focus on the two low-energy broad peaks. One of them is at ultralow frequency (<50 cm^{-1}). As temperature decreases from 8 K to 2 K, the spectral weight from the broad CDW amplitude mode (40 cm^{-1} at 2 K) evolves to the sharp SC mode (19 cm^{-1} at 2 K). In the meantime, The SC mode blue-shifts with decreasing temperature whereas the energy of the CDW mode stays almost constant in parallel polarization and slightly shifts to lower energy (~3 cm^{-1}) in cross polarization. Crucially, the SC and CDW modes develop in an opposite way: When superconductivity is gradually destroyed, the SC mode intensity collapses while the CDW mode intensity recovers [97]. The CDW amplitude mode is found to be layer-sensitive: It blue-shifts by >10 cm^{-1} from bulk to monolayer, suggesting the increasing electron–phonon coupling strength with decreasing thickness. This finding probably provides the major reason for stronger CDW in atomically thin $NbSe_2$ [93].

The other broad peak is centred at 180 cm^{-1} with the full width at half-maximum of 50 cm^{-1} at room temperature, which is much bigger than that of the other phonons (Fig. 2.11a). This peak at 180 cm^{-1} is called the soft mode, which just appears above the CDW phase transition temperature. Below the transition

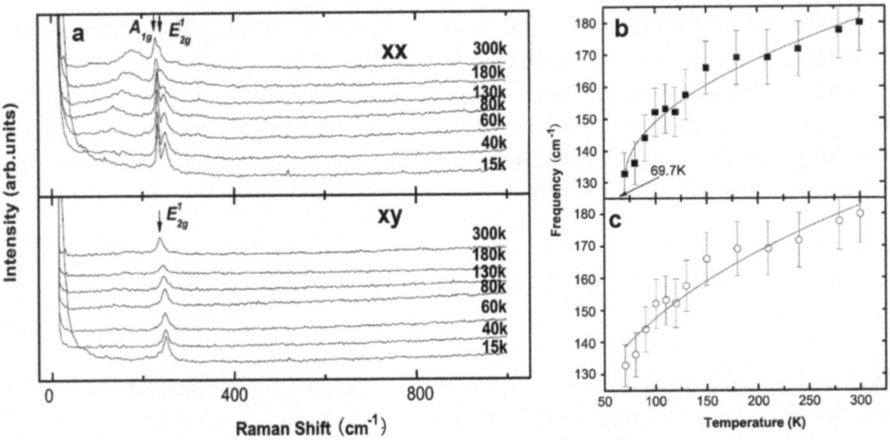

Fig. 2.11 (**a**) Raman spectra of the 2H–$NbSe_2$ single crystal measured from 15 K to 300 K with *xx* and *xy* polarization. The spectra have a systematic offset to arrange one above another to give a legible comparison at different temperatures. The peaks are marked by an arrow in the spectra measured at 300 K. (**b** and **c**) Temperature dependence of the soft mode. The solid line in (**b**) corresponds to Eq. (2.5) and the solid line in (**b**) corresponds to Eq. (2.6). The error bar = 5%. (Reproduced with permission from Ref. [94], © IOP PUBLISHING 2008)

temperature, the soften mode is frozen [93]. It is considered that the soften mode is not a simple one-phonon mode. As there is no magnetic order in $NbSe_2$, it rules out the possibility of magnetic origin for the peak [94]. Tsang et al. suggested that the broad peak may be associated with the scattering by two branches of LA soft phonons with opposite momentum [95].

Raman spectroscopy measurements show that the temperature-dependent Raman shift of the soft mode is different from other phonon modes. The A_{1g} and E_{2g}^1 modes stiffens with decreasing temperature, while the soft mode softens, as shown in Fig. 2.11b, c. Based on mean field theory, the energy change in the second-order soft mode can be described as [98]:

$$\Delta\omega = B\omega_0|T - T_C|^{1/2} \tag{2.5}$$

where B is related to the electron-phonon coupling constant, ω_0 is the high-temperature phonon frequency and T_c is the CDW transition temperature. However, the obtained T_c of 69.7 K from fitting is much higher than the measured CDW transition temperature of 33.5 K. Generally, the two-phonon Raman process is associated with scattering by two phonons from the opposite sides of the BZ, and the k-space vectors' relationship of the phonons is $\overrightarrow{k_1} + \overrightarrow{k_2} \approx 0$. Under the constraint of momentum conservation, it is reasonable to assume that the soft mode is contributed by a soft phonon mode plus a normal one. As reported by neutron diffraction studies [99], the soft mode can be considered as a weak first order phonon. Then the energy of the soft mode can be described as [94]:

$$\Delta\omega(T) = \omega_0 - A'\left(\frac{2}{e^{\hbar\omega_0/2k_BT} - 1}\right) + B'\omega_0'|T - T_C|^{1/2} \tag{2.6}$$

where ω_0 is the low-temperature phonon frequency of the normal mode and ω_0' is the high-temperature phonon frequency of the soft mode. The experimental results fit well with the calculation. When the CDW transition temperature of the sample is 33.5 K, one can get the ω_0 of the normal phonon to be 120 cm^{-1}. This mode is not scattered from the centre of the Brillouin zone and cannot be detected from normal Raman scattering [94].

The emerging possibilities to use BN as an ultra-thin insulator separating graphene layers, a spacer layer between TMDs and metal film (to prevent quenched photoluminescence), and an intermediate layer between SiO_2/Si and the TMD material (to observe intrinsic linewidth on spectroscopy) have drawn lots of interests to the mono- and few-layer BN insulator. Figure 2.11 shows the Raman spectra taken from atomically thin and bulk BN. BN exhibits a characteristic peak at ~1366 cm^{-1} that is due to the vibration of E_{2g} phonon mode. This peak is analogous to the G peak in graphene [100, 102]. As shown in Fig. 2.11a, the peak at ~1366 cm^{-1} becomes weaker as N decreases. In monolayer BN, the intensity of this peak is ~50 times smaller compared to the G peak in graphene under the same measurement conditions. The integrated intensity for the BN peak is proportional to N with high accuracy from 1 L to 6 L (inset in Fig. 2.11a), which can be exploited to distinguish one, two and more BN layers. However, Fig. 2.12b shows that mono- and bilayers exhibit unexpectedly strong variations in peak position, whereas these are essentially absent for crystals thicker than 5L. The maximum blue shift of monolayers (~4 cm^{-1}) is in agreement with the theory [100], and the typical shift is between 2 cm to 4 cm^{-1}. Gorbachev et al. suggests that it is strain that causes the sample variation on Raman shift.

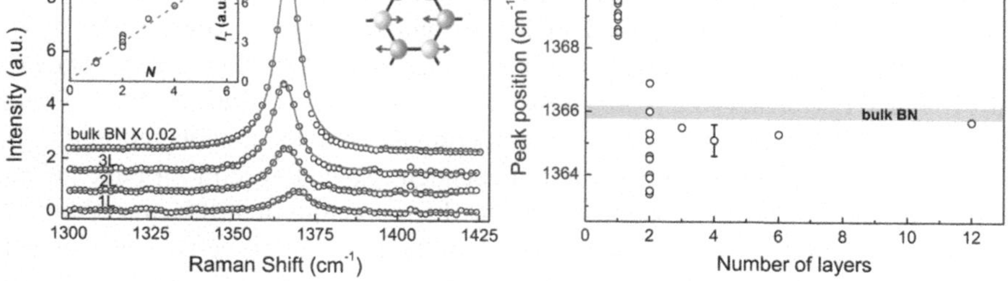

Fig. 2.12 (**a**) Raman spectra of atomically thin BN. The left inset show changes in integrated intensity with the number of layers N. The right picture illustrates the phonon mode responsible for the Raman peak. (**b**) Position of the Raman peak for different values of N. In mono- and bilayer BN, the peak position is sample-dependent and varies by as much as ±2 cm^{-1}. The dashed line is the Raman shift predicted for monolayer BN, compared to the bulk value (horizontal bar) [100]. The error bar indicates the typical accuracy of determining the peak position. (Reproduced with permission from Ref. [101], © John Wiley & Sons, Inc. 2011)

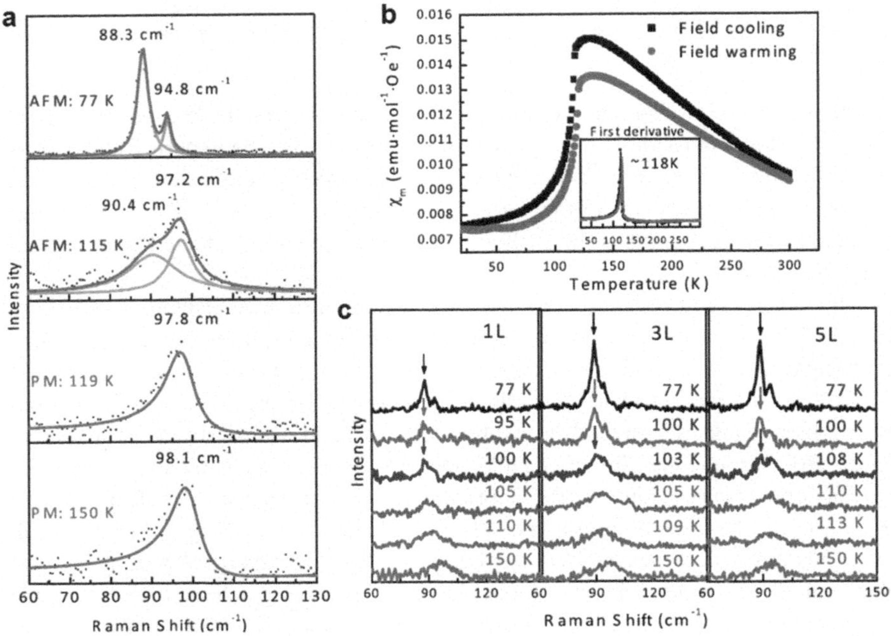

Fig. 2.13 (**a**) Temperature-dependent Raman spectra of bulk $FePS_3$ sample obtained with an excitation wavelength of 532 nm and power~0.5 mW around Néel temperature. The line shapes of fitting curves are Fano asymmetric and two-Lorentzian peaks above and below Néel temperature, respectively. (**b**) Plot of magnetic susceptibility of $FePS_3$ bulk crystal versus temperature measured in (filled squares) cooling and (filled circles) heating process. The first derivation of susceptibility was shown in the inset whose inflection points are both around 118 K. (**c**) Temperature-dependent Raman spectra of mono-, tri- and five-layer $FePS_3$ obtained with an excitation wavelength of 532 nm and power~0.5 mW. (Reproduced with permission from Ref. [105], © IOP Publishing 2016)

Recently, 2D magnetic layered materials have attracted lots of interests. Ferromagnetism has been discovered in mono- and few-layer samples [103, 104]. Meanwhile, through the observation of spin-order-induced Raman modes, Wang et al. [105] and Lee et al. [106] reported that iron phosphorus trisulfide ($FePS_3$) crystal exhibits an Ising-type antiferromagnetic ordering down to the monolayer limit.

Figure 2.13a shows the temperature-dependent Raman spectra on bulk $FePS_3$ crystal around the Néel temperature. The Néel temperature of 118 K was obtained from the magnetic susceptibility carried out in the physical property measurement system (Fig. 2.13b). At 77 K, two strong and symmetric peaks located at 88 and 95 cm^{-1} were resolved in the antiferromagnetic (AFM) phase whereas when the temperature was increased to 150 K, the two peaks disappeared accompanied by the appearance of a broad and asymmetric peak at around 100 cm^{-1} in the paramagnetic (PM) phase. First-principle calculations did not show any Raman-active modes in the frequency range of 0–130 cm^{-1} at the Γ point, which suggests that the peaks (88, 95, and 100 cm^{-1}) are spin-dependent. Temperature-dependent measurements show that the spin-order-induced modes at 88 and 95 cm^{-1} are still robust down to monolayer (Fig. 2.13c). This result confirms that the spin-order-induced Raman modes are attributed to the Brillouin-zone folding in the *ab* plane [107], because monolayer $FePS_3$ only has a layer of iron atoms, which is impossible to have double-sized magnetic cell along the *c* axis. The systematic layer-dependent and temperature-dependent measurements show that the Néel temperature decreases from 118 K in bulk to 104 K in monolayers.

In conclusion, with the focus on isotropic two-dimensional layered materials, we have reviewed the phonon modes in TMDs MX_2 (M = Mo, W; X = S, Te, Te), Bi_2X_3 (X = Se, Te), alloy, $NbSe_2$, h-BN, and magnetic $FePS_3$ crystals in this chapter. Dependence on thickness and the mechanism which leads to the thickness-evolutional trend have been discussed in detail. While temperature dependence, composition (doping) dependence, and the soft mode which is related to CDW have also been presented in this chapter.

Acknowledgments This work is supported by National Basic Research Program of China (2017YFA0303401), Strategic Priority Research Program of Chinese Academy of Sciences (Grant No. XDB28000000) LU JIAXI International team program. J. Z. also acknowledges support from National Young 1000 Talent Plan of China.

References

1. Q.H. Wang, K. Kalantar-Zadeh, A. Kis, J.N. Coleman, M.S. Strano, Electronics and optoelectronics of two-dimensional transition metal dichalcogenides. Nat. Nanotech. **7**, 699–712 (2012)
2. M. Chhowalla, H.S. Shin, G. Eda, L.J. Li, K.P. Loh, H.Z. Zhang, The chemistry of two-dimensional layered transition metal dichalcogenide nanosheets. Nat. Chem. **5**, 263–275 (2013)
3. A.K. Geim, I.V. Van der Grigorieva, Waals heterostructures. Nature **499**, 419–425 (2013)
4. K.S. Novoselov, D. Jiang, F. Schedin, T.J. Booth, V.V. Khotkevich, S.V. Morozov, A.K. Geim, Two-dimensional atomic crystals. Proc. Natl. Acad. Sci. U. S. A. **102**, 10451–10453 (2005)
5. X. Xu, W. Yao, D. Xiao, T.F. Heinz, Spin and pseudospins in layered transition metal dichalcogenides. Nat. Phys. **10**, 343–350 (2014)
6. K.F. Mak, K. He, J. Shan, T.F. Heinz, Control of valley polarization in monolayer MoS_2 by optical helicity. Nat. Nanotech. **7**, 494–498 (2012)
7. H. Zeng, J. Dai, W. Yao, D. Xiao, X. Cui, Valley polarization in MoS_2 monolayers by optical pumping. Nat. Nanotech. **7**, 490–493 (2012)
8. T. Cao, G. Wang, W. Han, H. Ye, C. Zhu, J. Shi, Q. Niu, P. Tan, E. Wang, B. Liu, J. Feng, Valley-selective circular dichroism of monolayer molybdenum disulphide. Nat. Commun. **3**, 887 (2012)
9. K.F. Mak, K.L. McGill, J. Park, P.L. McEuen, The valley hall effect in MoS_2 transistors. Science **344**, 1489–1492 (2014)
10. S.F. Wu, J.S. Ross, G.B. Liu, G. Aivazian, A. Jones, Z.Y. Fei, W.G. Zhu, D. Xiao, W. Yao, D. Cobden, X.D. Xu, Electrical tuning of valley magnetic moment through symmetry control in bilayer MoS_2. Nat. Phys. **9**, 149–153 (2013)
11. J. Lee, K.F. Mak, J. Shan, Electrical control of the valley hall effect in bilayer MoS_2 transistors. Nat. Nanotech. **11**, 421–425 (2016)

12. K.F. Mak, C. Lee, J. Hone, J. Shan, T.F. Heinz, Atomically thin MoS_2: a new direct-gap semiconductor. Phys. Rev. Lett. **105**, 136805 (2010)
13. A. Splendiani, L. Sun, Y. Zhang, T. Li, J. Kim, C.-Y. Chim, G. Galli, F. Wang, Emerging photoluminescence in monolayer MoS_2. Nano Lett. **10**, 1271–1275 (2010)
14. X. Lu, X. Luo, J. Zhang, S.Y. Quek, Q.H. Xiong, Lattice vibrations and Raman scattering in two-dimensional layered materials beyond graphene. Nano Res. **9**, 3559–3597 (2016)
15. X. Qian, J. Liu, L. Fu, J. Li, Quantum spin hall effect in two-dimensional transition metal dichalcogenides. Science **346**, 1344–1347 (2014)
16. B. Radisavljevic, A. Radenovic, J. Brivio, V. Giacometti, A. Kis, Single-layer MoS2 transistors. Nat. Nanotech. **6**, 147–150 (2011)
17. X. Zhang, X.-F. Qiao, W. Shi, J.-B. Wu, D.-S. Jiang, P.-H. Tan, Phonon and Raman scattering of two-dimensional transition metal dichalcogenides from monolayer, multilayer to bulk material. Chem. Soc. Rev. **44**, 2757–2785 (2015)
18. X. Lu, M.I.B. Utama, J. Lin, X. Luo, Y. Zhao, J. Zhang, S.T. Pantelides, W. Zhou, S.Y. Quek, Q.H. Xiong, Rapid and nondestructive identification of polytypism and stacking sequences in few-layer molybdenum diselenide by Raman spectroscopy. Adv. Mater. **27**, 4502–4508 (2015)
19. R. Suzuki, M. Sakano, Y.J. Zhang, R. Akashi, D. Morikawa, A. Harasawa, K. Yaji, K. Kuroda, K. Miyamoto, T. Okuda, K. Ishizaka, R. Arita, Y. Iwasa, Valley-dependent spin polarization in bulk MoS_2 with broken inversion symmetry. Nat. Nanotech. **9**, 611–617 (2014)
20. D.H. Keum, S. Cho, J.H. Kim, D.-H. Choe, H.-J. Sung, M. Kan, H. Kang, J.-Y. Hwang, S.W. Kim, H. Yang, K.J. Chang, Y.H. Lee, Bandgap opening in few-layered monoclinic $MoTe_2$. Nat. Phys. **11**, 482–U144 (2015)
21. Q.J. Song, Q.H. Tan, X. Zhang, J.B. Wu, B.W. Sheng, Y. Wan, X.Q. Wang, L. Dai, P.H. Tan, Physical origin of Davydov splitting and resonant Raman spectroscopy of Davydov components in multilayer $MoTe_2$. Phys. Rev. B **93**, 115409 (2016)
22. G. Froehlicher, E. Lorchat, F. Fernique, C. Joshi, A. Molina-Sanchez, L. Wirtz, S. Berciaud, Unified description of the optical phonon modes in N-layer $MoTe_2$. Nano Lett. **15**, 6481–6489 (2015)
23. C. Ruppert, O.B. Aslan, T.F. Heinz, Optical properties and band gap of single- and few-layer $MoTe_2$ crystals. Nano Lett. **14**, 6231–6236 (2014)
24. Y. Kim, Y.I. Jhon, J. Park, J.H. Kim, S. Lee, Y.M. Jhon, Anomalous Raman scattering and lattice dynamics in mono- and few-layer WTe_2. Nanoscale **8**, 2309–2316 (2016)
25. Y.C. Jiang, J. Gao, L. Wang, Raman fingerprint for semi-metal WTe_2 evolving from bulk to monolayer. Sci. Rep. **6**, 19624 (2016)
26. Y. Zhao, X. Luo, H. Li, J. Zhang, P.T. Araujo, C.K. Gan, J. Wu, H. Zhang, S.Y. Quek, M.S. Dresselhaus, Q.H. Xiong, Interlayer breathing and shear modes in few-trilayer MoS_2 and WSe_2. Nano Lett. **13**, 1007–1015 (2013)
27. N. Wakabayashi, H.G. Smith, R.M. Nicklow, Lattice-dynamics of hexagonal MoS_2 studied by neutron-scattering. Phys. Rev. B **12**, 659–663 (1975)
28. A. Molina-Sanchez, L. Wirtz, Phonons in single-layer and few-layer MoS_2 and WS_2. Phys. Rev. B **84**, 155413 (2011)
29. J.L. Verble, T.J. Wieting, Lattice mode degeneracy in MoS_2 and other layer compounds. Phys. Rev. Lett. **25**, 362 (1970)
30. P.N. Ghosh, C.R. Maiti, Interlayer force and Davydov splitting in 2H-MoS_2. Phys. Rev. B **28**, 2237–2239 (1983)
31. T.J. Wieting, J.L. Verble, Infrared and Raman studies of long-wavelength optical phonons in hexagonal MoS_2. Phys. Rev. B **3**, 4286 (1971)
32. C. Lee, H. Yan, L.E. Brus, T.F. Heinz, J. Hone, S. Ryu, Anomalous lattice vibrations of single- and few-layer MoS_2. ACS Nano **4**, 2695–2700 (2010)
33. X. Luo, Y. Zhao, J. Zhang, Q. Xiong, S.Y. Quek, Anomalous frequency trends in MoS_2 thin films attributed to surface effects. Phys. Rev. B **88**, 075320 (2013)
34. H. Zeng, B. Zhu, K. Liu, J. Fan, X. Cui, Q.M. Zhang, Low-frequency Raman modes and electronic excitations in atomically thin MoS_2 films. Phys. Rev. B **86**, 241301(R) (2012)

35. H. Li, Q. Zhang, C.C.R. Yap, B.K. Tay, T.H.T. Edwin, A. Olivier, D. Baillargeat, From bulk to monolayer MoS_2: evolution of Raman scattering. Adv. Funct. Mater. **22**, 1385–1390 (2012)
36. W. Zhao, Z. Ghorannevis, K.K. Amara, J.R. Pang, M. Toh, X. Zhang, C. Kloc, P.H. Tan, G. Eda, Lattice dynamics in mono- and few-layer sheets of WS_2 and WSe_2. Nanoscale **5**, 9677–9683 (2013)
37. P. Tonndorf, R. Schmidt, P. Boettger, X. Zhang, J. Boerner, A. Liebig, M. Albrecht, C. Kloc, O. Gordan, D.R.T. Zahn, S.M. de Vasconcellos, R. Bratschitsch, Photoluminescence emission and Raman response of monolayer MoS_2, $MoSe_2$, and WSe_2. Opt. Express **21**, 4908–4916 (2013)
38. J.-H. Fan, P. Gao, A.-M. Zhang, B.-R. Zhu, H.-L. Zeng, X.-D. Cui, R. He, Q.-M. Zhang, Resonance Raman scattering in bulk 2H-MX_2 (M = Mo, W; X = S, Se) and monolayer MoS_2. J. Appl. Phys. **115**, 053527 (2014)
39. H. Richter, Z.P. Wang, L. Ley, The one phonon Raman-spectrum in microcrystalline silicon. Solid State Commun. **39**, 625–629 (1981)
40. R.P. Wang, G.W. Zhou, Y.L. Liu, S.H. Pan, H.Z. Zhang, D.P. Yu, Z. Zhang, Raman spectral study of silicon nanowires: high-order scattering and phonon confinement effects. Phys. Rev. B **61**, 16827–16832 (2000)
41. K.W. Adu, H.R. Gutierrez, U.J. Kim, P.C. Eklund, Inhomogeneous laser heating and phonon confinement in silicon nanowires: a micro-Raman scattering study. Phys. Rev. B **73**, 155333 (2006)
42. K.W. Adu, Q. Xiong, H.R. Gutierrez, G. Chen, P.C. Eklund, Raman scattering as a probe of phonon confinement and surface optical modes in semiconducting nanowires. Appl. Phys. A Mater. Sci. Process. **85**, 287–297 (2006)
43. Y. Lin, X. Ling, L. Yu, S. Huang, A.L. Hsu, Y.-H. Lee, J. Kong, M.S. Dressehaus, T. Palacios, Dielectric screening of excitons and trions in single-layer MoS_2. Nano Lett. **14**, 5569–5576 (2014)
44. A. Molina-Sanchez, K. Hummer, L. Wirtz, Vibrational and optical properties of MoS_2: from monolayer to bulk. Surf. Sci. Rep. **70**, 554–586 (2015)
45. J. Yan, J. Xia, X. Wang, L. Liu, J.-L. Kuo, B.K. Tay, S. Chen, W. Zhou, Z. Liu, Z.X. Shen, Stacking-dependent interlayer coupling in Trilayer MoS_2 with broken inversion symmetry. Nano Lett. **15**, 8155–8161 (2015)
46. J.-U. Lee, K. Kim, S. Han, G.H. Ryu, Z. Lee, H. Cheong, Raman signatures of polytypism in molybdenum disulfide. ACS Nano **10**, 1948–1953 (2016)
47. A.A. Puretzky, L. Liang, X. Li, K. Xiao, K. Wang, M. Mahjouri-Samani, L. Basile, J.C. Idrobo, B.G. Sumpter, V. Meunier, D.B. Geohegan, Low-frequency Raman fingerprints of two-dimensional metal dichalcogenide layer stacking configurations. ACS Nano **9**, 6333–6342 (2015)
48. X.-F. Qiao, J.-B. Wu, L. Zhou, J. Qiao, W. Shi, T. Chen, X. Zhang, J. Zhang, W. Ji, P.-H. Tan, Polytypism and unexpected strong interlayer coupling in two-dimensional layered ReS_2. Nanoscale **8**, 8324–8332 (2016)
49. R. He, J.-A. Yan, Z. Yin, Z. Ye, G. Ye, J. Cheng, J. Li, C.H. Lui, Coupling and stacking order of ReS_2 atomic layers revealed by ultralow-frequency Raman spectroscopy. Nano Lett. **16**, 1404–1409 (2016)
50. Y. Yoo, Z.P. Degregorio, Y. Su, S.J. Koester, J.E. Johns, In-plane 2H-1T' $MoTe_2$ homojunctions synthesized by flux-controlled phase engineering. Adv. Mater. **29**, 1605461 (2017)
51. J.J. Wu, M.J. Liu, K. Chatterjee, K.P. Hackenberg, J.F. Shen, X.L. Zou, Y. Yan, J. Gu, Y.C. Yang, J. Lou, P.M. Ajayan, Exfoliated 2D transition metal disulfides for enhanced electrocatalysis of oxygen evolution reaction in acidic medium. Adv. Mater. Interfaces **3**, 1500669 (2016)
52. B.R. Carvalho, L.M. Malard, J.M. Alves, C. Fantini, M.A. Pimenta, Symmetry-dependent exciton-phonon coupling in 2D and bulk MoS_2 observed by resonance Raman scattering. Phys. Rev. Lett. **114**, 136403 (2015)

53. L. Sun, J. Yan, D. Zhan, L. Liu, H. Hu, H. Li, B.K. Tay, J.-L. Kuo, C.-C. Huang, D.W. Hewak, P.S. Lee, Z.X. Shen, Spin-orbit splitting in single-layer MoS_2 revealed by triply resonant Raman scattering. Phys. Rev. Lett. **111**, 126801 (2013)
54. Y. Wang, C. Cong, C. Qiu, T. Yu, Raman spectroscopy study of lattice vibration and crystallographic orientation of monolayer MoS_2 under uniaxial strain. Small **9**, 2857–2861 (2013)
55. B. Chakraborty, A. Bera, D.V.S. Muthu, S. Bhowmick, U.V. Waghmare, A.K. Sood, Symmetry-dependent phonon renormalization in monolayer MoS_2 transistor. Phys. Rev. B **85**, 161403(R) (2012)
56. R.S. Yan, J.R. Simpson, S. Bertolazzi, J. Brivio, M. Watson, X.F. Wu, A. Kis, T.F. Luo, A.R.H. Walker, H.G. Xing, Thermal conductivity of monolayer molybdenum disulfide obtained from temperature-dependent Raman spectroscopy. ACS Nano **8**, 986–993 (2014)
57. S. Sahoo, A.P.S. Gaur, M. Ahmadi, M.J.F. Guinel, R.S. Katiyar, Temperature-dependent Raman studies and thermal conductivity of few-layer MoS_2. J. Phys. Chem. C **117**, 9042–9047 (2013)
58. N.A. Lanzillo, A.G. Birdwell, M. Amani, F.J. Crowne, P.B. Shah, S. Najmaei, Z. Liu, P.M. Ajayan, J. Lou, M. Dubey, S.K. Nayak, T.P. O'Regan, Temperature-dependent phonon shifts in monolayer MoS_2. Appl. Phys. Lett. **103**, 093102 (2013)
59. A. Taube, J. Judek, C. Jastrzebski, A. Duzynska, K. Switkowski, M. Zdrojek, Temperature-dependent nonlinear phonon shifts in a supported MoS_2 monolayer. ACS Appl. Mater. Interfaces **6**, 8959–8963 (2014)
60. A.A. Balandin, S. Ghosh, W.Z. Bao, I. Calizo, D. Teweldebrhan, F. Miao, C.N. Lau, Superior thermal conductivity of single-layer graphene. Nano Lett. **8**, 902–907 (2008)
61. S. Najmaei, Z. Liu, P.M. Ajayan, J. Lou, Thermal effects on the characteristic Raman spectrum of molybdenum disulfide (MoS_2) of varying thicknesses. Appl. Phys. Lett. **100**, 013106 (2012)
62. C.R. Zhu, G. Wang, B.L. Liu, X. Marie, X.F. Qiao, X. Zhang, X.X. Wu, H. Fan, P.H. Tan, T. Amand, B. Urbaszek, Strain tuning of optical emission energy and polarization in monolayer and bilayer MoS_2. Phys. Rev. B **88**, 121301(R) (2013)
63. C.-H. Chang, X. Fan, S.-H. Lin, J.-L. Kuo, Orbital analysis of electronic structure and phonon dispersion in MoS_2, $MoSe_2$, WS_2, and WSe_2 monolayers under strain. Phys. Rev. B **88**, 125120 (2013)
64. M. Thripuranthaka, D.J. Late, Temperature dependent phonon shifts in single-layer WS_2. ACS Appl. Mater. Interfaces **6**, 1158–1163 (2014)
65. Y.Q. Cai, J.H. Lan, G. Zhang, Y.W. Zhang, Lattice vibrational modes and phonon thermal conductivity of monolayer MoS_2. Phys. Rev. B **89**, 035438 (2014)
66. X.-L. Qi, S.-C. Zhang, Topological insulators and superconductors. Rev. Mod. Phys. **83**, 1057–1110 (2011)
67. M.Z. Hasan, C.L. Kane, Colloquium: topological insulators. Rev. Mod. Phys. **82**, 3045–3067 (2010)
68. J.E. Moore, The birth of topological insulators. Nature **464**, 194–198 (2010)
69. H. Zhang, C.-X. Liu, X.-L. Qi, X. Dai, Z. Fang, S.-C. Zhang, Topological insulators in Bi_2Se_3, Bi_2Te_3 and Sb_2Te_3 with a single Dirac cone on the surface. Nat. Phys. **5**, 438–442 (2009)
70. Y.L. Chen, J.G. Analytis, J.H. Chu, Z.K. Liu, S.K. Mo, X.L. Qi, H.J. Zhang, D.H. Lu, X. Dai, Z. Fang, S.C. Zhang, I.R. Fisher, Z. Hussain, Z.X. Shen, Experimental realization of a three-dimensional topological insulator, Bi_2Te_3. Science **325**, 178–181 (2009)
71. A. Soni, Y. Zhao, L. Yu, M.K.K. Aik, M.S. Dresselhaus, Q. Xiong, Enhanced thermoelectric properties of solution grown $Bi_2Te_{3-x}Se_x$ Nanoplatelet composites. Nano Lett. **12**, 1203–1209 (2012)
72. J. Zhang, Z. Peng, A. Soni, Y. Zhao, Y. Xiong, B. Peng, J. Wang, M.S. Dresselhaus, Q.H. Xiong, Raman spectroscopy of few-quintuple layer topological insulator Bi_2Se_3 nanoplatelets. Nano Lett. **11**, 2407–2414 (2011)
73. G.L. Frey, R. Tenne, M.J. Matthews, M.S. Dresselhaus, G. Dresselhaus, Raman and resonance Raman investigation of MoS_2 nanoparticles. Phys. Rev. B **60**, 2883–2892 (1999)

74. Q.-C. Sun, D. Mazumdar, L. Yadgarov, R. Rosentsveig, R. Tenne, J.L. Musfeldt, Spectroscopic determination of phonon lifetimes in rhenium-doped MoS_2 nanoparticles. Nano Lett. **13**, 2803–2808 (2013)
75. C. Wang, X. Zhu, L. Nilsson, J. Wen, G. Wang, X. Shan, Q. Zhang, S. Zhang, J. Jia, Q. Xue, In situ Raman spectroscopy of topological insulator Bi_2Te_3 films with varying thickness. Nano Res. **6**, 688–692 (2013)
76. Y. Zhao, X. Luo, J. Zhang, J. Wu, X. Bai, M. Wang, J. Jia, H. Peng, Z. Liu, S.Y. Quek, Q. Xiong, Interlayer vibrational modes in few-quintuple-layer Bi_2Te_3 and Bi_2Se_3 two-dimensional crystals: Raman spectroscopy and first-principles studies. Phys. Rev. B **90**, 245428 (2014)
77. A.D. Laforge, A. Frenzel, B.C. Pursley, T. Lin, X. Liu, J. Shi, D.N. Basov, Optical characterization of Bi_2Se_3 in a magnetic field: Infrared evidence for magnetoelectric coupling in a topological insulator material. Phys. Rev. B **81**, 125120 (2010)
78. P. Di Pietro, M. Ortolani, O. Limaj, A. Di Gaspare, V. Giliberti, F. Giorgianni, M. Brahlek, N. Bansal, N. Koirala, S. Oh, P. Calvani, S. Lupi, Observation of Dirac plasmons in a topological insulator. Nat. Nanotech. **8**, 556–560 (2013)
79. C. Chen, Z. Xie, Y. Feng, H. Yi, A. Liang, S. He, D. Mou, J. He, Y. Peng, X. Liu, Y. Liu, L. Zhao, G. Liu, X. Dong, J. Zhang, L. Yu, X. Wang, Q. Peng, Z. Wang, S. Zhang, F. Yang, C. Chen, Z. Xu, X.J. Zhou, Tunable Dirac fermion dynamics in topological insulators. Sci. Rep. **3**, 2411 (2013)
80. M.V. Costache, I. Neumann, J.F. Sierra, V. Marinova, M.M. Gospodinov, S. Roche, S.O. Valenzuela, Fingerprints of inelastic transport at the surface of the topological insulator Bi_2Se_3: role of Electron-phonon coupling. Phys. Rev. Lett. **112**, 086601 (2014)
81. C.H. Zhang, Z.P. Peng, Z. Li, L.G. Yu, K.A. Khor, Q.H. Xiong, Controlled growth of bismuth antimony telluride $Bi_xSb_{2-x}Te_3$ nanoplatelets and their bulk thermoelectric nanocomposites. Nano Energy **15**, 688–696 (2015)
82. M. Zhang, J.X. Wu, Y.M. Zhu, D.O. Dumcenco, J.H. Hong, N.N. Mao, S.B. Deng, Y.F. Chen, Y.L. Yang, C.H. Jin, S.H. Chaki, Y.S. Huang, J. Zhang, L.M. Xie, Two-dimensional molybdenum tungsten diselenide alloys: photoluminescence, Raman scattering, and electrical transport. ACS Nano **8**, 7130–7137 (2014)
83. S. Tongay, D.S. Narang, J. Kang, W. Fan, C. Ko, A.V. Luce, K.X. Wang, J. Suh, K.D. Patel, V.M. Pathak, J. Li, J. Wu, Two-dimensional semiconductor alloys: monolayer $Mo_{1-x}W_xSe_2$. Appl. Phys. Lett. **104**, 012101 (2014)
84. Y.F. Chen, W. Wen, Y.M. Zhu, N.N. Mao, Q.L. Feng, M. Zhang, H.P. Hsu, J. Zhang, Y.S. Huang, L.M. Xie, Temperature-dependent photoluminescence emission and Raman scattering from $Mo_{1-x}W_xS_2$ monolayers. Nanotechnology **27**, 445705 (2016)
85. X.F. Qiao, X.L. Li, X. Zhang, W. Shi, J.B. Wu, T. Chen, P.H. Tan, Substrate-free layer-number identification of two-dimensional materials: a case of $Mo_{0.5}W_{0.5}S_2$ alloy. Appl. Phys. Lett. **106**, 223102 (2015)
86. S. Zheng, L. Sun, T. Yin, A.M. Dubrovkin, F. Liu, Z. Liu, Z.X. Shen, H.J. Fan, Monolayers of $W_xMo_{1-x}S_2$ alloy heterostructure with in-plane composition variations. Appl. Phys. Lett. **106**, 063113 (2015)
87. Q.L. Feng, N.N. Mao, J.X. Wu, H. Xu, C.M. Wang, J. Zhang, L.M. Xie, Growth of $MoS_{2(1-x)}Se_{2x}$ ($x = 0.41$–1.00) monolayer alloys with controlled morphology by physical vapor deposition. ACS Nano **9**, 7450–7455 (2015)
88. W.T. Zhang, X.D. Li, T.T. Jiang, J.L.Q. Song, Y. Lin, L.X. Zhu, X.L. Xu, CVD synthesis of $Mo_{(1-x)}WxS_2$ and $MoS_{2(1-x)}Se_{2x}$ alloy monolayers aimed at tuning the bandgap of molybdenum disulfide. Nanoscale **7**, 13554–13560 (2015)
89. Y.F. Chen, D.O. Dumcenco, Y.M. Zhu, X. Zhang, N.N. Mao, Q.L. Feng, M. Zhang, J. Zhang, P.H. Tan, Y.S. Huang, L.M. Xie, Composition-dependent Raman modes of $Mo_{1-x}W_xS_2$ monolayer alloys. Nanoscale **6**, 2833–2839 (2014)
90. X. Lu, M.I.B. Utama, J. Lin, X. Gong, J. Zhang, Y. Zhao, S.T. Pantelides, J. Wang, Z. Dong, Z. Liu, W. Zhou, Q.H. Xiong, Large-area synthesis of monolayer and few-layer $MoSe_2$ films on silicon substrates. Nano Lett. **14**, 2419–2425 (2014)

91. X. Luo, Y. Zhao, J. Zhang, M. Toh, C. Kloc, Q. Xiong, S.Y. Quek, Effects of lower symmetry and dimensionality on Raman spectra in two-dimensional WSe_2. Phys. Rev. B **88**, 195313 (2013)
92. J.A. Wilson, F.J. Di Salvo, S. Mahajan, Charge-density waves and superlattices in the metallic layered transition metal dichalcogenides (reprinted from *Adv. Phys*. 1974, *32*, 882). Adv. Phys. **50**, 1171–1248 (2001)
93. X.X. Xi, L. Zhao, Z.F. Wang, H. Berger, L. Forro, J. Shan, K.F. Mak, Strongly enhanced charge-density-wave order in monolayer $NbSe_2$. Nat. Nanotech. **10**, 765–770 (2015)
94. Y. Wu, M. An, R. Xiong, J. Shi, Q.M. Zhang, Raman scattering spectra in the normal phase of 2H-$NbSe_2$. J. Phys. D. Appl. Phys. **41**, 175408 (2008)
95. J.C. Tsang, J.E. Smith, M.W. Shafer, Raman-spectroscopy of soft modes at charge-density-wave phase-transition in 2h-$NbSe_2$. Phys. Rev. Lett. **37**, 1407–1410 (1976)
96. A. Mialitsin, Fano line shape and anti-crossing of Raman active E_{2g} peaks in the charge density wave state of $NbSe_2$. J. Phys. Chem. Solids **72**, 568–571 (2011)
97. M.A. Measson, Y. Gallais, M. Cazayous, B. Clair, P. Rodiere, L. Cario, A. Sacuto, Amplitude Higgs mode in the 2H-$NbSe_2$ superconductor. Phys. Rev. B **89**, 060503(R) (2014)
98. J.F. Scott, Soft-mode spectroscopy: experimental studies of structural phase-transitions. Rev. Mod. Phys. **46**, 83–128 (1974)
99. D.E. Moncton, J.D. Axe, F.J. Disalvo, Study of superlattice formation in 2H-$NbSe_2$ and 2H-$TaSe_2$ by neutron-scattering. Phys. Rev. Lett. **34**, 734–737 (1975)
100. R. Arenal, A.C. Ferrari, S. Reich, L. Wirtz, J.Y. Mevellec, S. Lefrant, A. Rubio, A. Loiseau, Raman spectroscopy of single-wall boron nitride nanotubes. Nano Lett. **6**, 1812–1816 (2006)
101. R.V. Gorbachev, I. Riaz, R.R. Nair, R. Jalil, L. Britnell, B.D. Belle, E.W. Hill, K.S. Novoselov, K. Watanabe, T. Taniguchi, A.K. Geim, P. Blake, Hunting for monolayer boron nitride: optical and Raman signatures. Small **7**, 465–468 (2011)
102. D. Golberg, Y. Bando, Y. Huang, T. Terao, M. Mitome, C.C. Tang, C.Y. Zhi, Boron nitride nanotubes and nanosheets. ACS Nano **4**, 2979–2993 (2010)
103. C. Gong, L. Li, Z. Li, H. Ji, A. Stern, Y. Xia, T. Cao, W. Bao, C. Wang, Y. Wang, Z.Q. Qiu, R.J. Cava, S.G. Louie, J. Xia, X. Zhang, Discovery of intrinsic ferromagnetism in two-dimensional van der Waals crystals. Nature **546**, 265–269 (2017)
104. B. Huang, G. Clark, E. Navarro-Moratalla, D.R. Klein, R. Cheng, K.L. Seyler, D. Zhong, E. Schmidgall, M.A. McGuire, D.H. Cobden, W. Yao, D. Xiao, P. Jarillo-Herrero, X.D. Xu, Layer-dependent ferromagnetism in a van der Waals crystal down to the monolayer limit. Nature **546**, 270–273 (2017)
105. X.Z. Wang, K.Z. Du, Y.Y.F. Liu, P. Hu, J. Zhang, Q. Zhang, M.H.S. Owen, X. Lu, C.K. Gan, P. Sengupta, C. Kloc, Q.H. Xiong, Raman spectroscopy of atomically thin two-dimensional magnetic iron phosphorus trisulfide (FePS3) crystals. 2D Materials **3**, 031009 (2016)
106. J.U. Lee, S. Lee, J.H. Ryoo, S. Kang, T.Y. Kim, P. Kim, C.H. Park, J.G. Park, H. Cheong, Ising-type magnetic ordering in atomically thin $FePS_3$. Nano Lett. **16**, 7433–7438 (2016)
107. T. Sekine, M. Jouanne, C. Julien, M. Balkanski, Light-scattering study of dynamic behavior of antiferromagnetic spins in the layered magnetic semiconductor $FePS_3$. Phys. Rev. B **42**, 8382–8393 (1990)

Chapter 3
Raman Spectroscopy of Anisotropic Two-Dimensional Materials

Juanxia Wu, Shishu Zhang, Lianming Tong, and Jin Zhang

Abstract Due to the in-plane structural anisotropy, two-dimensional (2D) layered materials with low symmetry exhibit unique crystalline-axis dependent properties, including the optical, mechanical and electrical properties. Raman spectroscopy, in particular, polarized Raman spectroscopy, has been used as a rapid and non-invasive technique to study the composition, structure and symmetry of 2D anisotropic layered materials. In this chapter, the recent advances on the Raman spectroscopic studies of anisotropic 2D materials are summarized. The Raman selection rules and the structural symmetry will be discussed, followed by the overview of the polarized Raman scattering studies of anisotropic 2D materials cataloged by crystal symmetries.

3.1 Introduction

Two-dimensional (2D) materials have attracted intense interest due to the unique physical and chemical properties and the tremendous potential applications since graphene was discovered in 2004 [1]. During the past decade, more and more members were discovered in the 2D material family, including planar hexagonal boron nitride (h-BN), [2] sandwiched transition metal dichalcogenides (TMDs), [3] puckered black phosphorus (BP), [4, 5] buckled silicene [6–8] and germanene, [9] and so on. Due to the unique structure, these 2D materials exhibit distinct mechanical, thermal, optical and electric properties compared to the bulk phase. At the same time, the composition, [10–12] layer number, [13, 14] stacking order [15–17] and other structural difference will have a great impact on their properties.

For 2D materials with high symmetry, such as graphene, h-BN and MoS_2, the mechanical, thermal, optical and electric properties are isotropic in the planes of the materials [18]. However, the reduced lattice symmetry of layered materials

J. Wu · S. Zhang · L. Tong (✉) · J. Zhang (✉)
Center for Nanochemistry, College of Chemistry and Molecular Engineering, Peking University, Beijing, China
e-mail: tonglm@pku.edu.cn; jinzhang@pku.edu.cn

P.-H. Tan (ed.), *Raman Spectroscopy of Two-Dimensional Materials*, Springer Series in Materials Science 276, https://doi.org/10.1007/978-981-13-1828-3_3

could induce interesting in-plane anisotropic optical, electrical, thermoelectric and mechanical properties, that is, these properties show obvious crystalline orientation dependence. In fact, recent works have revealed the in-plane anisotropic electric conductivity, [19] Raman scattering, [20, 21] photoluminescence emission, [22] optical absorption, [19] strain response [23, 24] and photocurrent response [25] of 2D layered materials with low symmetry, such as orthorhombic BP and tin selenide (SnSe), [26, 27] monoclinic gallium telluride (GaTe) [28, 29] and 1 T′-phase molybdenum ditelluride ($MoTe_2$), [30] triclinic rhenium disulfide (ReS_2) [31, 32] and rhenium diselenide ($ReSe_2$), [33] etc. Therefore, uncovering the anisotropic nature of these materials and relating these properties to the crystalline orientation are key issues for the understanding of such materials and the potential electric/opto-electric applications.

Raman spectroscopy is a rapid and nondestructive method in materials characterization and has been employed to investigate their structure, physical and chemical properties. It can also be used to study the electronic band structure, phonon energy dispersion and electron-phonon interactions in 2D materials [34]. The Raman peak positions, intensities and full widths at half maxima (FWHM) can directly reflect the layer number, [14] stacking order, [35] crystalline orientation [21] and applied strain [24] of the 2D materials. Similarly, Raman spectroscopy has also been used to study the anisotropic in-plane structure and optical properties of 2D materials with low symmetry [20, 21, 31, 36]. The anisotropic thermal, [37] mechanical [24] and electric [38] properties can be reflected in the anisotropic variation of Raman frequency and intensity. Therefore, Raman spectroscopy is a reliable characterization tool to reveal the anisotropic nature of anisotropic 2D materials and to relate the anisotropic properties with their crystalline orientation.

This chapter focuses on the Raman spectroscopic studies of anisotropic 2D materials. Firstly, the structural symmetry of crystals and Raman selection rules are introduced, and the relationship between Raman scattering efficiency and crystalline orientation is discussed. According to the Raman selection rules, all the 2D materials belonging to triclinic, monoclinic and orthorhombic crystal systems have obvious crystalline-orientation dependent Raman scattering efficiency in the plane of the material. Secondly, on the basis of crystalline orientation dependent Raman scattering efficiency, polarized Raman spectroscopy can be used to identify the crystalline orientation of few-layered anisotropic 2D materials. Thirdly, recent advances on the Raman spectroscopy of anisotropic 2D materials is reviewed from the orthorhombic (e.g., BP, SnSe), monoclinic (e.g., GaTe) and triclinic (e.g., ReS_2, $ReSe_2$) crystal systems. A summary is given in the last section.

3.2 The Raman Selection Rules and Structural Symmetry

At the crystal surface, the lattice periodicity is maintained in the *xy* plane, same as that in the bulk phase, while it is broken along the *z* direction. In other words, the base vectors of unit cell in the crystal surface can be represented by the two

Table 3.1 Crystal systems and corresponding unit cells, Bravais lattices and point groups in three-dimensional space

Crystal system	Axial lengths and angles	Bravais lattice	Point group
Triclinic	$a \neq b \neq c, \alpha \neq \beta \neq \gamma \neq 90°$	Primitive	C_1, C_i
Monoclinic	$a \neq b \neq c, \alpha = \gamma = 90° \neq \beta$	Primitive, base-centered	C_2, C_s, C_{2h}
Orthorhombic	$a \neq b \neq c, \alpha = \beta = \gamma = 90°$	Primitive, base-centered, body-centered, face-centered	D_2, C_{2v}, D_{2h}
Tetragonal	$a = b \neq c, \alpha = \beta = \gamma = 90°$	Primitive, body-centered	C_4, S_4, C_{4h}
			C_{4v}, D_4, D_{2d}, D_{4h}
Trigonal	$a = b = c$, $\alpha = \beta = \gamma < 120°, \neq 90°$	Rhombohedrally-centered	C_3, C_{3i}
			D_3, C_{3v}, D_{3d}
Hexagonal	$a = b \neq c$, $\alpha = \beta = 90°, \gamma = 120°$	Primitive	C_6, C_{3h}, C_{6h}
			D_6, C_{6v}, D_{3h}, D_{6h}
Cubic	$a = b = c, \alpha = \beta = \gamma = 90°$	Primitive, body-centered, face-centered	T, T_h, O, T_d, O_h

fundamental vectors that are parallel to the surface plane. Depending on the relative lengths of the two vectors and the angles between them in the unit cells, there are 4 crystal systems in 2D space, containing 5 Bravais lattices, that is, oblique, rectangular, centered rectangular, hexagonal and square.

Differently, there are 14 Bravais lattices in three-dimensional (3D) space, grouped into 7 crystal systems and 32 point groups. Although there are innumerable kinds of crystals in nature, anyone can be classified to one of these 14 Bravais lattices. Table 3.1 lists the Bravais lattices and the corresponding point groups in the 7 crystal systems. Although this chapter focuses on the Raman spectroscopy of anisotropic 2D materials, for a full understanding, the Raman selection rules and the relationships between Raman scattering efficiency and crystalline orientation will be represented in 3D space in the following content in this section.

When a molecule is placed in an electric field (***E***) of incident light, the electrons in the molecule will deviate from the nucleus to produce a dipole moment (***P***). For a first-order Raman scattering process, ***P*** and ***E*** have a linear relationship: $\boldsymbol{P} = \boldsymbol{\alpha} \cdot \boldsymbol{E}$, where $\boldsymbol{\alpha}$ is the molecular polarizability. Generally, ***P*** and ***E*** are vectors along different directions, and $\boldsymbol{\alpha}$ is a second-order tensor as following

$$\boldsymbol{\alpha} = \begin{pmatrix} \alpha_{xx} & \alpha_{xy} & \alpha_{xz} \\ \alpha_{yx} & \alpha_{yy} & \alpha_{yz} \\ \alpha_{zx} & \alpha_{zy} & \alpha_{zz} \end{pmatrix} \tag{3.1}$$

Therefore, the polarization states of incident and scattered light are correlated through the polarizability tensor. For a crystal under excitation of a linearly polarized laser, the polarization state of the scattered light is determined by the relative values of the elements in the polarizability tensor, which in turn depend on the symmetry of the vibrational mode.

The polarizability tensor of the first-order Raman mode and the Raman tensor are correlated by the following relationship: [39].

$$\alpha_{zy,x} = -\frac{e^2}{m^2\omega_i^2\hbar^2 d} R^x_{yz}(-\omega_i, \omega_i, 0) \tag{3.2}$$

where e and m are the unit charge and mass of an electron, respectively, ω_i is the frequency of incident photon, $\hbar = 6.58 \times 10^{-14}$ eV·s is the reduced Planck constant, and d is lattice constant. Accordingly, $\boldsymbol{P}$ can be described by the Raman tensors, and then reflect in the Raman intensity. For a crystal with known structure, the point group can be determined according to the allowed symmetric operations. Then, the irreducible representation can be obtained based on the group theory analysis, and the Raman active modes can be determined according to the corresponding basic functions. Moreover, the Raman tensor of a Raman-active mode is only related to the point group and the symmetry of the mode, which does not depend on the specific material.

Generally, for a given Raman mode, the Raman scattering efficiency is related to the form of Raman tensor and scattering geometry of Raman system, which can be expressed by the following equation [39]:

$$S = A\left[\sum_{k,l=x,y,z} e_i^k R_{kl} e_s^l\right]^2 = A|e_i \cdot R \cdot e_s|^2 \tag{3.3}$$

where e_i and e_s are both unit vectors and represent the polarization directions of incident and scattering light, respectively. For doubly or triply degenerated Raman modes, the total scattering efficiency should be the sum of contributions of the two or three Raman tensors [39]. According to Eq. (3.3), Raman intensity of a Raman mode is only determined by the diagonal elements in the Raman tensor under parallel polarization configuration, while it is only related to the off-diagonal elements under cross polarization configuration. The Raman signals can be detected only when the corresponding elements in Raman tensor have non-zero values.

The Raman tensor is given when the axes of crystal are parallel or perpendicular to a symmetric axis or mirror surface, that is, the polarization directions of incident (e_i) and scattering light (e_s) are both along the crystalline orientation of the crystal. However, the crystalline orientation is generally unknown when the Raman measurements are carried out, so the generalized form of Raman tensor should be transformed from the crystalline orientation of crystal to the coordinate frame in the geometry of Raman system. Here, we only consider the transformation in xy plane (the incident and scattering light are both propagated along to the z axis) which is generally parallel to the surface of sample, and the angle between crystalline orientation of 2D materials and e_i is defined as θ. The transformation of Raman tensors from the crystalline coordinate to the measurement coordinate is given by $R' = rRr^t$, [40] where r is the transformation matrix

$$r = \begin{pmatrix} \cos\theta & \sin\theta & 0 \\ -\sin\theta & \cos\theta & 0 \\ 0 & 0 & 1 \end{pmatrix} \quad r^t = \begin{pmatrix} \cos\theta & -\sin\theta & 0 \\ \sin\theta & \cos\theta & 0 \\ 0 & 0 & 1 \end{pmatrix} \tag{3.4}$$

Take the orthorhombic 2D materials as an example, the Raman tensors can be expressed as

$$R\left(\mathrm{A_g}\right) = \begin{pmatrix} a & 0 & 0 \\ 0 & b & 0 \\ 0 & 0 & c \end{pmatrix} \quad R\left(\mathrm{B_{2g}}\right) = \begin{pmatrix} 0 & e & 0 \\ e & 0 & 0 \\ 0 & 0 & 0 \end{pmatrix} \tag{3.5}$$

$$\begin{aligned} R'\left(\mathrm{A_g}\right) &= \begin{pmatrix} a\cos^2\theta + b\sin^2\theta & (b-a)\sin\theta\cos\theta & 0 \\ (b-a)\sin\theta\cos\theta & a\sin^2\theta + b\cos^2\theta & 0 \\ 0 & 0 & c \end{pmatrix} \\ R'\left(\mathrm{B_{2g}}\right) &= \begin{pmatrix} e\sin 2\theta & e\cos 2\theta & 0 \\ e\cos 2\theta & -e\sin 2\theta & 0 \\ 0 & 0 & 0 \end{pmatrix} \end{aligned} \tag{3.6}$$

Based on Eq. (3.3), the Raman scattering efficiency of A_g and B_{2g} modes can be expressed as

$$\begin{aligned} S_{XX}^{\mathrm{A_g}} &\propto \left(a\cos^2\theta + b\sin^2\theta\right)^2 \quad S_{XX}^{\mathrm{B_{2g}}} \propto e^2\sin^2 2\theta \\ S_{XY}^{\mathrm{A_g}} &\propto (b-a)^2\sin^2 2\theta/4 \quad S_{XY}^{\mathrm{B_{2g}}} \propto e^2\cos^2 2\theta \end{aligned} \tag{3.7}$$

The subscripts XX denote that e_i and e_s are both along to x axis, and XY denote that e_i and e_s are along to x and y axes, respectively. It can be seen that the Raman intensity is closely depended on the angle θ (crystalline orientation of 2D materials) under both parallel and cross polarization configurations. In other words, all the orthorhombic 2D materials show the same anisotropy in Raman scattering.

Similar analysis can be carried out for 2D materials belonging to other crystal systems, and the Raman scattering efficiency for triclinic, monoclinic, orthorhombic and tetragonal 2D materials is listed in Table 3.2. It is found that all the 2D materials belonging to triclinic, monoclinic, orthorhombic and tetragonal crystal system show polarization-dependent Raman scattering. It is important to point out that the anisotropy here refers to the in-plane (xy plane) anisotropic structures of 2D materials. So far, there are two methods to study the in-plane anisotropic Raman scattering of 2D materials: (1) a polarizer is placed in the incident light path, and Raman spectra are detected by rotating the polarizer or sample; [22] (2) a polarizer and an analyzer are placed in the incident and scattering light path, respectively, and Raman measurements are carried out by rotating the sample when e_i and e_s are parallel or perpendicular to each other [20, 21]. On the other hand, for thin samples of only a few layers, the anisotropic Raman scattering can be used to identify the

Table 3.2 Raman scattering efficiency for triclinic, monoclinic, orthorhombic and tetragonal 2D materials

Crystal system	Point group, scattering geometry	Raman modes and Raman scattering efficiency				
Triclinic	$Z(XX)\overline{Z}$	$(a\cos^2\theta + b\sin^2\theta - d\sin 2\theta)^2$				
	$Z(XY)\overline{Z}$	$[(a\text{-}b)\sin\theta\cos\theta + d\cos 2\theta]^2$				
	C_1	A				
	C_i	A_g				
Monoclinic	$Z(XX)\overline{Z}$	$(a\cos^2\theta + b\sin^2\theta)^2$	$e^2\sin^2 2\theta$			
	$Z(XY)\overline{Z}$	$(b\text{-}a)^2\sin^2\theta\cos^2\theta$	$e^2\cos^2 2\theta$			
	C_2	A(*y*)	B(*x*, *z*)			
	C_s	A′(*x*, *z*)	A′′(*y*)			
	C_{2h}	A_g	B_g			
Orthorhombic	$Z(XX)\overline{Z}$	$(a\cos^2\theta + b\sin^2\theta)^2$	$d^2\sin^2 2\theta$	*0*	*0*	
	$Z(XY)\overline{Z}$	$(b\text{-}a)^2\sin^2\theta\cos^2\theta$	$d^2\cos^2 2\theta$	*0*	*0*	
	D_2	A	$B_1(z)$	$B_2(y)$	$B_3(x)$	
	C_{2v}	$A_1(z)$	A_2	$B_1(x)$	$B_2(y)$	
	D_{2h}	A_g	B_{1g}	B_{2g}	B_{3g}	
Tetragonal	$Z(XX)\overline{Z}$	a^2	$(c\cos 2\theta + d\sin 2\theta)^2$	*0*	*0*	
	$Z(XY)\overline{Z}$	*0*	$(c\sin 2\theta\text{-} d\cos 2\theta)^2$	*0*	*0*	
	C_4	A(z)	B	E(x)	E(y)	
	S_4	A	B(z)	E(x)	E(−y)	
	C_{4h}	A_g	B_g	E_g	E_g	
	$Z(XX)\overline{Z}$	a^2	$c^2\cos^2 2\theta$	$d^2\sin^2 2\theta$	*0*	*0*
	$Z(XY)\overline{Z}$	*0*	$c^2\sin^2 2\theta$	$d^2\cos^2 2\theta$	*0*	*0*
	C_{4v}	$A_1(z)$	B_1	B_2	E(*x*)	E(*y*)
	D_4	A_1	B_1	B_2	E(−*y*)	E(*x*)
	D_{2d}	A_1	B_1	$B_2(z)$	E(*y*)	E(*x*)
	D_{4h}	A_{1g}	B_{1g}	B_{2g}	E_g	E_g

crystalline orientation of anisotropic 2D materials according to the angle-dependent Raman scattering efficiency listed in Table 3.2.

3.3 Identifying the Crystalline Orientation

In this section, we will discuss the capability of identifying the crystalline orientation of 2D materials using polarized Raman scattering. As shown in Table 3.2, the Raman intensity of all the Raman modes in triclinic, monoclinic, orthorhombic and tetragonal 2D materials shows 90° or 180° periodic variation with θ under both parallel and cross polarization configurations, that is, the intensity reaches maximum or minimum when the crystalline orientation is parallel to e_i ($\theta = 0°/90°$) or they have an angle of 45° ($\theta = 45°$). Combining other structural characterization method, such as TEM, angle-resolved conductance, the crystalline orientation of these 2D materials can be accurately identified [20, 21].

However, not all the Raman modes can be used to identify the crystalline orientation. For triclinic 2D materials, the Raman intensity of Raman modes generally has 180° periodic variation with θ under both parallel and cross polarization configurations, which certainly can identify the crystalline orientation. For monoclinic and orthorhombic 2D materials, the Raman intensity of totally symmetric Raman modes can also identify the crystalline orientation of anisotropic 2D materials under parallel polarization configuration. Under cross polarization configuration, the Raman intensity has 90° periodic variation, and the maximum intensity should be corresponding to the crystalline orientations, however, if there is no difference in the maxima, it is incapable to distinguish the specific crystalline orientation. Similar situation can be observed for the non-totally symmetric Raman modes, which are also not possible to identify the crystalline orientation. In a word, the specific crystalline orientation of triclinic, monoclinic and orthorhombic 2D materials can be identified generally through the periodically changing Raman intensity of totally symmetric Raman modes. For tetragonal 2D materials, totally symmetric mode can be detected with no crystalline orientation dependence under parallel polarization configuration, while it is forbidden under cross polarization configuration. Raman intensity of non-totally symmetric mode (non-degenerated B, B_g, B_1, B_2, B_{1g} and B_{2g} modes) show 90° periodic variation so that the specific crystalline orientation cannot be distinguished. Therefore, only the crystalline orientation of triclinic, monoclinic and orthorhombic 2D materials can be identified using polarized Raman spectroscopy.

Besides, studies have suggested that periodic variation of Raman intensity of layered materials is also affected by photon-phonon interactions, and is related to the sample thickness and excitation energy [47]. Further, optical birefringence in anisotropic materials should also be considered in the polarized Raman scattering studies of anisotropic 2D materials [41]. For the polarized Raman spectroscopy of optical anisotropic crystal, the polarization states of incident and scattering light in Raman scattering process will both be altered due to the birefringence effect [20, 42]. The influence of birefringence effect on the Raman selection rule varies with the specific materials, which will be discussed in the following sections.

3.4 Orthorhombic Crystal System

3.4.1 Black Phosphorus

BP is a semiconductor with a direct bandgap, which changes from 0.3 eV to 2.0 eV at Γ point from bulk phase to monolayer [43]. BP has a folded honeycomb structure, as shown in Fig. 3.1a. The top view is shown in Fig. 3.1b, where the two in-plane

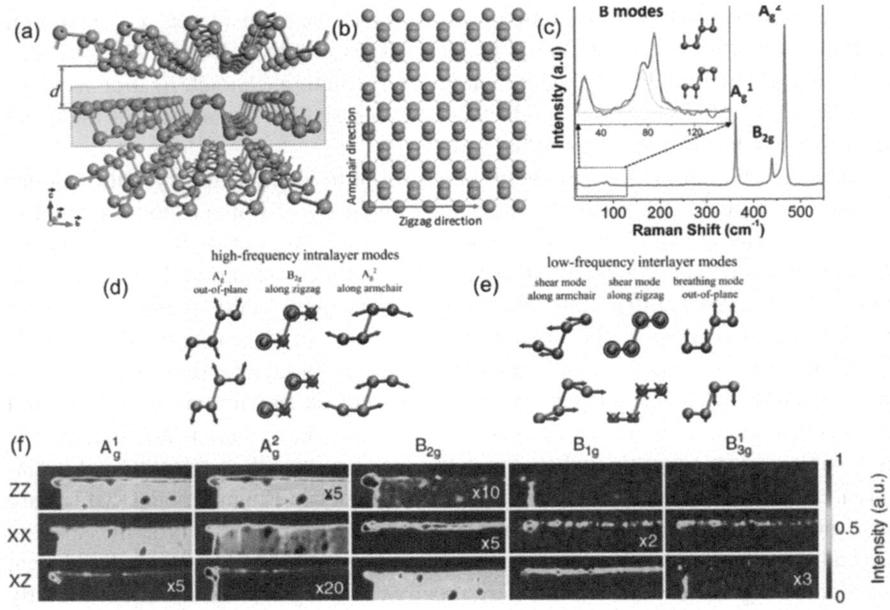

Fig. 3.1 (**a**) Structural illustration of BP. (**b**) Top view of BP, with the zigzag and armchair directions indicated by arrows [45]. (**c**) Typical Raman spectra of few-layer BP. The low-frequency range is zoomed-in in the inset. (**d–e**) Illustrations of the vibrational motions in BP, including the high-frequency intralayer vibrations (**d**) and low-frequency interlayer vibrations (**e**) [44]. (**f**) Hyperspectral Raman intensity images of all observed modes: $A_g{}^1$, $A_g{}^2$, B_{2g}, B_{1g} and B_{3g} (columns), for the different polarization configurations indicated on the left side [46]

crystalline directions, the armchair and zigzag directions, are orthogonal to each other. BP belongs to the orthorhombic crystal system, and bulk BP belongs to $D_{2h}{}^{18}$ point group and *Cmce* (No.64) space group [5]. According to the group theory analysis, there are 4 atoms in a unit cell, so at the center of the Brillouin zone there are 12 phonon modes:

$$\Gamma = 2A_g + B_{1g} + B_{2g} + 2B_{3g} + A_{1u} + 2B_{1u} + 2B_{2u} + B_{3u} \tag{3.8}$$

Six of them are Raman-active modes, including $2A_g$, B_{1g}, B_{2g}, $2B_{3g}$, five of them are infrared-active, including $2B_{1u}$, $2B_{2u}$, B_{3u}, and A_u is optically inactive mode [21, 23]. Although there are six Raman-active modes, only $A_g{}^1$ (362 cm^{-1}), B_{2g} (439 cm^{-1}), $A_g{}^2$ (468 cm^{-1}) modes are allowed (Fig. 3.1c) in the back scattering geometry where e_i and e_s are both in the plane of the sample [44]. The $A_g{}^1$ is an out-of-plane mode, with the P atoms in the two sub-layers vibrating in opposite directions. The B_{2g} and $A_g{}^2$ modes are the in-plane vibrations in zigzag and armchair directions, as shown in Fig. 3.1d. The low frequency Raman modes correspond to the relative motions between the adjacent BP layers, as shown in

Fig. 3.1e. These modes are related to the interlayer coupling and sample thickness, and the Raman frequencies are usually below 100 cm^{-1} due to the weak van der Waals interaction. Because the shear modes have B_{1g} and B_{3g} symmetry, only the breathing modes (A_g mode) can be detected under the normal incidence condition. Ling et al. [44] reported the low-frequency Raman spectra in few layer BP sample. As shown in Fig. 3.1c, three breathing modes located in 26.2 cm^{-1}, 75.6 cm^{-1} and 85.6 cm^{-1} can be observed with *N*-layers BP ($N > 6$). The breathing modes also exhibit polarization dependence. The number and peak positions of the breathing modes have obvious thickness dependence and the variation can be explained by the linear chain model [45].

3.4.1.1 Edge Modes of Black Phosphorus

At the edges of the BP samples, new peaks can appear in polarized Raman scattering depending on the lattice termination mode (zigzag or armchair). Ribeiro et al. [46] measured the polarized Raman hyperspectral images of a BP sample, as shown in Fig. 3.1f. In the center of the BP sample, the ${A_g}^1$ and ${A_g}^2$ modes only appear in ZZ and XX configuration, while the B_{2g} only appears in XZ configuration. For the edge of the BP flakes, in the XZ configuration, both the ${A_g}^1$ and ${A_g}^2$ modes are visible. The ${A_g}^1$ mode has higher intensities for the zigzag edges, and the ${A_g}^2$ mode are more significant for the armchair edges. The B_{1g} and ${B_{3g}}^1$ modes are forbidden in the XZ plane backscattering geometry. However, the B_{1g} mode is also significant for the zigzag edges in XZ configuration, and a weak peak can be observed in the XX configuration. In ZZ configuration, the mode is visible for the armchair edges. The ${B_{3g}}^1$ mode is observed at the armchair edges in cross polarization configuration and at the zigzag edges in XX configuration.

The new edge modes of BP can be explained by the lattice rearrangement at the edges. According to the *ab* initio DFT calculations, the lattice rearrangements along armchair and zigzag directions are different. For the armchair edges, the edge atoms within the same layer move close together. However, for the zigzag edges, the stretching and contraction of the puckered structure lead to a distortion of the edge. Due to the lattice rearrangement at the edges, the Raman tensor of the A_g mode has off-diagonal elements and the B_{2g} mode has diagonal components. Besides, the forbidden B_{1g} and ${B_{3g}}^1$ modes have XX, XZ, and ZZ elements. Hence, polarized Raman scattering is an effective tool to identify the different edges of the BP flakes.

3.4.1.2 Polarized Raman Spectra of BP

Because of the in-plane anisotropy in BP, the high-frequency Raman modes have obvious polarization dependence. The Raman intensities are decided by the polarization directions of the incident and scattered light relative to the crystalline axes. The Raman intensities of different modes show different polarization dependence. Several groups [20, 21, 47–49] have reported such effects, and it turns out that the

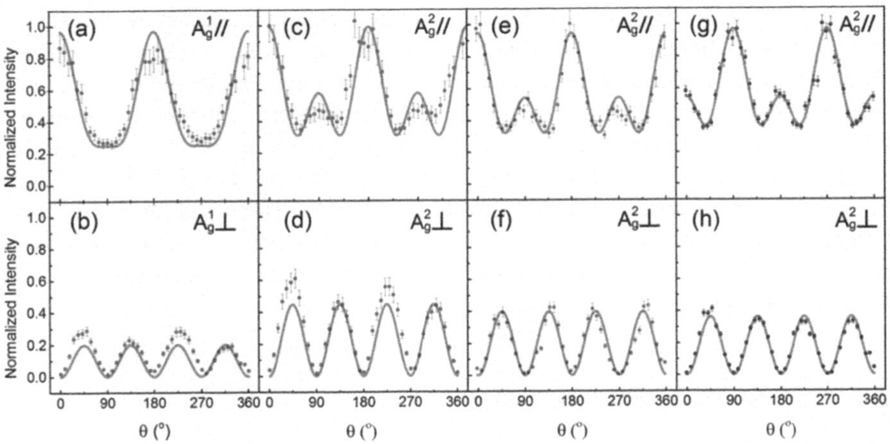

Fig. 3.2 Angular dependence of the Raman intensities measured with the (**a–d**) 633 nm, (**e, f**) 532 nm and (**g, h**) 488 nm laser excitation. The polarization configurations (parallel or cross) and the different Raman modes are indicated in each panel. Dots are experimental data, and solid curves correspond to the best fits to the data [49]

classical Raman selection rules cannot explain the polarization-dependence for all the three modes.

Ribeiro et al. [49] measured the angle-dependence of Raman intensities under parallel and cross polarization configurations with different laser wavelengths as shown in Fig. 3.2. With different configuration and excitation wavelength, the angle-dependence of B_{2g} mode can be well fitted according to the classical Raman selection rules. For the A_g modes, under cross polarization configuration, both the ${A_g}^1$ (362 cm^{-1}) and ${A_g}^2$ (468 cm^{-1}) modes have 90° variation periods and can also be fitted by the classical Raman selection rules. The angles of the maximum intensities are coincident with the zigzag or the armchair direction. However, under parallel polarization configuration, the ${A_g}^1$ and ${A_g}^2$ modes have a 180° variation periods, at the same time, secondary maximums appear, which cannot be explained by the classical Raman selections rules. For the classical Raman tensor, the elements are real values. However, for absorptive materials, the dielectric functions are complex with real and imaginary parts. Since Raman tensor is the derivative of the dielectric function tensor, the Raman tensor elements can also be expressed as complex values with real and imaginary parts. By considering the linear dichroism and modifying the Raman tensors accordingly, the Raman tensor of A_g mode is rewritten as

$$R\left(\mathrm{A_g}\right) = \begin{pmatrix} |a|\,e^{i\varphi_a} & 0 & 0 \\ 0 & |b|\,e^{i\varphi_b} & 0 \\ 0 & 0 & |c|\,e^{i\varphi_c} \end{pmatrix} \tag{3.9}$$

The Raman intensities of $A_g{}^1$ and $A_g{}^2$ modes under parallel polarization configuration are

$$I_{A_g}^{//} = \left(|a|\sin^2\theta + |c|\cos\varphi_{ca}\cos^2\theta\right)^2 + |c|^2\sin^2\varphi_{ca}\cos^4\theta \tag{3.10}$$

where the $\varphi_{ca} = \varphi_c - \varphi_a$, and θ is the angle between e_i and the zigzag direction. Using Eq. 3.10, the angle-dependence can be well fitted both in the variation and the extreme values.

From Eq. 3.10, it can be seen that the angle of the maximum intensity corresponds to either the armchair and zigzag direction depending on the value of $|c/a|$. When $|c/a| > 1$, the maximum corresponds to the zigzag direction and the secondary maximum corresponds to the armchair direction. When $|c/a| < 1$, the maximum corresponds to the armchair direction and the secondary maximum corresponds to the zigzag direction. The value of φ_{ca} determines the magnitude of the secondary maximum. The values of both $|c/a|$ and φ_{ca} are dependent on the sample thickness, excitation laser wavelength and the Raman modes.

Kim et al. and Ling et al. measured the angle-resolved polarized Raman spectra of BP flakes with different sample thickness and excitation laser wavelength [47, 48]. It can be observed that the maximum angles can locate at different direction with the same crystalline orientation under parallel configuration, when the sample thickness and excitation laser wavelength change both for $A_g{}^1$ and $A_g{}^2$ modes. So, using the maximum angles to identifying the crystalline orientation should consider the sample thickness, excitation laser wavelength and the Raman frequency. After correction for the interference effect, the values of $|c/a|$ show increasing trend with the thickness increasing and the φ_{ca} also change with the thickness increasing under 633 nm laser wavelength excitation.

Black phosphorus is an anisotropic layered material, and the refractive indexes along the armchair and zigzag directions are expected to be different, so both the polarization of the incident and scattered light suffer from the birefringence effect. Mao et al. [20] explained the abnormal polarized Raman scattering of BP by considering the birefringence effect. The electric fields of the incident and scattered light are modified by introducing a phase factor δ. As a result, the Raman intensity for the A_g Raman modes in BP crystal under parallel polarization configuration can be given in a corrected model by

$$I_{A_g}^{//} = a^2\sin^4\theta + c^2\cos^4\theta + 2ac\sin^2\theta\cos^2\theta\cos 2\delta \tag{3.11}$$

where the Raman tensor of A_g modes is $\begin{pmatrix} a & 0 & 0 \\ 0 & b & 0 \\ 0 & 0 & c \end{pmatrix}$, and δ is the phase factor due to the birefringence effect. The experimental data can now be fitted according to the birefringence-directed Raman selection rules.

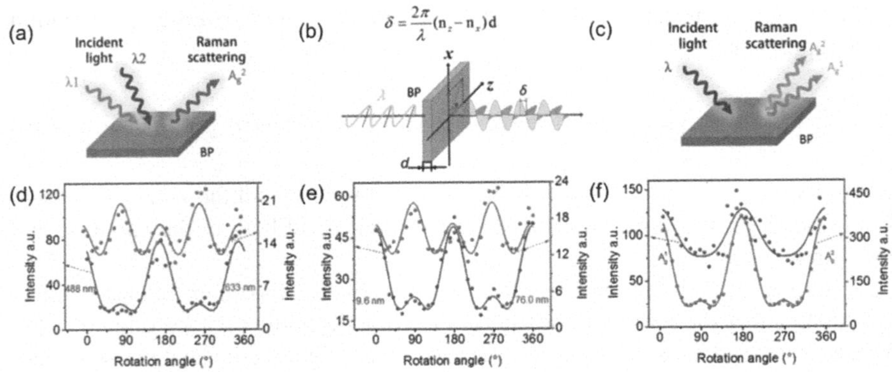

Fig. 3.3 Birefringence-directed Raman selection rules of the BP crystal. (**a**, **b**, **c**) Schematic illustration for the three key factors determining the birefringence effect in polarized Raman scattering of BP: (**a**) the incident wavelength (λ, *n*), (**b**) BP thickness and (**c**) the Raman modes (λ, *n*). (**d**, **e**, **f**) The angle-dependent Raman intensity of A_g modes under parallel polarization: (**d**) A_g^2 mode for thick BP (76 nm) with excitation of 488.0 nm (lower line) and 632.8 nm (upper line) laser; (**e**) A_g^2 mode for the thin (9.6 nm, lower line) and thick (76 nm, upper line) BP regions with excitation of 632.8 nm laser; (**f**) Both of A_g^1 (upper line) and A_g^2 (lower line) Raman modes for thin BP region with excitation of 514.5 nm laser. The solid lines are the fitting results [20]

According to the birefringence-corrected model, the phase shift is a key factor that influences the angle-dependence of the polarized Raman scattering. The phase difference between the two components along the armchair and zigzag directions due to birefringence can be estimated as

$$\delta = \frac{2\pi}{\lambda}(n_z - n_x)\,\Delta d \tag{3.12}$$

where n_x, n_z are the refractive indexes along the armchair and zigzag direction, λ is the wavelength and Δd is the sample thickness. Although, it is hard to quantitatively calculate the phase difference, a semi-quantitative description can be used to verify the determinant factors of the birefringence effect.

As shown in Fig. 3.3, the excitation laser wavelength, the sample thickness and the frequency of Raman modes determine the angular dependence of the Raman intensity. As shown in Fig. 3.3d, with 488 nm laser wavelength and 633 nm laser wavelength, the maximum angle located at armchair or zigzag direction determined by the different anisotropic factor *a*/*c*. The phase factors are different for different laser wavelengths, following the trend as $\delta_{514.5nm} > \delta_{488.0nm} > \delta_{632.8nm}$. With 633 nm excitation wavelength, the BP flakes with different thicknesses show different maximum angles, and the phase difference for 76 nm thickness is slightly larger than that for 9.6 nm. The anisotropic factor, the relative ratio between Raman tensor elements *a*/*c*, is also influenced by the sample thickness. The Raman modes have different frequencies, so the wavelength of the scattered light is different, although for simplicity, it is ignored in the present model, but the phase factors are not same.

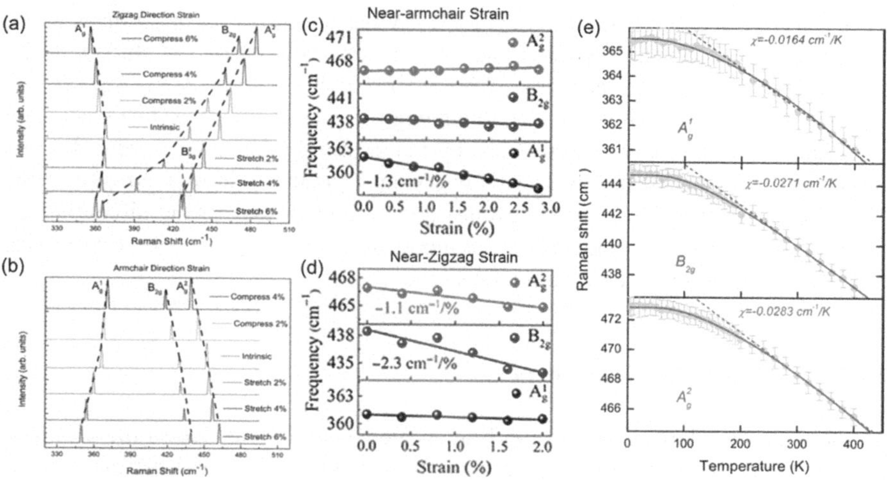

Fig. 3.4 (**a–b**) Raman spectra of BP under uniaxial strain along zigzag (**a**) and armchair (**b**). The peak height is in the logarithmic scale. The peak has a 4 cm^{-1} Gaussian smearing [23]. (**c**) Fitted Raman frequencies as functions of armchair strain and (**d**) zigzag strain [24]. (**e**) Temperature dependence of BP Raman modes in the range of 4–400 K [37]

The anisotropic factor *a*/*c* is dependent on the Raman frequency, so the angle-dependence of $A_g{}^1$ and $A_g{}^2$ modes show different patterns.

3.4.1.3 Strain Effect in BP Monitored by Raman Scattering

Strain is an effective tool to engineer the energy band structures of the materials. Monolayer black phosphorus can sustain tensile strain to 30% according to DFT calculations [50]. Owing to its unique honeycomb structure, it has obvious anisotropic mechanical properties, and the strain along different crystalline axes can be well monitored by Raman scattering measurements.

Fei et al. [23] have calculated the lattice vibrational modes and Raman spectra of strained black phosphorus using the first-principle linear response method. As shown in Fig. 3.4a and b, under strain along zigzag, different Raman modes exhibit different response. The B_{2g} mode and $A_g{}^2$ mode shifts monotonically. When the BP flake is stretched, they are red shifted, and when the BP flake is compressed, they are blue shifted. But for the $A_g{}^1$ mode, no matter the sample is stretched or compressed, the Raman frequency is red shifted. Under strain along the armchair direction, all the three modes show monotonic frequency shifts. The B_{2g} and $A_g{}^2$ modes red shift under compression and blue shift under stretching, while the $A_g{}^1$ mode shows opposite shift.

Wang et al. [24] measured the Raman spectra of the strained black phosphorus, as shown in Fig. 3.4c, d. They also found that the $A_g{}^1$ mode are sensitive to the armchair-strain, and it is linearly red shifted at the rate of -1.3 cm^{-1}/%, while the shifts of B_{2g} mode and $A_g{}^2$ are negligible. When the direction of strain is along the zigzag direction, the B_{2g} mode and $A_g{}^2$ mode show linear red shift with the stretching strain, and the rates are $-2.3\ cm^{-1}/\%$ and $-\ 1.1\ cm^{-1}/\%$, respectively. The $A_g{}^1$ mode is insensitive to the strain along the zigzag direction. The results are coincident with the theory calculations by Fei et al. [23] The difference in Raman frequency shifts can be briefly viewed in a simple picture of atomic arrangements. The displacement of atoms under strains along armchair and zigzag directions are different, so that the in-plane vibration modes and the out-plane vibration modes have different responses to the uniaxial strain.

3.4.1.4 Temperature Dependent

The temperature dependence of the Raman modes is obvious due to the electron-phonon interactions, inharmonic phonon-phonon interactions and thermal expansions. The temperature dependence of the Raman modes shifts is described by the Gruneisen model [51]:

$$\omega(T) = \omega_0 + \chi T \tag{3.13}$$

where ω_0 is the Raman peak position of black phosphorus at the temperature of zero Kelvin, and χ is the first-order temperature coefficient of the same mode.

Lapinska et al. [37] measured the temperature dependence of BP flakes with 6 nm thickness from 4 K to 400 K. As shown in Fig. 3.4e, above 250 K, the temperature dependence of the three Raman modes is linear and follows the Gruneisen model. The χ of $A_g{}^1$, B_{2g} and $A_g{}^2$ modes are $-0.0164\ cm^{-1}/K$, $-0.0271\ cm^{-1}/K$ and $-\ 0.0283\ cm^{-1}/K$, respectively. Below 250 K, the temperature dependence is not linear owing to that the optical phonon decays into two or three acoustic phonons. The FWHM and the intensity ratio of different Raman modes both show temperature dependence. Late et al. [52] measured the temperature dependence of BP from 78 K to 573 K and also obtained the first-order temperature coefficients of the three Raman modes.

3.4.2 Other Orthorhombic 2D Materials

Qiu et al [53] measured the angle-resolved Raman spectra of $ZrTe_5$. $ZrTe_5$ has attracted much attention owing to its unique properties [54–57]. It is a Dirac semimetal, which is similar to graphene and exhibits topologic properties. The structure of $ZrTe_5$ is shown in Fig. 3.5a and b. It can be seen that the Zr chains are along the crystalline *a*-axis, and the *c*-axis is perpendicular to the *a*-axis. The 2D planes piled up form the crystal along the *b*-axis. $ZrTe_5$ belongs to the $D_{12}{}^7$ point space and *Cmcm* space group and there are 12 atoms in a unit cell, so at

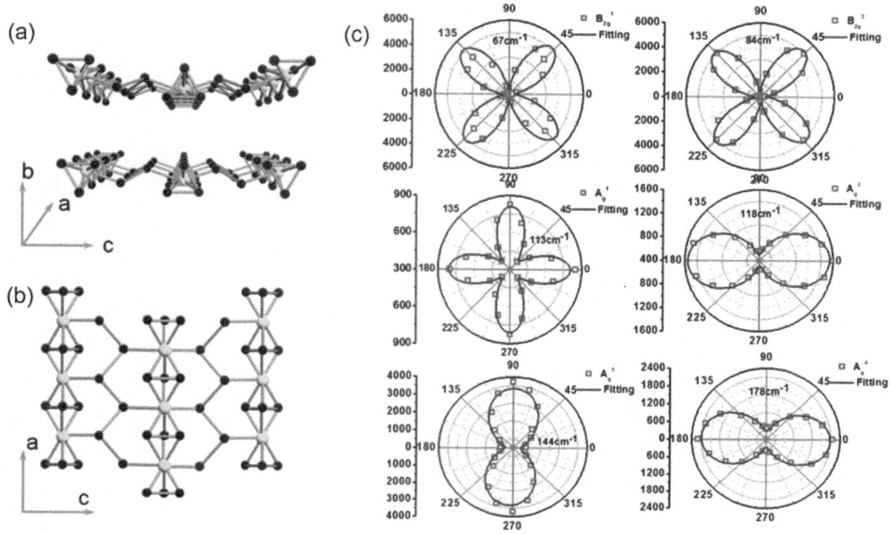

Fig. 3.5 (**a**) Crystal structure of $ZrTe_5$. (**b**) Top view of monolayer $ZrTe_5$. (**c**) The angle-resolved polarized Raman spectra of $B_{2g}{}^1$, $B_{2g}{}^2$, $A_g{}^1$, $A_g{}^2$, $A_g{}^3$, and $A_g{}^4$ Raman modes which are located at 67, 84, 113, 118, 144, and 178 cm^{-1} [53]

the Γ point, it has 36 phonon modes. Although there are 18 Raman-active modes, only a few of them can be detected including the A_g modes and B_{2g} modes under the backscattering geometry. As shown in Fig. 3.5c, all the A_g modes and B_{2g} modes exhibit polarized Raman responses under parallel configuration, and can be explained by the Raman selection rules.

WTe_2 is a semi-metallic layered material, which has a distorted 1 T phase, and belongs to the orthorhombic crystal system [58, 59]. The reduced symmetry leads to the in-plane anisotropy. WTe_2 belongs to the $C_{2v}{}^7$ point space and *Pmn2₁* space group, so it has 33 optical phonon modes at the Brillion zone center:

$$\Gamma_{bulk} = 11A_1 + 6A_2 + 5B_1 + 11B_2 \tag{3.14}$$

which are all Raman-active modes. However, under the backscattering geometry, only the A_1 and A_2 modes can be detected. All the A_1 and A_2 modes showed angle-dependence that can be explained by the Raman selection rules by considering the complex Raman tensor elements. The polar pattern of the Raman intensity is determined by the real parts and the imaginary parts of Raman tensor elements.

A family of Group IV monochalcogenides layered materials, including SnSe, [60, 61] SnS, [62] GeSe and GeS, [63] has the similar crystal structure, which is familiar with black phosphorus. T_d-$MoTe_2$ [64, 65] also has similar crystal structure as WTe_2 with distorted 1 T phase.

3.5 Monoclinic Crystal System

3.5.1 GaTe

GaTe is a layered material with monoclinic crystal structure. As shown in Fig. 3.6a and b, there are six Ga atoms and six Te atoms in a unit cell. GaTe crystal has C_{2h}^3 symmetry. At the center of the Brillion zone, there are 36 phonon modes, and 18 modes of them are Raman-active modes. Huang et al. [28] reported the calculated and experimental Raman scattering of GaTe. Most of the Raman-active modes can be observed, as shown in Fig. 3.6c, including the peaks at 107 cm^{-1}, 115 cm^{-1}, 126 cm^{-1}, 142 cm^{-1}, 161 cm^{-1}, 208 cm^{-1}, 268 cm^{-1}, and 280 cm^{-1} can be observed. According to the DFPT calculation, the Raman peaks of 107 cm^{-1}, 115 cm^{-1}, 208 cm^{-1}, 268 cm^{-1}, and 280 cm^{-1} Raman peaks are assigned to the A_g modes, the 161 cm^{-1} peak belongs to the B_g mode, and the 126 cm^{-1}, 142 cm^{-1} peaks are of double frequencies two modes between 60~70 cm^{-1}.

The angle-resolved polarized Raman spectra of GaTe was measured under 633 nm laser wavelength excitation as shown in Fig. 3.6d. The angle 0° is coincident

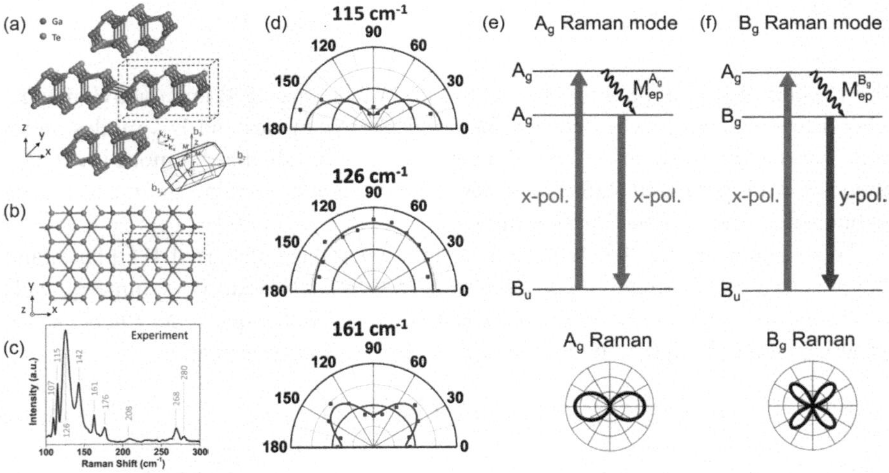

Fig. 3.6 (**a**) Crystal structure of bulk GaTe. (**b**) Top view of monolayer GaTe. The black dashed boxes in (**a**) and (**b**) correspond to the same group of atoms. Inset of (**a**): Brillouin zone for the primitive unit cell of bulk GaTe. (**c**) Experimental Raman spectrum of a 125 nm thick GaTe flake at room temperature and under vacuum (10^{-5} mbar). The measurement was performed with 532 nm laser excitation. (**d**) The Raman intensity vs. polarization angle with 633 nm excitation wavelength for three typical Raman modes: A_g mode (115 cm^{-1}), double-resonant mode (126 cm^{-1}) and B_g mode (161 cm^{-1}). The squares are experimental values and the curves are numerical fittings. (**e, f**) One of the expected transitions for Raman scattering and polarization dependence for the A_g (**e**) and B_g (**f**) modes at the Z or P point in Brillouin zone. M_{ep}^{Ag} (M_{ep}^{Bg}) indicates the electron−phonon interaction emitting an A_g (B_g) phonon. The x-pol. and y-pol. arrows indicate the transition activated by x- and y-polarized light, respectively [28]

with the x-direction. The 115 cm^{-1} Raman peak has a 180° variation period, indicating that it belongs to the A_g mode, and the 161 cm^{-1} has a 90° variation period, so it belongs to the B_g mode. The experimental results are similar with the DFPT calculations. The 126 cm^{-1} Raman peak does not show obvious polarization dependence, because the double-resonance Raman modes are contributed by the combination of different processes.

The angle resolved Raman spectra of different thicknesses under different photon energies were also measured. The A_g modes under 633 nm and 785 nm excitation laser wavelength have a 180° variation period, and all the maximum angles locate at 0°. However, under 532 nm excitation laser wavelength, the maximum angles of A_g modes locate at 0° or 90°. The secondary maximums are owing to the dichroism and the birefringence effect. The B_g mode (161 cm^{-1}) has a 90° variation period in all situations.

The photon energy dependence is explained by the optical transitions selection rules. The Raman scattering process involves three sub-processes: (1) photon absorption, (2) electron-phonon interaction, and (3) photon emission. The optical transitions selection rules can be expressed as [66]:

$$I = \left| \sum\nolimits_{i,m,m'} \frac{\langle f|\nabla|m'\rangle\langle m'|H_{el-ph}|m\rangle\langle m|\nabla|i\rangle}{(E_L - \Delta E_{mi})(E_L - h\omega_v - \Delta E_{m'i})} \right|^2 \quad (3.15)$$

$\langle f|\nabla|m'\rangle$ corresponds to the electro-photon interaction of the photon emission and the $\langle m|\nabla|i\rangle$ corresponds to the electron-photon interaction of the photon absorption. For the A_g modes, both these two processes have the same polarization dependence, so they have a 180° variation period as shown in Fig. 3.6e. For the B_g mode, the two processes have different symmetry in Fig. 3.6f. The matrix element of the light absorption has x-polarization and the matrix element of the light emission has y-polarization, so the B_g mode have a 90° variation period. The A_g modes with 532 nm laser wavelength excitation can exhibit different polarization dependence, because the high photon energy can excite the electron to several energy bands with different symmetry.

3.5.2 *Layered Transition Metal Trichalcogenides (TMTCs)*

Layered transition metal trichalcogenides (TMTCs), a new family of two-dimensional materials, exhibit quasi-1D properties owing to its high in-plane anisotropy.

Transition metal trichalcogenides can be represented by the chemical formula of MX_3, where M is the group IV transition metal atom and X is a chalcogen element. There are 2 transition metal atoms and 6 chalcogen atoms in a unit cell. As shown in Fig. 3.7 [67]. Some M-X bonds are longer than the others to form trigonal prismatic configuration and the metal chain is along the b-axis.

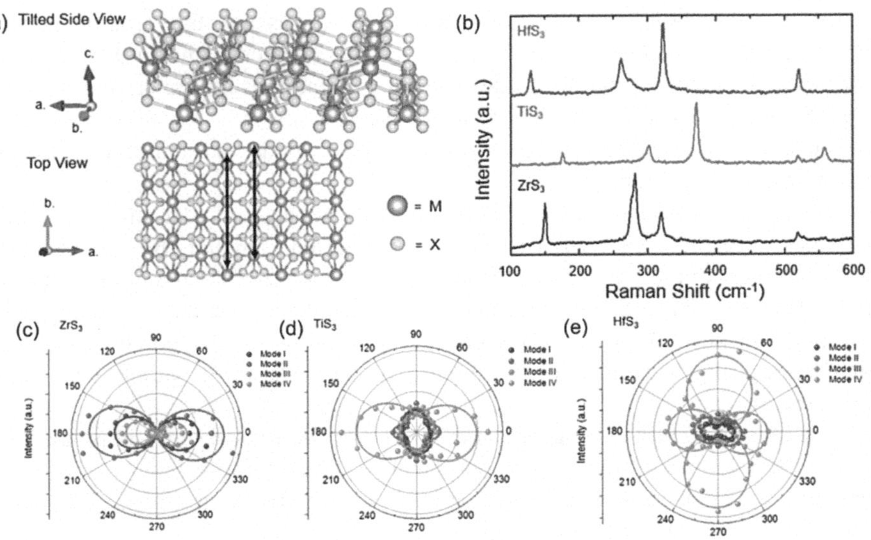

Fig. 3.7 (**a**) Tilted side view of a typical MX_3 crystal structure. A top view is shown with the black arrows indicating adjacent chain-like structures. The "M" atoms represent the transition metals while the "X" atoms represent the chalcogen atoms. (**b**) Raman spectra obtained for TiS_3, ZrS_3, and HfS_3 bulk flakes using a 488 nm laser. (**c**) Polar plot for all the modes of ZrS_3 with mode III corresponding to 320 cm^{-1}. (**d**) Polar plot of all the modes of TiS_3 modes with mode III at 372 cm^{-1}. (**e**) Polar plot of all the modes of HfS_3 with mode III at 321 cm^{-1} [67]

Kong et al. [67] reported the Raman spectra of TiS_3, ZrS_3 and HfS_3. Both the three materials belong to monoclinic crystal system and $p2_{1/m}$ space group. There are 24 vibration modes at the Γ point, including 21 optical branches and 3 acoustic branches, according to the DFT calculations. However, only a few Raman peaks can be observed. For TiS_3, only the Raman peaks at 172, 299, 372 and 557 cm^{-1} can be observed, and the four peaks of 151, 281, 320, 528 cm^{-1} for ZrS_3, and the 128, 260, 321, 524 cm^{-1} modes for HfS_3 can be observed. All the four Raman-active modes are A_g modes and are defined as Mode I, II, III, IV from lowest to highest frequencies.

The angle-resolved polarized Raman spectra of TiS_3, ZrS_3 and HfS_3 are also measured. θ is defined as the angle between the *b*-axis and the incident (scattered) light. Under parallel configuration, all the three modes in ZrS_3 have 180° variation period and the maximum Raman intensities appear at 0° and 180°, while the minimums appear at 90° and 270°. However, the modes of these three materials show different behaviors. In TiS_3, Mode III is more sensitive than the other three modes when the polarization angle is changed. The maximum is at 0° (180°) and the minimum at 90° (270°). In HfS_3, Mode III and Mode IV are both sensitive to the polarization, the maximum angles of Mode III are 0° and 180°, but the minimum angles of Mode IV are 90° and 270°. In the three materials, the bond lengths are different, so that the sensitivity of the modes of the same symmetry to the change of polarization angle differs. For the TMTCs, angle-resolved polarized Raman scattering is very effective to identify the crystalline orientation.

3.5.3 1T′-MoTe$_2$

The 2H-$MoTe_2$ and 1T′-$MoTe_2$ have different crystal symmetry. 2H-$MoTe_2$ belongs to the hexagonal crystal system and it is an in-plane isotropic material. Instead, bulk 1T′-$MoTe_2$ is a monoclinic crystal. It belongs to $C_{2h}{}^2$ point group and $p2_{1/m}$ space group. Below 250 K, the 1T′-$MoTe_2$ can convert into T_d-$MoTe_2$, which is a semi-metallic layered material and belongs to the orthorhombic crystal system. The optical anisotropy of the 1T′-$MoTe_2$ can be revealed by Raman scattering. Beams et al [68] reported the second harmonic generation and Raman scattering of 1T′-$MoTe_2$. For N-layer 1T′-$MoTe_2$, when N is odd, it belongs to $C_{2h}{}^2$ point group and there are 9N Raman-active modes including 6N A_g modes and 3N B_g modes. When N is even, it belongs to the $C_s{}^1$ point group and has 6N A′ and 3N A″ Raman-active modes. All the 9 Raman peaks can be measured under 532 nm laser wavelength excitation and show up in the range of 70∼260 cm^{-1}. However, when measuring few-layer 1T′-$MoTe_2$, the samples are easily damaged by the laser. The angle-resolved polarized Raman scattering was also measured under both parallel and cross polarization configurations. The Raman intensities of the A_g (A′) modes exhibit polarization dependence and can be explained by the Raman selection rules by considering the Raman tensor elements as complex values due to anisotropic absorption.

3.6 Triclinic Crystal System

3.6.1 ReS$_2$

ReS_2 has the lowest in-plane symmetry in the TMDs family [69, 70]. The Re atoms move away from the metal sites compared to MoS_2 or WS_2, leading to the distorted 1T′ phase [71]. Except the Re-S bonds, Re-Re bonds also exist in ReS_2. Four Re atoms constitute a Re_4 parallelogram. The zigzag chain of Re is defined as *b*-axis, as shown in Fig. 3.8. The reduced symmetry of ReS_2 leads to the in-plane anisotropy of electrical and optical properties [32, 36, 72, 73].

Bulk ReS_2 belongs to the C_i point group and *P-1(2)* space group, which has the lowest symmetry. There are 12 atoms in a unit cell owing to the Re_4 parallelogram, so at the center of the Brillouin zone, there are 18 A_g and 18 A_u modes [74]. All the 18 A_g modes are Raman-active modes and the 18 A_u modes include 15

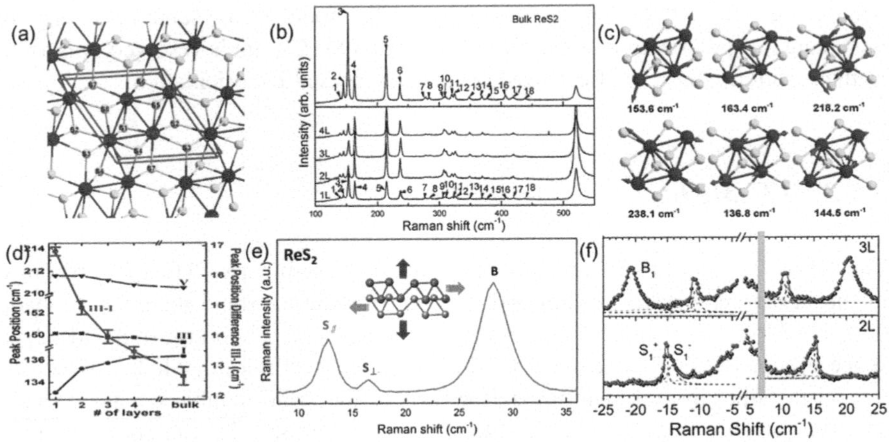

Fig. 3.8 (**a**) The lattice structure of bulk ReS_2. The parallelepiped is the primitive cell of bulk ReS_2. (**b**) Raman spectrum of bulk ReS_2 and few-layer ReS_2. The arrows mark the positions of the 18 Raman active modes. (**c**) The six vibrational modes at 136.8, 144.5, 153.6, 163.4, 218.2, and 238.1 cm^{-1} for bulk ReS_2 [74]. (**d**) Frequencies for modes I, III, and V of ReS_2, as a function of thickness (1L, 2L, 3L, 4L, and bulk) on the left vertical axis with the frequency difference between modes I and III on the right [31]. (**e**) The Raman spectrum of bilayer (2L) ReS_2 of stacking mode I [75]. (**f**) The Raman spectra of bilayer (2L) and trilayer (3L) ReS_2 of stacking mode 2 [76]

infrared-active modes and 3 optical inactive modes. The Raman tensor of the A_g mode can be represented as $\begin{pmatrix} u & v \\ v & w \end{pmatrix}$ in two-dimensions.

Feng et al. [74] reported the Raman vibrations of bulk ReS_2. According to their calculation results, the 18 Raman peaks of ReS_2 can be divided into two parts. The Raman peaks below 250 cm^{-1} belong to the vibrations modes involving the Re atoms and the Raman peaks above 250 cm^{-1} correspond to vibrations involving the S atoms. Some of the modes are shown in Fig. 3.8c. The 136.8 cm^{-1} and 144.5 cm^{-1} peaks involve the out-plane Re atoms vibrations and the 153.6 cm^{-1}, 163.4 cm^{-1}, 218.2 cm^{-1} and 238.1 cm^{-1} are contributed by the in-plane Re atoms vibrations. The calculated Raman shifts are slightly different from the experimental data. The Raman-active modes are defined as Mode I, Mode II, Mode III, Mode IV etc. from lowest to highest Raman frequency. The experimental Raman spectra are shown in Fig. 3.8b.

Chenet et al. [31] measured the Raman spectra of 1L, 2L, 3L, 4L and bulk ReS_2 as shown in Fig. 3.8d. With the thickness increasing, the peak position of the Mode I blue-shifts from 133.1 cm^{-1} for monolayer to 136.2 cm^{-1} for 4L. Mode III and Mode V red-shift when the layer number is increased. The frequency difference between mode III and mode I is 16.8 cm^{-1} for monolayer, 14.9 cm^{-1} for bilayer, 14.0 cm^{-1} for trilayer, 13.5 cm^{-1} for tetralayer and 12.7 cm^{-1} for bulk. The Raman peak positions can be used to identify the thickness in few layers.

3.6.2 Ultralow-Frequency Raman Spectra of ReS_2

The low-frequency Raman spectroscopy is an effective method to reveal the interlayer coupling and the stacking order of atomically thin ReS_2. He *et al* [75] and Lorchat et al. [76] measured two kinds of low-frequency Raman spectra of few-layered ReS_2 owing to the two different interlayer stacking. The bilayer and few-layers ReS_2 exhibit shear modes and breathing modes below 50 cm^{-1}. The behaviors of breathing modes are similar with other TMDS, and can be explained by a linear chain model. There are two shear modes which are non-degenerated owing to the anisotropic crystal structure. The low-frequency Raman spectra of stacking mode 1 are shown in Fig. 3.8e. The two shear modes of 2 L ReS_2 can be resolved and are separated by 3.5 cm^{-1}. The interlayer modes show obvious polarization dependence and the three modes have different angular dependence. Furthermore, the Raman intensities of the shear modes are weak than the breathing modes. This is quite different from other TMDs because of the distorted and enlarged unit cell of ReS_2. The low-frequency Raman spectra of stacking mode 2 are shown in Fig. 3.8f. Only two interlayer modes are can be observed because the two non-degenerated shear modes are closed to each other and cannot be resolved in the Raman spectra. The linear chain model can be used to fit the Raman frequencies and interlayer force constant. The force constants of the two shear modes are 17.8×10^{18} and 19.4×10^{18} N/m^3, and the force constant of the breathing modes are 69×10^{18} N/m^3, which is comparable to MoS_2, indicating rather strong interlayer coupling. Qiao et al. [77] and Nagler et al [78] also systematically studied the ultralow frequency Raman spectra of ReS_2.

3.6.3 Polarized Raman Spectra of ReS_2

Due to the reduced symmetry, ReS_2 shows strong in-plane anisotropy. Chenet et al. [31] measured the angle-resolved Raman spectra of ReS_2 with 532 nm laser excitation wavelength for 1–4 L. The polarization dependence of the Raman intensity in Mode III and Mode V are shown in Fig. 3.9a and b. Under parallel configuration, both the Mode III and Mode V exhibit 180° variation period, which is coincident with the Raman selection rule. The maximum angles of different Raman-active modes are determined by the absolute values of the Raman tensor elements. For Mode III of 1–4L samples, the angles of the maximum Raman intensities are almost the same. In contrast, the angle of Mode V for monolayer is 83° for 1L, 90° for 2L, and 91° for 3L and 4L, which may be caused by the strain, interlayer coupling or substrate effect. TEM characterization shows that the maximum axis of Mode V in few layers is parallel to the zigzag Re-Re chain (*b*-axis).

Compared to other TMDs, ReS_2 has the lowest (triclinic) symmetry and only has an inversion center. Turning over ReS_2 flake is not a symmetry operation, however, the orientation in the vertical direction is reverted. This is unique for ReS_2

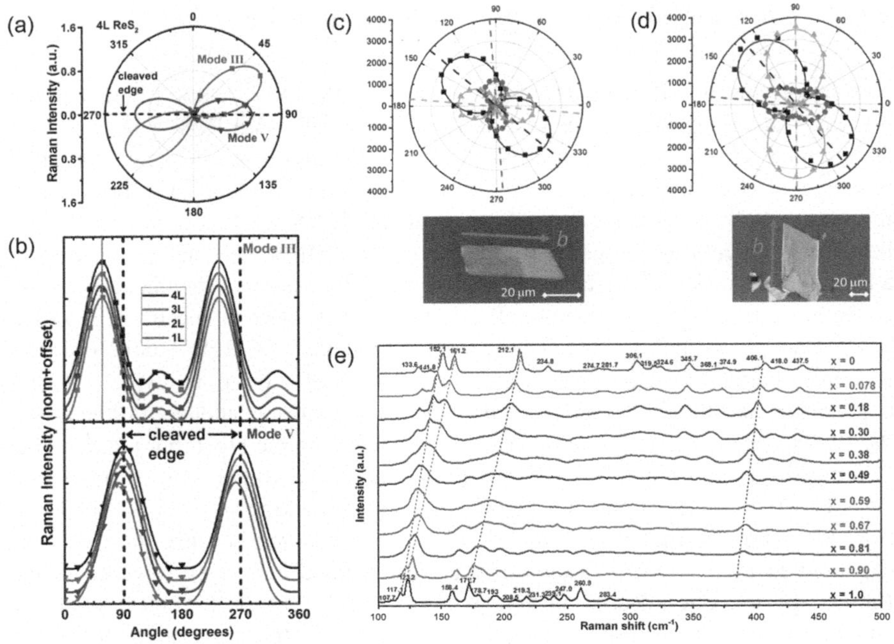

Fig. 3.9 (**a**) Polar plots of Raman intensities of modes III (squares) and V (triangles) for a 4L sample of ReS_2. (**b**) Variation of the intensities of modes III and Mode V for samples of 1L–4L thickness. The data for each thickness are normalized in the plot. The gray solid lines represent the angle of maximum intensities [31]. (**c, d**) Intensities as a function of angle with respect to the laboratory *x*-axis of the three peaks in the Raman spectra of ReS_2 flakes (thickness >100 nm) which were cleaved from the same crystal but with opposite vertical orientations: (**c**) "upward" and (**d**) "downward". The insets are the optical images of the ReS_2 flakes with *b*-axis indicated by the arrow. The chosen peaks are from modes III−V, that is, the 150, 160, and 211 cm^{-1} peaks (black squares, circles, and triangles), respectively [79]. (**e**) Composition-dependent Raman spectra of $ReS_{2(1-x)}Se_{2x}$ monolayers [80]

and does not hold for materials of higher symmetry. Hart et al. [79] reported the identification of the vertical orientation by Raman spectra. The Raman tensor of the Raman-active modes in ReS_2 is $\begin{pmatrix} u & v \\ v & w \end{pmatrix}$. If the ReS_2 flake is flipped upside down, the Raman tensor changes to $\begin{pmatrix} u & -v \\ -v & w \end{pmatrix}$. According to the Raman selection rules, when under a polarized incident light, the Raman intensity is $|R \bullet e_i|^2$. There are two kinds of angle-dependence which are determined by the vertical orientation: (1) $I_s(\theta) \propto u^2cos^2\theta + w^2sin^2\theta + v^2 + 2v(u + w)sin\theta cos\theta$ and (2) $I_s(\theta) \propto u^2cos^2\theta + w^2sin^2\theta + v^2 - 2v(u + w)sin\theta cos\theta$. As shown in Fig. 3.9c, d, for flake 1, the difference of the maximum angles with Mode III and Mode V is −45° and the difference of the maximum angles with Mode IV and Mode V is 39°. For flake 2, the differences become 45° and −36° respectively. These results are in good agreement with the Raman tensor analysis and the method can be used to identify the vertical orientation of ReS_2 flakes.

3.6.4 $ReS_{2(1-x)}Se_{2x}$ Alloys

$ReSe_2$ has the same crystal structure with ReS_2, so it is possible to alloy these two materials and the properties can be engineered by change the compositions of S and Se atoms in the alloy. Wen et al [80] fabricated $ReS_{2(1-x)}Se_{2x}$ alloys with different compositions. Raman scattering characterizations were performed on samples with different x values. The Raman peaks of $ReSe_2$ are below 300 cm^{-1}, while the ReS_2 are in the range of 100~450 cm^{-1}. With the increase of Se atoms in the alloy, the Raman peaks are red shifted as shown in Fig. 3.9e. It is seen that, in the low-frequency range, the Raman peaks still exist with different Se compositions, since these vibration modes are mostly originated from the vibrations of the Re atoms. In the middle-frequency range (160–300 cm^{-1}), most peaks were observed with high Se compositions. In the high-frequency range (300–450 cm^{-1}), the peaks disappeared with Se composition increases. The Raman spectra of $ReS_{2(1-x)}Se_{2x}$ alloys also exhibit polarization dependence owing to the reduced symmetry.

3.7 Summary

With the discovery of emerging new members in the 2D materials family, the Raman characterization of the structures and properties of such materials remains essential. The in-plane symmetry of 2D layered materials adds an important measure to the characterization of the materials. In this chapter, we discussed the Raman selection rules of polarized Raman scattering of anisotropic 2D layered materials, the identification of crystalline orientation using angle resolved polarized Raman scattering, and summarized the recent advances of Raman scattering studies on anisotropic 2D materials by the order of symmetry reduction. As a matter of fact, if we only consider the in-plane symmetry, only a 2 by 2 Raman tensor is used in the Raman selection rules, and all the anisotropic 2D layered materials can be cataloged into the three symmetries: orthorhombic, monoclinic and triclinic. Thus this chapter could be useful for the Raman scattering studies of any new 2D materials of low symmetry.

Acknowledgments This work was supported by the Ministry of Science and Technology of China (2016YFA0200100 and 2015CB932400) and the National Natural Science Foundation of China (51432002, 51720105003, 21790052, 11374355 and 21573004).

References

1. K.S. Novoselov, A.K. Geim, S.V. Morozov, D. Jiang, Y. Zhang, S.V. Dubonos, I.V. Grigorieva, A.A. Firsov, Electric field effect in atomically thin carbon films. Science **306**(5696), 666–669 (2004). https://doi.org/10.1126/science.1102896
2. Y. Kubota, K. Watanabe, O. Tsuda, T. Taniguchi, Deep ultraviolet light-emitting hexagonal boron nitride synthesized at atmospheric pressure. Science **317**(5840), 932–934 (2007). https://doi.org/10.1126/science.1144216
3. B. Radisavljevic, A. Radenovic, J. Brivio, V. Giacometti, A. Kis, Single-layer MoS2 transistors. Nat. Nanotechnol. **6**(3), 147–150 (2011). https://doi.org/10.1038/nnano.2010.279
4. H.O.H. Churchill, P. Jarillo-Herrero, Two-dimensional crystals: Phosphorus joins the family. Nat. Nanotechnol. **9**(5), 330–331 (2014). https://doi.org/10.1038/nnano.2014.85
5. L.K. Li, Y.J. Yu, G.J. Ye, Q.Q. Ge, X.D. Ou, H. Wu, D.L. Feng, X.H. Chen, Y.B. Zhang, Black phosphorus field-effect transistors. Nat. Nanotechnol.. **9**(5), 372–377 (2014). https://doi.org/10.1038/nnano.2014.35
6. L. Meng, Y.L. Wang, L.Z. Zhang, S.X. Du, R.T. Wu, L.F. Li, Y. Zhang, G. Li, H.T. Zhou, W.A. Hofer, H.J. Gao, Buckled silicene formation on Ir(111). Nano Lett.. **13**(2), 685–690 (2013). https://doi.org/10.1021/nl304347w
7. A. Fleurence, R. Friedlein, T. Ozaki, H. Kawai, Y. Wang, Y. Yamada-Takamura, Experimental evidence for epitaxial silicene on diboride thin films. Phys. Rev. Lett.. **108**(24), 245501 (2012). https://doi.org/10.1103/PhysRevLett.108.245501
8. P. Vogt, P. De Padova, C. Quaresima, J. Avila, E. Frantzeskakis, M.C. Asensio, A. Resta, B. Ealet, G. Le Lay, Silicene: compelling experimental evidence for graphenelike two-dimensional silicon. Phys. Rev. Lett.. **108**(15), 155501 (2012). https://doi.org/10.1103/PhysRevLett.108.155501
9. L.F. Li, S.Z. Lu, J.B. Pan, Z.H. Qin, Y.Q. Wang, Y.L. Wang, G.Y. Cao, S.X. Du, H.J. Gao, Buckled germanene formation on Pt(111). Adv. Mater.. **26**(28), 4820–4824 (2014). https://doi.org/10.1002/adma.201400909
10. Q.L. Feng, Y.M. Zhu, J.H. Hong, M. Zhang, W.J. Duan, N.N. Mao, J.X. Wu, H. Xu, F.L. Dong, F. Lin, C.H. Jin, C.M. Wang, J. Zhang, L.M. Xie, Growth of large-area 2D $MoS_{2(1-x)}Se_{2x}$ semiconductor alloys. Adv. Mater. **26**(17), 2648–2653 (2014). https://doi.org/10.1002/adma.201306095
11. D.O. Dumcenco, K.Y. Chen, Y.P. Wang, Y.S. Huang, K.K. Tiong, Raman study of 2H-Mo1-xWxS2 layered mixed crystals. J. Alloys Compd. **506**(2), 940–943 (2010). https://doi.org/10.1016/j.jallcom.2010.07.120
12. M. Zhang, J.X. Wu, Y.M. Zhu, D.O. Dumcenco, J.H. Hong, N.N. Mao, S.B. Deng, Y.F. Chen, Y.L. Yang, C.H. Jin, S.H. Chaki, Y.S. Huang, J. Zhang, L.M. Xie, Two-dimensional molybdenum tungsten diselenide alloys: photoluminescence, Raman scattering, and electrical transport. ACS Nano **8**(7), 7130–7137 (2014). https://doi.org/10.1021/nn5020566
13. C. Lee, H. Yan, L.E. Brus, T.F. Heinz, J. Hone, S. Ryu, Anomalous lattice vibrations of single- and few-layer MoS2. ACS Nano **4**(5), 2695–2700 (2010). https://doi.org/10.1021/nn1003937
14. Z.H. Ni, Y.Y. Wang, T. Yu, Z.X. Shen, Raman spectroscopy and imaging of graphene. Nano Res. **1**(4), 273–291 (2008). https://doi.org/10.1007/s12274-008-8036-1
15. J.M.B.L. dos Santos, N.M.R. Peres, A.H. Castro, Graphene bilayer with a twist: electronic structure. Phys. Rev. Lett. **99**(25), 256802 (2007). https://doi.org/10.1103/PhysRevLett.99.256802
16. J.G. He, K. Hummer, C. Franchini, Stacking effects on the electronic and optical properties of bilayer transition metal dichalcogenides MoS2, MoSe2, WS2, and WSe2. Phys. Rev. B **89**(7), 075409 (2014). https://doi.org/10.1103/PhysRevB.89.075409
17. S.X. Yang, J. Kang, Q. Yue, K. Yao, Vapor phase growth and imaging stacking order of bilayer molybdenum disulfide. J. Phys. Chem. C **118**(17), 9203–9208 (2014). https://doi.org/10.1021/jp500050r

18. Y.L. Wang, C.X. Cong, C.Y. Qiu, T. Yu, Raman spectroscopy study of lattice vibration and crystallographic orientation of monolayer MoS2 under uniaxial strain. Small **9**(17), 2857–2861 (2013). https://doi.org/10.1002/smll.201202876
19. F.N. Xia, H. Wang, Y.C. Jia, Rediscovering black phosphorus as an anisotropic layered material for optoelectronics and electronics. Nat. Commun. **5**, 4458 (2014). https://doi.org/10.1038/ncomms5458
20. N.N. Mao, J.X. Wu, B.W. Han, J.J. Lin, L.M. Tong, J. Zhang, Birefringence-directed Raman selection rules in 2D black phosphorus crystals. Small **12**(19), 2627–2633 (2016). https://doi.org/10.1002/smll.201600295
21. J.X. Wu, N.N. Mao, L.M. Xie, H. Xu, J. Zhang, Identifying the crystalline orientation of black phosphorus using angle-resolved polarized Raman spectroscopy. Angew. Chem.Int. Ed **54**(8), 2366–2369 (2015). https://doi.org/10.1002/anie.201410108
22. X.M. Wang, A.M. Jones, K.L. Seyler, V. Tran, Y.C. Jia, H. Zhao, H. Wang, L. Yang, X.D. Xu, F.N. Xia, Highly anisotropic and robust excitons in monolayer black phosphorus. Nat. Nanotechnol. **10**(6), 517–521 (2015). https://doi.org/10.1038/nnano.2015.71
23. R.X. Fei, L. Yang, Lattice vibrational modes and Raman scattering spectra of strained phosphorene. Appl. Phys. Lett. **105**, 083120 (2014). https://doi.org/10.1063/1.4894273
24. Y.L. Wang, C.X. Cong, R.X. Fei, W.H. Yang, Y. Chen, B.C. Cao, L. Yang, T. Yu, Remarkable anisotropic phonon response in uniaxially strained few-layer black phosphorus. Nano Res. **8**(12), 3944–3953 (2015). https://doi.org/10.1007/s12274-015-0895-7
25. T. Hong, B. Chamlagain, W.Z. Lin, H.J. Chuang, M.H. Pan, Z.X. Zhou, Y.Q. Xu, Polarized photocurrent response in black phosphorus field-effect transistors. Nanoscale **6**(15), 8978–8983 (2014). https://doi.org/10.1039/c4nr02164a
26. J. Xia, X.Z. Li, X. Huang, N.N. Mao, D.D. Zhu, L. Wang, H. Xu, X.M. Meng, Physical vapor deposition synthesis of two-dimensional orthorhombic SnS flakes with strong angle/temperature-dependent Raman responses. Nanoscale **8**(4), 2063–2070 (2016). https://doi.org/10.1039/c5nr07675g
27. L.D. Zhao, S.H. Lo, Y.S. Zhang, H. Sun, G.J. Tan, C. Uher, C. Wolverton, V.P. Dravid, M.G. Kanatzidis, Ultralow thermal conductivity and high thermoelectric figure of merit in SnSe crystals. Nature **508**(7496), 373–377 (2014). https://doi.org/10.1038/nature13184
28. S.X. Huang, Y. Tatsumi, X. Ling, H.H. Guo, Z.Q. Wang, G. Watson, A.A. Puretzky, D.B. Geohegan, J. Kong, J. Li, T. Yang, R. Saito, M.S. Dresselhaus, In-plane optical anisotropy of layered gallium telluride. ACS Nano **10**(9), 8964–8972 (2016). https://doi.org/10.1021/acsnano.6b05002
29. Z.X. Wang, K. Xu, Y.C. Li, X.Y. Zhan, M. Safdar, Q.S. Wang, F.M. Wang, J. He, Role of Ga vacancy on a multilayer GaTe phototransistor. ACS Nano **8**(5), 4859–4865 (2014). https://doi.org/10.1021/nn500782n
30. D.H. Keum, S. Cho, J.H. Kim, D.H. Choe, H.J. Sung, M. Kan, H. Kang, J.Y. Hwang, S.W. Kim, H. Yang, K.J. Chang, Y.H. Lee, Bandgap opening in few-layered monoclinic MoTe2. Nat. Phys. **11**(6), 482–486 (2015). https://doi.org/10.1038/nphys3314
31. D.A. Chenet, O.B. Aslan, P.Y. Huang, C. Fan, A.M. van der Zande, T.F. Heinz, J.C. Hone, In-plane anisotropy in mono- and few-layer ReS2 probed by Raman spectroscopy and scanning transmission Electron microscopy. Nano Lett. **15**(9), 5667–5672 (2015). https://doi.org/10.1021/acs.nanolett.5b00910
32. S. Tongay, H. Sahin, C. Ko, A. Luce, W. Fan, K. Liu, J. Zhou, Y.S. Huang, C.H. Ho, J.Y. Yan, D.F. Ogletree, S. Aloni, J. Ji, S.S. Li, J.B. Li, F.M. Peeters, J.Q. Wu, Monolayer behaviour in bulk ReS2 due to electronic and vibrational decoupling. Nat. Commun. **5**, 3252 (2014). https://doi.org/10.1038/ncomms4252
33. D. Wolverson, S. Crampin, A.S. Kazemi, A. Ilie, S.J. Bending, Raman spectra of monolayer, few-layer, and bulk ReSe2: an anisotropic layered semiconductor. ACS Nano **8**(11), 11154–11164 (2014). https://doi.org/10.1021/nn5053926
34. J.A. Yan, W.Y. Ruan, M.Y. Chou, Phonon dispersions and vibrational properties of monolayer, bilayer, and Trilayer graphene: density-functional perturbation theory. Phys. Rev. B **77**(12), 125401 (2008). https://doi.org/10.1103/PhysRevB.77.125401

35. J.X. Wu, H. Xu, W.H. Mu, L.M. Xie, X. Ling, J. Kong, M.S. Dresselhaus, J. Zhang, Observation of low-frequency combination and overtone Raman modes in Misoriented graphene. J. Phys. Chem. C **118**(7), 3636–3643 (2014). https://doi.org/10.1021/jp411573c
36. O.B. Aslan, D.A. Chenet, A.M. van der Zande, J.C. Hone, T.F. Heinz, Linearly polarized excitons in single- and few-layer ReS2 crystals. ACS Photonics **3**(1), 96–101 (2016). https://doi.org/10.1021/acsphotonics.5b00486
37. A. Lapinska, A. Taube, J. Judek, M. Zdrojek, Temperature evolution of phonon properties in few-layer black phosphorus. J. Phys. Chem. C **120**(9), 5265–5270 (2016). https://doi.org/10.1021/acs.jpcc.6b01468
38. F. Ahmed, Y.D. Kim, M.S. Choi, X. Liu, D.S. Qu, Z. Yang, J.Y. Hu, I.P. Herman, J. Hone, W.J. Yoo, High electric field carrier transport and power dissipation in multilayer black phosphorus field effect transistor with dielectric engineering. Adv. Funct. Mater. **27**(4), 1604025 (2017). https://doi.org/10.1002/adfm.201604025
39. R. Loudon, The Raman effect in crystals. Adv. Phys. **50**(7), 813–864 (2001). https://doi.org/10.1080/00018730110101395
40. J.B. Bates, A.S. Quist, Polarized Raman spectra of Beta-quartz. J. Chem. Phys. **56**(4), 1528–1533 (1972). https://doi.org/10.1063/1.1677402
41. K. Khaliji, A. Fallahi, L. Martin-Moreno, T. Low, Tunable Plasmon-enhanced birefringence in ribbon array of anisotropic two-dimensional materials. Phys. Rev. B **95**(20), 201401 (2017). https://doi.org/10.1103/PhysRevB.95.201401
42. C. Kranert, C. Sturm, R. Schmidt-Grund, M. Grundmann, Raman tensor formalism for optically anisotropic crystals. Phys. Rev. Lett. **116**(12), 127401 (2016). https://doi.org/10.1103/PhysRevLett.116.127401
43. V. Tran, R. Soklaski, Y.F. Liang, L. Yang, Layer-controlled band gap and anisotropic excitons in few-layer black phosphorus. Phys. Rev. B **89**(23), 235319 (2014). https://doi.org/10.1103/PhysRevB.89.235319
44. X. Ling, L.B. Liang, S.X. Huang, A.A. Puretzky, D.B. Geohegan, B.G. Sumpter, J. Kong, V. Meunier, M.S. Dresselhaus, Low-frequency interlayer breathing modes in few-layer black phosphorus. Nano Lett. **15**(6), 4080–4088 (2015). https://doi.org/10.1021/acs.nanolett.5b01117
45. X. Luo, X. Lu, G.K.W. Koon, A.H.C. Neto, B. Ozyilmaz, Q.H. Xiong, S.Y. Quek, Large frequency change with thickness in interlayer breathing mode-significant interlayer interactions in few layer black phosphorus. Nano Lett. **15**(6), 3931–3938 (2015). https://doi.org/10.1021/acs.nanolett.5b00775
46. H.B. Ribeiro, C.E.P. Villegas, D.A. Bahamon, D. Muraca, A.H.C. Neto, E.A.T. de Souza, A.R. Rocha, M.A. Pimenta, C.J.S. de Matos, Edge phonons in black phosphorus. Nat. Commun. **7**, 12191 (2016). https://doi.org/10.1038/ncomms12191
47. X. Ling, S.X. Huang, E.H. Hasdeo, L.B. Liang, W.M. Parkin, Y. Tatsumi, A.R.T. Nugraha, A.A. Puretzky, P.M. Das, B.G. Sumpter, D.B. Geohegan, J. Kong, R. Saito, M. Drndic, V. Meunier, M.S. Dresselhaus, Anisotropic electron-photon and electron-phonon interactions in black phosphorus. Nano Lett. **16**(4), 2260–2267 (2016). https://doi.org/10.1021/acs.nanolett.5b04540
48. J. Kim, J.U. Lee, J. Lee, H.J. Park, Z. Lee, C. Lee, H. Cheong, Anomalous polarization dependence of Raman scattering and crystallographic orientation of black phosphorus. Nanoscale **7**(44), 18708–18715 (2015). https://doi.org/10.1039/c5nr04349b
49. H.B. Ribeiro, M.A. Pimenta, C.J.S. de Matos, R.L. Moreira, A.S. Rodin, J.D. Zapata, E.A.T. de Souza, A.H.C. Neto, Unusual angular dependence of the Raman response in black phosphorus. ACS Nano **9**(4), 4270–4276 (2015). https://doi.org/10.1021/acsnano.5b00698
50. Q. Wei, X. Peng, Superior mechanical flexibility of phosphorene and few-layer black phosphorus. Appl. Phys. Lett. **104**(25), 251915 (2014). https://doi.org/10.1063/1.4885215
51. E.S. Zouboulis, M. Grimsditch, Raman scattering in diamond up to 1900 K. Phys. Rev. B **43**(15), 12490–12493 (1991). https://doi.org/10.1103/PhysRevB.43.12490
52. D.J. Late, Temperature dependent phonon shifts in few-layer black phosphorus. ACS Appl. Mater. Interfaces **7**(10), 5857–5862 (2015). https://doi.org/10.1021/am509056b

53. G. Qiu, Y.C. Du, A. Charnas, H. Zhou, S.Y. Jin, Z. Luo, D.Y. Zemlyanov, X.F. Xu, G.J. Cheng, P.D.D. Ye, Observation of optical and electrical in-plane anisotropy in high-mobility few-layer ZrTe5. Nano Lett. **16**(12), 7364–7369 (2016). https://doi.org/10.1021/acs.nanolett.6b02629
54. Y. Liu, X. Yuan, C. Zhang, Z. Jin, A. Narayan, C. Luo, Z. Chen, L. Yang, J. Zou, X. Wu, S. Sanvito, Z. Xia, L. Li, Z. Wang, F. Xiu, Zeeman splitting and dynamical mass generation in Dirac semimetal ZrTe5. Nat. Commun. **7**, 12516 (2016). https://doi.org/10.1038/ncomms12516
55. G. Zheng, J. Lu, X. Zhu, W. Ning, Y. Han, H. Zhang, J. Zhang, C. Xi, J. Yang, H. Du, K. Yang, Y. Zhang, M. Tian, Transport evidence for the three-dimensional Dirac semimetal phase inZrTe5. Phys. Rev. B **93**, 115414 (2016). https://doi.org/10.1103/PhysRevB.93.115414
56. H. Weng, X. Dai, Z. Fang, Transition-metal PentatellurideZrTe5andHfTe5: a paradigm for large-gap quantum spin hall insulators. Phys. Rev. X **4**, 011002 (2014). https://doi.org/10.1103/PhysRevX.4.011002
57. R. Wu, J.Z. Ma, S.M. Nie, L.X. Zhao, X. Huang, J.X. Yin, B.B. Fu, P. Richard, G.F. Chen, Z. Fang, X. Dai, H.M. Weng, T. Qian, H. Ding, S.H. Pan, Evidence for topological edge states in a large energy gap near the step edges on the surface ofZrTe5. Phys. Rev. X **6**, 021017 (2016). https://doi.org/10.1103/PhysRevX.6.021017
58. M. Kim, S. Han, J.H. Kim, J.U. Lee, Z. Lee, H. Cheong, Determination of the thickness and orientation of few-layer tungsten ditelluride using polarized Raman spectroscopy. 2d Materials **3**(3), 034004 (2016). https://doi.org/10.1088/2053-1583/3/3/034004
59. Q.J. Song, X.C. Pan, H.F. Wang, K. Zhang, Q.H. Tan, P. Li, Y. Wan, Y.L. Wang, X.L. Xu, M.L. Lin, X.G. Wan, F.Q. Song, L. Dai, The in-plane anisotropy of WTe2 investigated by angle-dependent and polarized Raman spectroscopy. Sci. Rep. **6**, 29254 (2016). https://doi.org/10.1038/srep29254
60. X.L. Xu, Q.J. Song, H.F. Wang, P. Li, K. Zhang, Y.L. Wang, K. Yuan, Z.C. Yang, Y. Ye, L. Dai, In-plane anisotropies of polarized Raman response and electrical conductivity in layered tin selenide. ACS Appl. Mater. Interfaces **9**(14), 12601–12607 (2017). https://doi.org/10.1021/acsami.7b00782
61. S.W. Luo, X. Qi, H. Yao, X.H. Ren, Q. Chen, J.X. Zhong, Temperature-dependent Raman responses of the vapor-deposited tin selenide ultrathin flakes. J. Phys. Chem. C **121**(8), 4674–4679 (2017). https://doi.org/10.1021/acs.jpcc.6b12059
62. Z. Tian, C.L. Guo, M.X. Zhao, R.R. Li, J.M. Xue, Two-dimensional SnS: a phosphorene analogue with strong in-plane electronic anisotropy. ACS Nano **11**(2), 2219–2226 (2017). https://doi.org/10.1021/acsnano.6b08704
63. D.Z. Tan, H.E. Lim, F.J. Wang, N.B. Mohamed, S. Mouri, W.J. Zhang, Y. Miyauchi, M. Ohfuchi, K. Matsuda, Anisotropic optical and electronic properties of two-dimensional layered germanium sulfide. Nano Res. **10**(2), 546–555 (2017). https://doi.org/10.1007/s12274-016-1312-6
64. Q.J. Song, Q.H. Tan, X. Zhang, J.B. Wu, B.W. Sheng, Y. Wan, X.Q. Wang, L. Dai, P.H. Tan, Physical origin of Davydov splitting and resonant Raman spectroscopy of Davydov components in multilayer MoTe2. Phys. Rev. B **93**(11), 115409 (2016). https://doi.org/10.1103/PhysRevB.93.115409
65. X.L. Ma, P.J. Guo, C.J. Yi, Q.H. Yu, A.M. Zhang, J.T. Ji, Y. Tian, F. Jin, Y.Y. Wang, K. Liu, T.L. Xia, Y.G. Shi, Q.M. Zhang, Raman scattering in the transition-metal dichalcogenides of 1T′-MoTe2, Td-MoTe2, and Td-WTe2. Phys. Rev. B **94**(21), 214105 (2016). https://doi.org/10.1103/PhysRevB.94.214105
66. A. Jorio, M. Dresselhaus, R. Saito, G. Dresselhaus, *Raman spectroscopy in graphene related systems* (Wiley VCH, Weinheim, 2011)
67. W. Kong, C. Bacaksiz, B. Chen, K.D. Wu, M. Blei, X. Fan, Y.X. Shen, H. Sahin, D. Wright, D.S. Narang, S. Tongay, Angle resolved vibrational properties of anisotropic transition metal trichalcogenide nanosheets. Nanoscale **9**(12), 4175–4182 (2017). https://doi.org/10.1039/c7nr00711f

68. R. Beams, L.G. Cancado, S. Krylyuk, I. Kalish, B. Kalanyan, A.K. Singh, K. Choudhary, A. Bruma, P.M. Vora, F. Tavazza, A.V. Da-Vydov, S.J. Stranick, Characterization of few-layer 1T′ MoTe2 by polarization-resolved second harmonic generation and Raman scattering. ACS Nano **10**(10), 9626–9636 (2016). https://doi.org/10.1021/acsnano.6b05127
69. K.F. Mak, C. Lee, J. Hone, J. Shan, T.F. Heinz, Atomically thin MoS2: A new direct-gap semiconductor. Phys. Rev. Lett. **105**(13), 136805 (2010). https://doi.org/10.1103/PhysRevLett.105.136805
70. H.S.S.R. Matte, A. Gomathi, A.K. Manna, D.J. Late, R. Datta, S.K. Pati, C.N.R. Rao, MoS2 and WS2 analogues of graphene. Angew. Chem.Int..Ed **49**(24), 4059–4062 (2010). https://doi.org/10.1002/anie.201000009
71. K. Friemelt, M.C. Luxsteiner, E. Bucher, Optical properties of the layered transition-metal-dichalcogenide ReS2: anisotropy in the van der waals plane. J. Appl. Phys. **74**(8), 5266–5268 (1993). https://doi.org/10.1063/1.354268
72. F.C. Liu, S.J. Zheng, X.X. He, A. Chaturvedi, J.F. He, W.L. Chow, T.R. Mion, X.L. Wang, J.D. Zhou, Q.D. Fu, H.J. Fan, B.K. Tay, L. Song, R.H. He, C. Kloc, P.M. Ajayan, Z. Liu, Highly sensitive detection of polarized light using anisotropic 2D ReS2. Adv. Funct. Mater. **26**(8), 1169–1177 (2016). https://doi.org/10.1002/adfm.201504546
73. E.F. Liu, Y.J. Fu, Y.J. Wang, Y.Q. Feng, H.M. Liu, X.G. Wan, W. Zhou, B.G. Wang, L.B. Shao, C.H. Ho, Y.S. Huang, Z.Y. Cao, L.G. Wang, A.D. Li, J.W. Zeng, F.Q. Song, X.R. Wang, Y. Shi, H.T. Yuan, H.Y. Hwang, Y. Cui, F. Miao, D.Y. Xing, Integrated digital inverters based on two-dimensional anisotropic ReS2 field-effect transistors. Nat. Commun. **6**, 6991 (2015). https://doi.org/10.1038/ncomms7991
74. Y.Q. Feng, W. Zhou, Y.J. Wang, J. Zhou, E.F. Liu, Y.J. Fu, Z.H. Ni, X.L. Wu, H.T. Yuan, F. Miao, B.G. Wang, X.G. Wan, D.Y. Xing, Raman vibrational spectra of bulk to monolayer ReS2 with lower symmetry. Phys. Rev. B **92**(5), 054110 (2015). https://doi.org/10.1103/PhysRevB.92.054110
75. R. He, J.A. Yan, Z.Y. Yin, Z.P. Ye, G.H. Ye, J. Cheng, J. Li, C.H. Lui, Coupling and stacking order of ReS2 atomic layers revealed by ultralow-frequency Raman spectroscopy. Nano Lett. **16**(2), 1404–1409 (2016). https://doi.org/10.1021/acs.nanolett.5b04925
76. E. Lorchat, G. Froehlicher, S. Berciaud, Splitting of interlayer shear modes and photon energy dependent anisotropic Raman response in N-layer ReSe2 and ReS2. ACS Nano **10**(2), 2752–2760 (2016). https://doi.org/10.1021/acsnano.5b07844
77. X.F. Qiao, J.B. Wu, L.W. Zhou, J.S. Qiao, W. Shi, T. Chen, X. Zhang, J. Zhang, W. Ji, P.H. Tan, Polytypism and unexpected strong interlayer coupling in two-dimensional layered ReS2. Nanoscale **8**(15), 8324–8332 (2016). https://doi.org/10.1039/c6nr01569g
78. P. Nagler, G. Plechinger, C. Schuller, T. Korn, Observation of anisotropic interlayer Raman modes in few-layer ReS2. Phys. Status Solidi Rapid Res. Lett **10**(2), 185–189 (2016). https://doi.org/10.1002/pssr.201510412
79. L. Hart, S. Dale, S. Hoye, J.L. Webb, D. Wolverson, Rhenium dichalcogenides: layered semiconductors with two vertical orientations. Nano Lett. **16**(2), 1381–1386 (2016). https://doi.org/10.1021/acs.nanolett.5b04838
80. W. Wen, Y.M. Zhu, X.L. Liu, H.P. Hsu, Z. Fei, Y.F. Chen, X.S. Wang, M. Zhang, K.H. Lin, F.S. Huang, Y.P. Wang, Y.S. Huang, C.H. Ho, P.H. Tan, C.H. Jin, L.M. Xie, Anisotropic spectroscopy and electrical properties of 2D ReS2(1-x)Se2x alloys with distorted 1T structure. Small **13**(12), 1603788 (2017). https://doi.org/10.1002/smll.201603788

Chapter 4
Raman Spectroscopy of van der Waals Heterostructures

C. H. Lui

4.1 Introduction to van der Waals Heterostructures

The research of two-dimensional (2D) atomic crystals has progressed rapidly since the isolation of graphene in 2004 [1, 2]. The family of 2D crystals now include many different types of materials, including metals (e.g. graphene, $NbSe_2$), semiconductors (e.g. phosphorene, MoS_2, WSe_2), insulators (e.g. BN), superconductors and charge-density-wave materials (e.g. $NbSe_2$ and $TiSe_2$). Alongside with the rapid development of individual 2D materials, the research frontier has also advanced to explore their hybrid systems [3, 4]. In particular, the flat and inert surfaces of 2D materials enable the construction of heterogeneous stacks of different 2D crystals with atomically sharp interfaces, coupled vertically only by van der Waals forces. These van der Waals heterostructures exhibit many unique properties that cannot be realized in individual 2D crystals [3, 4]. For instance, graphene on hexagonal boron nitride (BN) can exhibit the Hofstadter's butterfly phenomenon because of the nanoscale periodic interaction between the graphene and BN lattices [5–7]. Transition metal dichalcogenide (TMD) heterostructures can host long-lived interlayer excitons due to the staggered band alignment between different TMD layers [8]. Electronic and optoelectronic devices made from van der Waals heterostructures can exhibit performance superior to that of traditional devices with lateral 2D junctions [9–11]. More generally, the 2D building blocks can be combined to form more complex structures. By incorporating the unique properties of each class of 2D crystal (e.g., semiconducting TMDs, insulating BN and metallic graphene), integrated circuits can in principle be constructed entirely with 2D materials. Such 2D systems of electronics, once realized, could open a route to post-silicon technology.

C. H. Lui (✉)
Department of Physics and Astronomy, University of California, Riverside, CA, USA
e-mail: joshua.lui@ucr.edu

P.-H. Tan (ed.), *Raman Spectroscopy of Two-Dimensional Materials*, Springer Series in Materials Science 276, https://doi.org/10.1007/978-981-13-1828-3_4

The great promises of hybrid 2D materials have called for much effort to explore their rich physics. However, complete investigations of van der Waals heterostructures are rather challenging, given the plethora of 2D materials and the huge number of possible combined structures [3]. One effective approach is to first study several prototype structures, which feature specific characteristics of van der Waals heterostructures, and then use these primary units to construct and explore more complicated heterostructures. Therefore, the research of van der Waals heterostructures usually starts with a stack of only two layers of different 2D materials (called the heterobilayer) and proceeds to study structures with three and four layers. Despite their relative simplicity, the heterobilayers already exhibit many interesting properties, such as the emergence of interlayer vibrations, interlayer charge transfer, and interlayer correlated states (e.g. excitons). We will mainly focus on the investigation of heterobilayers in this chapter.

Raman spectroscopy has played a significant role in the research of van der Waals heterostructures. It is a convenient and non-destructive optical probe of many basic material characteristics. In this chapter, I will present some representative Raman studies of van der Waals heterostructures. We will start with a general introduction of heterostructure fabrication, and then present the investigation of interlayer charge transfer and interlayer vibration modes, and finally the study of the intralayer lattice distortion in the heterostructures.

4.2 Fabrication of van der Waals Heterostructures

The research of van der Waals heterostructures starts with the design and construction of a vertical stack of selected 2D materials. The heterostructures can generally be fabricated by two methods [4]. The first method is the mechanical co-lamination of 2D materials (Fig. 4.1). We first prepare individual 2D crystals, by either mechanical exfoliation or chemical vapor deposition (CVD), and then transfer and stack one 2D crystal on top of another one. The method of mechanical assembly is the most common one for scientific research, because it is versatile and relatively easy to implement, and the heterostructures thus fabricated can achieve very high quality. The other method is the direct growth of heterostructures by CVD or physical epitaxy (Fig. 4.1). This method of sample growth is more challenging because it requires sophisticated apparatus and specific expertise. However, this method can produce large-scale heterostructures, which are important for industrial applications.

The method of mechanical co-lamination of 2D materials was first developed by Dean et al. in 2010, who demonstrated very high performance of graphene devices deposited on BN substrates [12]. Their technique makes use of sacrificial membranes. As displayed in Fig. 4.2, a 2D crystal is first deposited on a double sacrificial layer. After dissolving the bottom layer, the top layer with the 2D crystal is lifted off. The crystal is then aligned in an upside-down way and placed on top of another 2D crystal. The bilayer structure is formed after dissolving the top

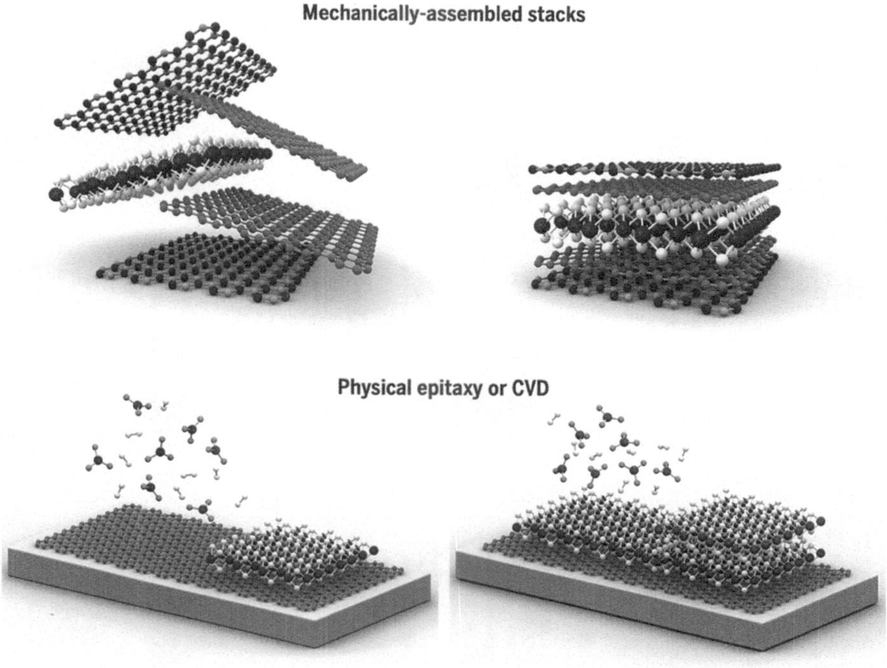

Fig. 4.1 Two ways to fabricate van der Waals heterostructures. (Top) mechanical assembly of of 2D crystals. (Bottom) Large-area growth of heterostructures by physical epitaxy or chemical vapor deposition (CVD). (The figure is adapted from Ref. [4])

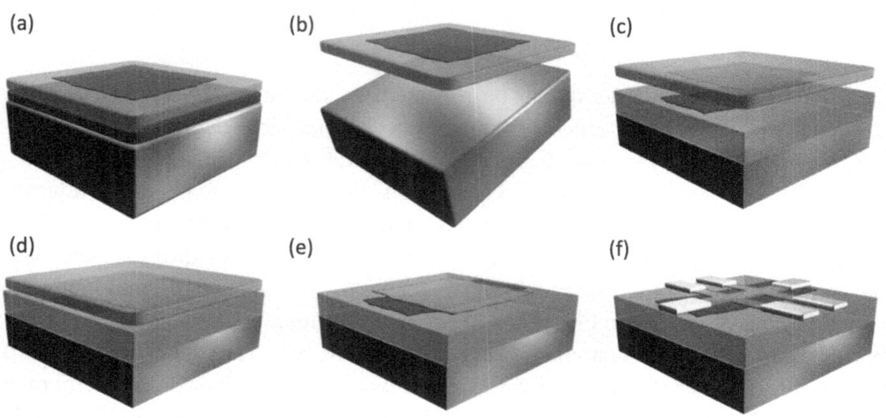

Fig. 4.2 Wet-transfer technique. A 2D crystal prepared on a double sacrificial layer (**a**) is lifted on the top layer by dissolving the bottom layer (**b**). The 2D crystal is then aligned upside-down (**c**) and placed (**d**) on top of another 2D crystal. Upon the removal of the top sacrificial membrane (**e**), the electrodes can be deposited to make it an electrical device (**f**). This process could be repeated to add more 2D crystals. (The figure is adapted from Ref. [4])

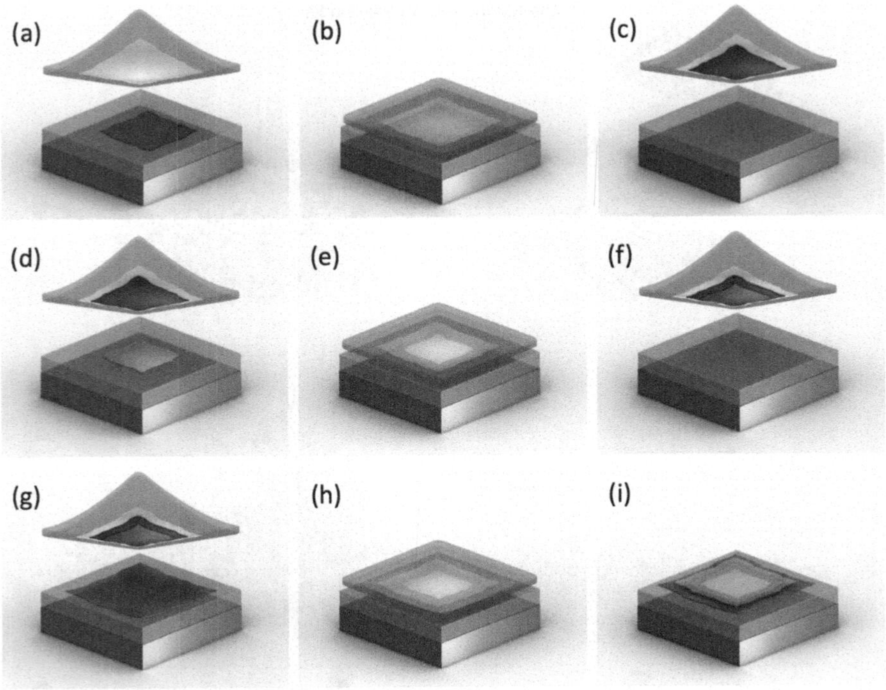

Fig. 4.3 The "pick-and-lift" technique. A 2D crystal on a membrane is aligned (**a**) and placed on top of another 2D crystal (**b**). Due to the adhesion between the two crystals, it is possible to lift both flakes using the same membrane (**c**). By repeating the process, we can pick up additional crystals (**d**–**f**). Finally, the whole stack is placed on the substrate and the top sacrificial membrane is dissolved to expose the entire stack (**g**–**i**). (The figure is adapted from Ref. [4])

sacrificial layer. This process can be repeated to add more 2D crystals. Finally, the heterostructure is patterned and made into an electrical device by depositing electrodes. There are other similar and improved versions of this technique, which involve different types of sacrificial layers and solvents. They are generally referred to as "wet-transfer" methods, because they involve chemical solutions.

After the successful implementation of the wet-transfer methods, researchers developed another, much cleaner, transfer method – the so-called "pick-and-lift" technique. As displayed in Fig. 4.3, a 2D crystal on a membrane is first aligned and brought into contact with another 2D crystal on a substrate. The two 2D crystals will attach to each other through van der Waals interactions. Afterward, the membrane is lifted off. In some cases, the van der Waals attraction between the two 2D crystals is stronger than the attraction between the bottom 2D crystal and the substrate. The bottom crystal will then be picked up by the membrane. This process can be repeated for several times to pick up more 2D crystals. Finally, the whole stack is placed on a substrate and the membrane on the top is dissolved. This technique is called the dry-transfer method because no chemical solutions are involved during the transfer process. The dry-transfer method is more effective than the wet-transfer method, because it is easier to implement and the heterostructures thus fabricated often have higher quality.

The quality of the interface between the 2D crystals is crucial to the performance of the heterostructure. In the transfer process, exposure to the ambient environment and chemical solvents can contaminate the interface between the 2D layers. However, the interfacial contaminants can usually be removed by annealing the samples in high temperature. Remarkably, some van der Waals heterostructures exhibit a "self-cleansing" mechanism, which spontaneously produces atomically clean interface [13]. This mechanism occurs when the affinity between the two 2D crystals is larger than the affinity between the crystals and the contaminants. In this situation, it is energetically more favorable to maximize the contact area between the 2D crystals, and the contaminants will be pushed away to form bubbles in some local areas. To further improve the interface quality, the whole exfoliation and transfer process can be carried out in a glovebox (or vacuum chamber) with low humidity and oxygen levels. A dry and inert environment is particularly important for assembling unstable 2D materials, such as $NbSe_2$, CrI_3 and phosphorene, which can degrade quickly in ambient conditions.

4.3 Interlayer Charge Transfer in Heterobilayers

Some of the most widely studied van der Waals heterostructures are the heterobilayers formed from two TMD monolayers, such as MoS_2, $MoSe_2$, WS_2 and WSe_2. High-quality monolayer TMD crystals can now be routinely grown through chemical vapor deposition. Their number of layers can be easily confirmed by optical contrast, photoluminescence (PL) and Raman spectroscopy, and atomic force microscopy (AFM). The single-crystal flakes with single-layer (1L) thickness usually exhibit triangular shape. Their crystalline orientation can be identified simply from their triangular shape, or more accurately, by second-harmonic optical spectroscopy. Chemical vapor deposition can produce numerous TMD flakes on one substrate. One can pick up all the flakes on the substrate together and transfer them onto another substrate with a different type of TMD flakes. In this way, many heterobilayer samples with various stacking orders can be obtained in one transfer. Here we will present the optical studies of two prototype TMD heterostructures, namely, the MoS_2/WSe_2 and $MoSe_2/MoS_2$ heterobilayers (see Fig. 4.4a–b for their lattice configuration). Figure 4.4c displays the optical image of a MoS_2/WSe_2 heterobilayer on a Si/SiO_2 substrate. The sample exhibits 1L MoS_2 area, 1L WSe_2 area, and their overlapping bilayer area.

Figure 4.4d–e display the representative PL spectra (in a logarithmic scale) of the MoS_2/WSe_2 and $MoSe_2/MoS_2$ bilayer stacks, in comparison with the spectra of monolayer areas of the same samples, under the excitation of continuous 532 nm laser. All 1L MoS_2, 1L WSe_2, and 1L $MoSe_2$ spectra exhibit strong PL signals, with their characteristic emission energies at ~1.85 eV, ~1.59 eV and ~1.56 eV,

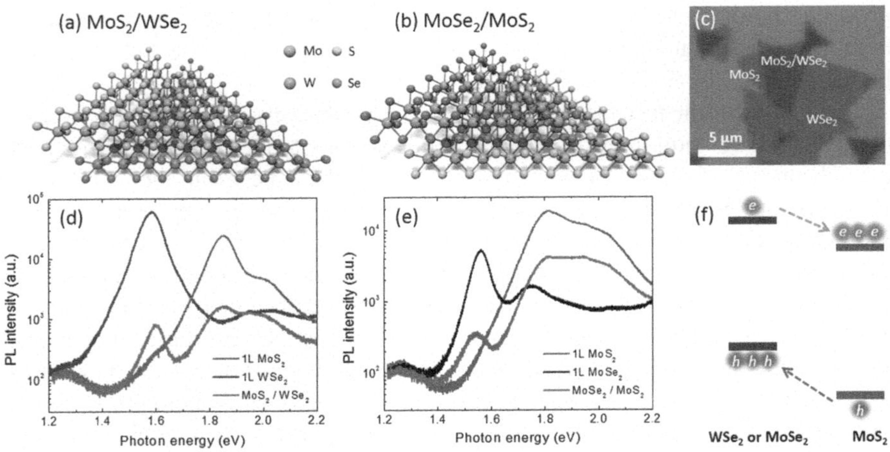

Fig. 4.4 (**a–b**) Schematic atomic configurations of the MoS_2/WSe_2 and $MoSe_2/MoS_2$ heterobilayers. (**c**) Optical image of a MoS_2/WSe_2 heterobilayer. (**d**) Photoluminescence (PL) spectrum of a MoS_2/WSe_2 heterobilayer in comparison with spectra of the 1L MoS_2 and 1L WSe_2 regions in the same sample under the same measurement conditions. (**e**) Similar PL spectra for a $MoSe_2/MoS_2$ heterobilayer. The excitation laser wavelength is 532 nm for all spectra. (**f**) Schematic of interlayer charge transfer in the heterostructure. The left two bars represent the conduction and valence band edges of 1L WSe_2 or $MoSe_2$. The right bars represent those of 1L MoS_2. They form type II heterostructures with staggered band gaps. The photoexcited electrons tend to flow from the WSe_2 ($MoSe_2$) layer to the MoS_2 layer, and the holes from the MoS_2 to WSe_2 ($MoSe_2$) layer. The resultant charge separation strongly suppresses the intralayer recombination processes and the PL. (The figure is adapted from Ref. [27]).

respectively [14, 15]. The observed PL comes from the recombination of excitons across the direct band gap at the K and K′ valleys of their respective band structures [14, 15]. Remarkably, the PL intensity is strongly reduced in the overlapping MoS_2/WSe_2 and $MoSe_2/MoS_2$ regions. The reduction of PL intensity can vary from a few tens of percent to two orders of magnitude for different heterobilayer samples due to their different quality. The PL suppression in the heterobilayers arises from the charge transfer between the two TMD layers. According to the calculated electronic band structure of different TMD materials, the conduction and valence band edges of 1L MoS_2 are lower than those of 1L WSe_2 and 1L $MoSe_2$, giving rise to type II band alignment with staggered gaps in their heterojunction (Fig. 4.4f) [16–18]. As a result, the photoexcited electrons in the WSe_2 ($MoSe_2$) layer tend to flow to the MoS_2 layer, and the holes in the MoS_2 layer tend to flow to the WSe_2 ($MoSe_2$) layer. The spatial separation of the electrons and holes therefore suppresses the intralayer optical recombination processes and thus the PL intensity in each individual layer. This strong PL suppression (up to two orders of magnitude) indicates that the charge transfer rate is close to the rate of exciton generation under continuous laser excitation. Such high transfer efficiency is consistent with the extremely fast charge transfer time (<50 fs) determined by ultrafast studies [19].

4.4 Interlayer Phonon Modes of van der Waals Heterostructures

(a) Observation of layer-breathing modes in heterobilayers

Few-layer graphene and TMDs possess a set of shear modes (SMs) and layer-breathing modes (LBMs) that involve lateral and vertical displacement of individual layers, respectively. These interlayer phonon modes have been extensively studied for individual 2D materials. Their properties are highly sensitive to the layer thickness, stacking order, and surface adsorbates of 2D materials [20–26]. Such interlayer phonons should also exist in van der Waals heterostructures, but their behavior is expected to be sensitive to the interfacial conditions and layer orientation of the heterostructure.

Lui et al. has carried out the first experimental investigation of interlayer phonon modes in TMD heterostructures by using ultralow-frequency Raman spectroscopy [27]. Figure 4.5 presents the Raman spectra of the MoS_2/WSe_2 and $MoSe_2/MoS_2$ heterobilayer samples. A pronounced Raman peak emerges at $\sim$32 cm^{-1} for both heterobilayers. This Raman feature is not observed in any TMD monolayers, indicating that it arises from the interaction between the two TMD layers (Fig. 4.5a, i). The same Raman feature has been observed in more than 70 heterobilayer samples, with frequencies varying between 30 and 35 cm^{-1} for both types of heterobilayers (Table 4.1; we will discuss the sample dependence in more detail later).

To understand the origin of the new Raman mode, Lui et al. make comparison with the Raman spectra of Bernal-stacked 2L MoS_2, 2L WSe_2 and 2L $MoSe_2$ samples (Fig. 4.5c–h). All these bilayer spectra exhibit two pronounced Raman peaks. The lower-frequency peaks (17–22 cm^{-1}) are the interlayer shear mode and the higher-frequency peaks (30–40 cm^{-1}) are the LBM. In addition, the line width of these interlayer modes generally increases with the increase of mode frequency. For instance, the full width at half maximum (FWHM) of the LBM increases from 3.5 to 9.5 cm^{-1} as its frequency increases from 2L WSe_2 (29.5 cm^{-1}) to 2L $MoSe_2$ (30.5 cm^{-1}) and to 2L MoS_2 (39.5 cm^{-1}), and the LBMs are also broader than the shear modes (Table 4.1). These observations reflect the increasing anharmonic decay rate of phonons at higher energy.

The frequency ($\sim$32 cm^{-1}) of the new Raman mode in the MoS_2/WSe_2 ($MoSe_2/MoS_2$) heterobilayer lies between the LBM frequencies of 2L MoS_2 and 2L WSe_2 (2L $MoSe_2$) (vertical dashed lines in Fig. 4.5). Also, the width of the new Raman mode (FWHM = 8 cm^{-1} and 5.5 cm^{-1} for MoS_2/WSe_2 and $MoSe_2/MoS_2$ heterobilayers, respectively) roughly matches those of the LBMs from Bernal bilayers (Table 4.1). Therefore, these observations indicate that the new Raman mode in the heterobilayers originates from the LBM vibration between the two TMD layers (Fig. 4.5b, j).

The LBM are observed in heterobilayers with different stacking configurations. This indicates that the formation of LBM is robust against the change of relative orientation (translation and rotation) between the two TMD layers. In contrast,

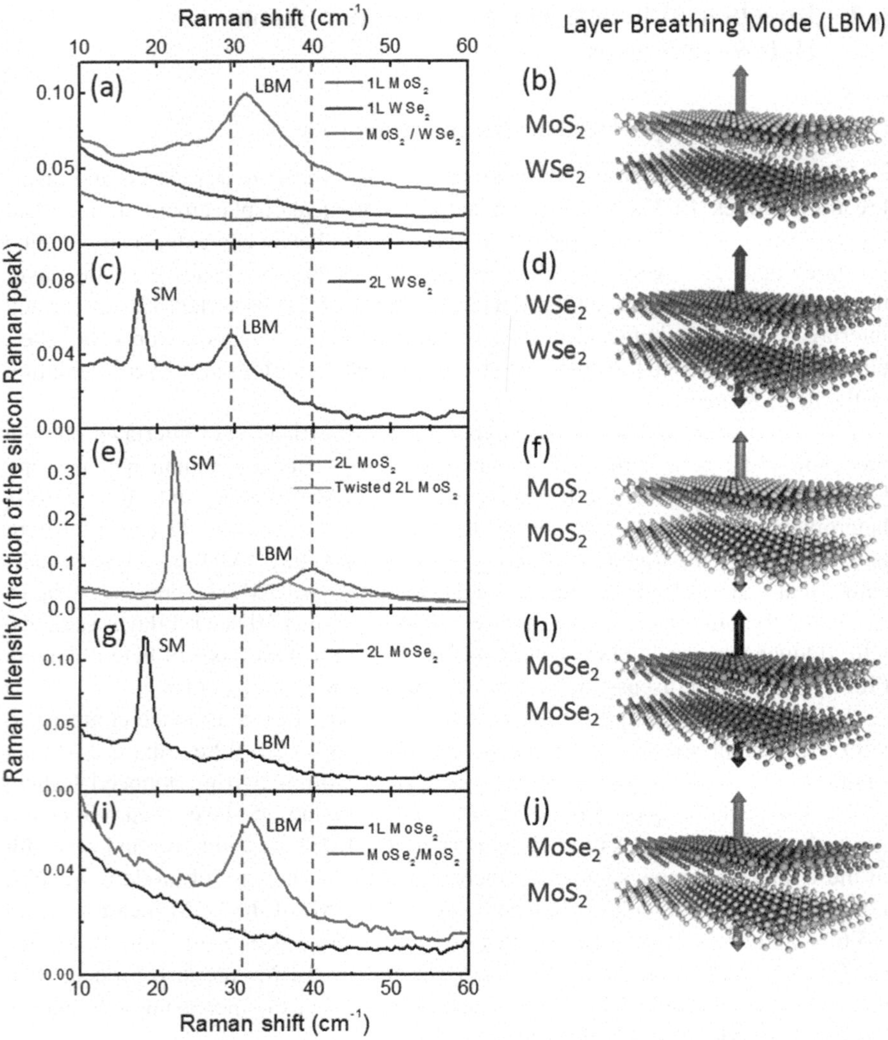

Fig. 4.5 Ultralow-frequency Raman spectra (left column) and schematics of layer breathing mode (LBM) vibrations (right column) for (**a**–**b**) the MoS_2/WSe_2 heterobilayer, 1L MoS_2 and 1L WSe_2; (**c**–**d**) Bernal-stacked 2L WSe_2; (**e**–**f**) Bernal-stacked and twisted 2L MoS_2; (**g**–**h**) Bernal-stacked 2L $MoSe_2$; (**i**–**j**) $MoSe_2/MoS_2$ heterobilayer and 1L $MoSe_2$. All spectra were measured with 532 nm laser excitation, and normalized with the Raman peak of the underlying silicon substrate (at 520 cm^{-1}). The shear mode (SM) and layer breathing mode (LBM) are denoted. The dashed vertical lines highlight the LBM position in Bernal-stacked 2L WSe_2, 2L MoS_2 and 2L $MoSe_2$. (The figure is adapted from Ref. [27])

Table 4.1 The frequency and full width at half maximum (FHWM) of the layer-breathing mode (LBM) and shear mode (SM) for different TMD bilayers. The uncertainty for all numbers is ±0.5 cm^{-1}. The table is adapted from Ref. [27]

	MoS_2/WSe_2	$MoSe_2/MoS_2$	Twisted 2L MoS_2	2L MoS_2	2L $MoSe_2$	2L WSe_2
LBM Frequency (FWHM) (cm^{-1})	30–35 (8)	30–35 (5.5)	32–39 (6.5)	39.5 (9.5)	30.5 (5)	29.5 (3.5)
SM Frequency (FWHM) (cm^{-1})	–	–	–	22 (1.5)	18.5 (1.5)	17.5 (1.5)

the TMD heterobilayers do not exhibit any Raman feature near the shear-mode range of Bernal bilayers (17–22 cm^{-1}), indicating the absence of shear mode in the heterostructures. We can understand these observations by considering the interlayer interactions in the heterostructures. As the MoS_2 and WSe_2 ($MoSe_2$) monolayers have ~4% mismatch of lattice constant and orient randomly with one another in the heterobilayers, the two lattices are generally incommensurate. Therefore, a lateral displacement of the two layers does not produce any overall restoring force, and hence no shear mode vibration can be generated in the heterostructures. Vertical displacement of the two layers, however, can create a finite restoring force due to the van der Waals interactions between them. This gives rise to the LBM vibration. As the overall strength of the van der Waals force depends predominantly on the interlayer distance, the formation of LBM vibration is stable regardless of the detailed lattice matching in the lateral dimension. This is supported by the similar LBM observed in trilayer graphene with ABA and ABC stacking order [21].

Lui et al. has also examined twisted MoS_2 bilayers, which are formed by randomly stacking two MoS_2 monolayers together. Figure 4.5e displays a representative Raman spectrum, which exhibits a pronounced Raman peak at 35 cm^{-1} with FWHM = 6.5 cm^{-1} (Table 4.1). The same Raman mode has been observed in more than 60 twisted 2L MoS_2 samples, with frequencies varying between 32 and 39 cm^{-1}. These results are comparable to the LBM in Bernal 2L MoS_2, which has slightly higher frequency (39.5 cm^{-1}) and correspondingly larger line width (FWHM = 9.5 cm^{-1}) (Table 4.1). The results indicate that LBM can be formed in twisted bilayer van der Waals materials [28–30]. The decreased LBM frequency in twisted bilayers is reasonable because the two layers are packed less efficiently than the Bernal stacking, resulting in weaker interlayer coupling. In addition, the shear mode is quenched in twisted bilayers due to the lateral lattice mismatch. These observations are consistent with the results and interpretation for the MoS_2/WSe_2 and $MoSe_2/MoS_2$ heterostructures.

(b) The influence of interface on the interlayer modes

As the LBM arises directly from the interlayer coupling, it is interesting to investigate its sensitivity to the interface conditions. To this end, Lui et al. have mapped the spatial profile of the LBM in an unannealed MoS_2/WSe_2 heterobilayer sample, which contains regions of 1L MoS_2, 1L WSe_2 and overlapping MoS_2/WSe_2 layers (Fig. 4.6a). As expected, the LBM is observed only in the area where the two TMD monolayers overlap (Fig. 4.6b). However, the LBM emerges only on the left side of the MoS_2/WSe_2 region. To understand the inhomogeneous LBM response, they have also mapped the intensity of the PL peaks of 1L MoS_2 and 1L WSe_2 at ~1.85 and ~1.59 eV, respectively (Figs. 4.4d–e, and 4.6c–d). Both PL intensities decrease in the MoS_2/WSe_2 region due to the interlayer electron-hole separation. The PL reduction is, however, much more severe on the left side than on the right side, indicating more efficient interlayer charge transfer and hence better layer-layer contact on the left side. This is further confirmed by AFM analysis. The AFM topographic image of the sample exhibits two distinct regions with different heights in the overlapping MoS_2/WSe_2 area (Fig. 4.6e–f). The left side of the area has a step height of ~0.7 nm, which is the same as the monolayer step height of MoS_2 crystal. The two TMD layers are therefore in direct contact in this region. In contrast, the right side has a larger interlayer separation of ~1.8 nm, which can be caused by unintentional residues trapped at the interface. The overall agreement between the LBM/PL intensity mapping and AFM scanning shows that the interlayer phonons can only be generated in heterostructure regions with direct interlayer contact and atomically clean interface. Such extreme sensitivity of LBM, even higher than that of the PL quenching, would enable ultra-sensitive characterization of the interface conditions in van der Waals materials. These results also indicate that the observed Raman signals are not from any interfacial molecules.

The interface of the heterobilayers can be improved by annealing the samples. Figure 4.7 shows the optical image, LBM and PL mappings of an annealed $MoSe_2$/MoS_2 heterobilayer sample. The LBM and the associated PL reduction are observed in most of the overlapping $MoSe_2$/MoS_2 areas, indicating an overall high-quality layer-layer contact in the bilayer region.

(c) The influence of layer orientation on the interlayer modes

As the LBM arises directly from interlayer coupling, it would be interesting to investigate its dependence on the relative orientation of the layers in the heterostructures. The CVD TMD monolayers usually appear to be flakes with triangular shapes, which have been shown to be single crystals that terminate with the transition-metal atoms (Mo, W) at zigzag edges [31, 32]. Therefore, it is convenient to determine the orientation of individual TMD flakes and the rotational angle (θ) between the top and bottom TMD flakes from their optical images (with uncertainty <5°). Lui et al. have measured a large collection of $MoSe_2$/MoS_2 heterobilayer samples, among which 58 samples exhibit strong LBM spectra and well-defined rotational angle between the $MoSe_2$ and MoS_2 layers. Figure 4.8a–f display the optical images and LBM spectra of three representative heterobilayers with $\theta \approx 45°$, 32° and 5°. Figure 4.8g shows the LBM frequency of all 58 samples as a function of θ. While the LBM frequency varies somewhat irregularly due to the different sample conditions (e.g. strain, defects, surface and interface conditions), the average value is found to decrease from 33.5 to 30.7 cm^{-1} as θ increases from 0° to 60°. The result clearly

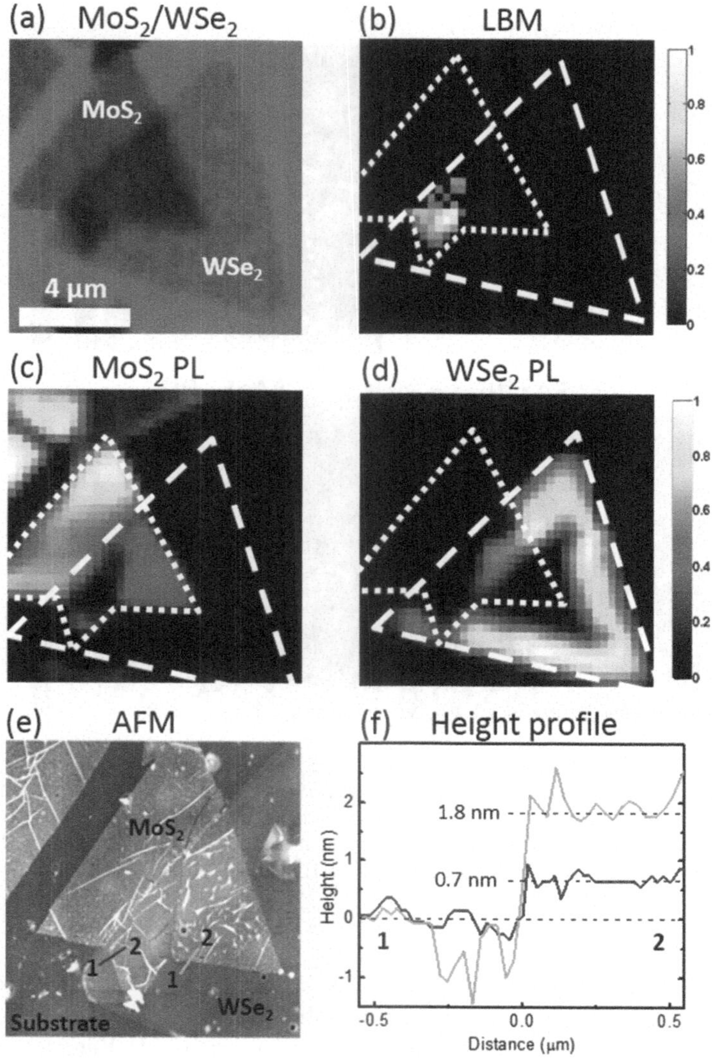

Fig. 4.6 (**a**) Optical image of an unannealed MoS_2/WSe_2 heterobilayer. (**b**) Raman intensity map of the LBM at $\sim 32\ cm^{-1}$ (see the Raman spectrum of MoS_2/WSe_2 in Fig. 4.5a). (**c**) Intensity map of the PL peak at 1.85 eV that corresponds to the main PL band of MoS_2 (see Fig. 4.4d). (**d**) Intensity map of the PL peak at 1.59 eV that corresponds to the main PL band of WSe_2 (see Fig. 4.4d). The PL is enhanced at the edge of the WSe_2 monolayer, as also reported in the literature [38]. The short and long dashed lines highlight the regions of 1L MoS_2 and 1L WSe_2, respectively. (**e**) AFM topographic image. (**f**) Height profiles of the two cutting lines across the boundary between the 1L WSe_2 region and the MoS_2/WSe_2 region in Panel (**e**). The green/upper and blue/lower lines are used to denote the right and left lines in (**e**) respectively, and the numbers 1 and 2 to denote the corresponding positions. (The figure is adapted from Ref. [27])

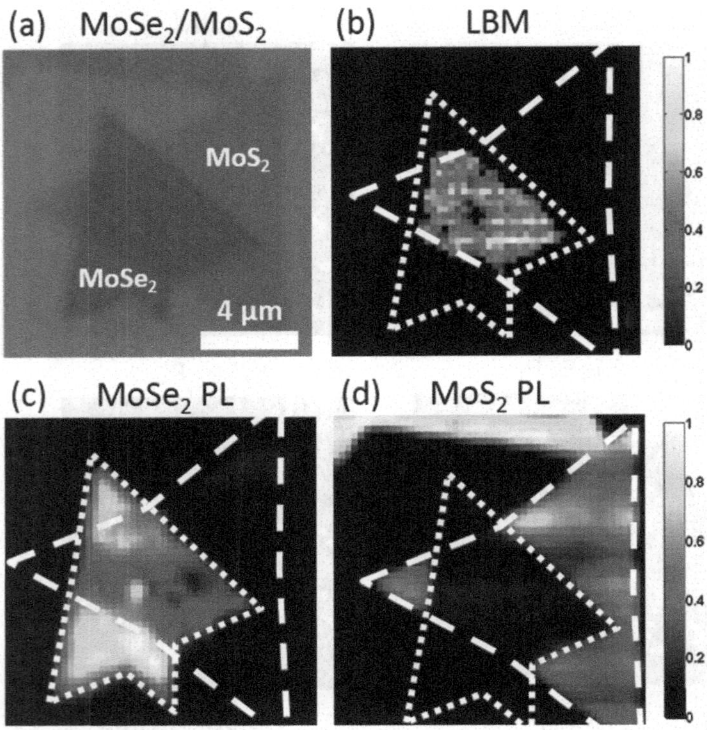

Fig. 4.7 (**a**) Optical image of an annealed $MoSe_2/MoS_2$ heterobilayer. (**b**) Raman intensity map of the LBM at $\sim$32 cm^{-1} (see the Raman spectrum of $MoSe_2/MoS_2$ in Fig. 4.5i). (**c**) Intensity map of the PL peak at 1.56 eV that corresponds to the main PL band of $MoSe_2$ (see Fig. 4.4e). (**d**) Intensity map of the PL peak at 1.85 eV that corresponds to the main PL band of MoS_2 (see Fig. 4.4e). The short and long dashed lines highlight the regions of 1L $MoSe_2$ and 1L MoS_2, respectively. (The figure is adapted from Ref. [27])

demonstrates the sensitivity of the LBM to the layer stacking. The softening of LBM implies that the interlayer coupling becomes slightly weaker as θ increases.

According to previous first-principles calculations by J. Kang et al. [33], $MoSe_2/MoS_2$ heterobilayers are expected to exhibit moiré patterns with a period (<10 nm) that varies with θ due to the $\sim$4% lattice mismatch between the $MoSe_2$ and MoS_2 layers. In addition, J. Kang et al. predicted an average interlayer distance $d = 3.345$ Å and interlayer adsorption energy $E_{ad} = 160$ meV for $\theta = 0°$, but a slightly larger interlayer distance $d = 3.368$ Å and smaller adsorption energy $E_{ad} = 157$ meV for $\theta = 60°$ in the $MoSe_2/MoS_2$ heterobilayer. These theoretical results agree qualitatively with the observed angle dependence of LBM frequency in the experiment.

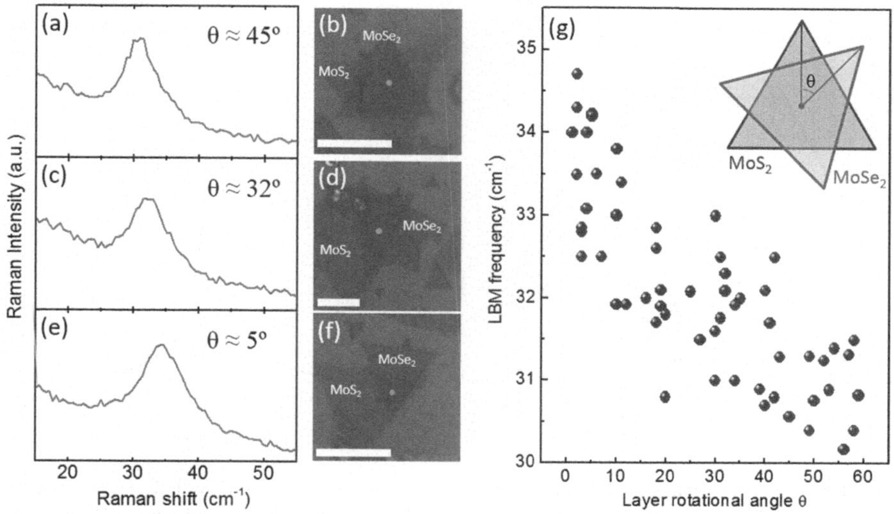

Fig. 4.8 **(a–f)** Layer-breathing mode (LBM) Raman spectra and the associated optical images of three representative $MoSe_2/MoS_2$ heterobilayer samples at different layer rotational angles (θ). The scale bars are 10 μm. The dots in the optical images denote the position of the laser spot in the measurements. **(g)** The LBM frequencies of 58 $MoSe_2/MoS_2$ heterobilayer samples as a function of the layer rotational angle (θ). The inset shows the schematic rotational angle between single-crystal MoS_2 and $MoSe_2$ monolayers. (The figure is adapted from Ref. [27])

4.5 Intralayer Lattice Distortion of van der Waals Heterostructures

When two 2D crystals are stacked together, the interlayer interactions can sometimes induce in-plane lattice strain within each 2D crystal. One example to demonstrate this effect is the graphene/BN heterostructure. BN is known to be an excellent substrate for graphene due to its flat and clean surface [12]. Graphene devices fabricated on BN substrates can exhibit ultrahigh electron mobility, comparable to the mobility of suspended graphene [34]. However, the interlayer coupling between graphene and BN is not negligible. In some situations, their mutual interaction can significantly modulate the electron and phonon properties of graphene. Both graphene and BN crystals have honeycomb lattice, but the lattice constant of BN is 1.8% larger than that of graphene. As a result, when graphene is deposited on BN, it will exhibit a hexagonal moiré pattern due to their lattice mismatch (Fig. 4.9) [35]. The wavelength (λ) of the moiré pattern can be expressed as [35]:

$$\lambda = \frac{(1+\delta)\,a}{\sqrt{2\,(1+\delta)\,(1-cos\phi)+\delta^2}}$$

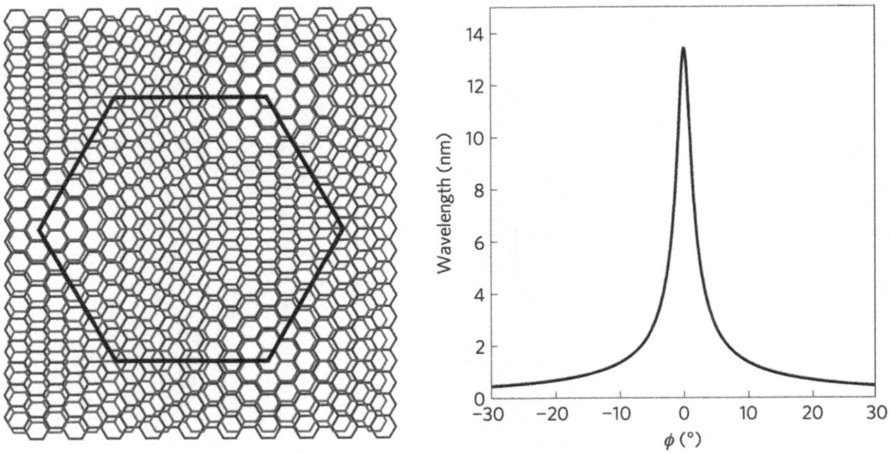

Fig. 4.9 (Left) Schematic representation of the moiré pattern of graphene on BN. The mismatch between the lattices is exaggerated (~10%). The black hexagon marks the moiré unit. (Right) The calculated wavelength of the moiré pattern as a function of rotational angle between the graphene and BN lattices. (The figure is adapted from Ref. [35, 36])

Here a is the graphene lattice constant. δ and ϕ are, respectively, the lattice mismatch and the rotational angle between graphene and BN. Figure 4.9 displays the moiré wavelength as a function of the rotational angle (ϕ) between graphene and BN. The moiré wavelength is highly sensitive to the rotational angle. It reaches a sharp maximum of 14 nm at $\phi = 0^{\circ}$ and decreases abruptly at $\phi > 0^{\circ}$. The graphene/BN moiré pattern imposes a periodic potential on graphene and makes it an effective superlattice. On the one hand, it can modulate the electronic band structure of graphene, giving rise to secondary Dirac cones, as reported in the literature [5–7]. On the other hand, it can also induce intralayer lattice strain of graphene in some situations [36].

The lattice properties of graphene/BN superlattice can be revealed by Raman spectroscopy [37]. Figure 4.10a shows the Raman 2D mode of graphene at three different rotational angles $\phi = 0^{\circ}$, 1° and $\phi > 2^{\circ}$ (misaligned). The full width at half maximum (FWHM) of the 2D mode is found to increase from 21 cm^{-1} for the misaligned sample to 34 cm^{-1} for $\phi = 1^{\circ}$ and 43 cm^{-1} for $\phi = 0^{\circ}$. Figure 4.10b shows such a change of line width for multiple samples with varying rotational angle. The 2D mode remains similar for all rotational angles larger than 2°. But its width increases dramatically when ϕ goes below 2° and rapidly approaches 43 cm^{-1} at $\phi = 0^{\circ}$. Such a critical behavior implies some significant change of the graphene properties when it is aligned well with BN. Figure 4.10c also shows the width of the 2D mode as a function of moiré wavelength at three different excitation photon energies (2.54, 2.41 and 1.96 eV). The 2D mode line width is found to increase linearly with the moiré period and is independent of the excitation photon energy. The insensitivity of the line width to the excitation photon energy shows that the broadening of the 2D mode does not arise from the coupling to the electronic band

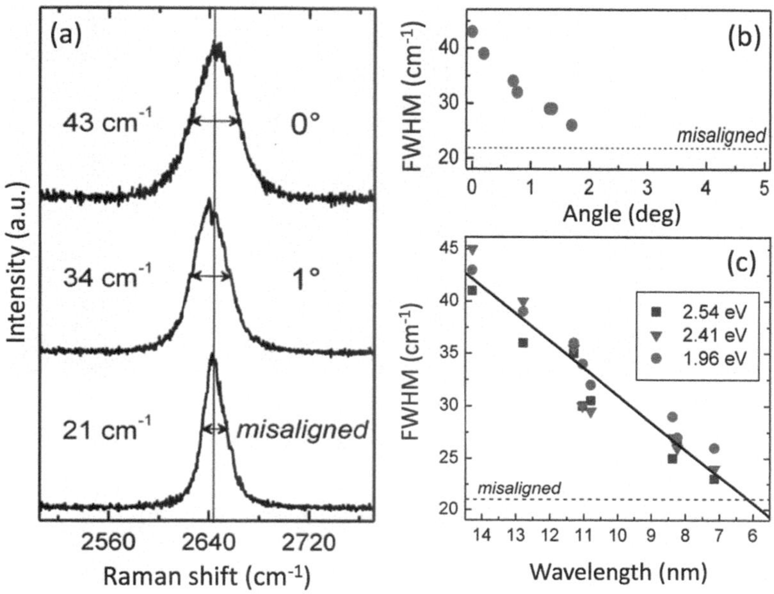

Fig. 4.10 (**a**) The Raman 2D mode of graphene deposited on BN with a mismatch angle of 0 and 1 degree, compared to that of a misaligned structure. (**b**) The FWHM of the 2D mode as a function of rotational angle between graphene and BN. The excitation photon energy is 1.96 eV for both (**a**) and (**b**). (**c**) The FWHM of the 2D mode as a function of moiré wavelength of graphene. Data for three excitation photon energies are presented (1.96, 2.41 and 2.54 eV). The straight line is a linear fit of the data. (The figure is adapted from Ref. [37])

structure through the double-resonance Raman mechanism. Indeed, the modulation of the electronic band structure due to the moiré pattern mainly occurs in energies (<300 meV) far below the excitation photon energy and hence should have little influence on the 2D mode.

The broadening of the 2D mode in graphene/BN sample has been argued to arise from the lattice distortion of graphene. When graphene is aligned with BN, it is energetically favorable for parts of its lattice to expand slightly to match the underlying BN lattice. Such a lattice distortion will cause an inhomogeneous distribution of strain in graphene and as a result, broaden the phonon modes. Woods et al. have investigated the strain distribution of graphene on BN substrate by using atomic force microscopy (AFM) [36]. By measuring the contact force between the AFM tip and the sample as a function of tip-sample distance, they can extract the local Young's modulus of the sample in the out-of-plane direction. Such Young's modulus depends on the local lattice matching between graphene and BN, and can be used to monitor the local strain in graphene. Figure 4.11 displays the Young's modulus distribution of two graphene/BN superlattices with moiré wavelengths of $\lambda = 8$ and 14 nm. Both images exhibit hexagonal patterns from the moiré interlayer coupling, but their Young's modulus distribution is quite different. The Young's modulus of the sample with $\lambda = 8$ nm (slightly misaligned) exhibits gradual spatial variation. This is reasonable because the local coupling strength between the

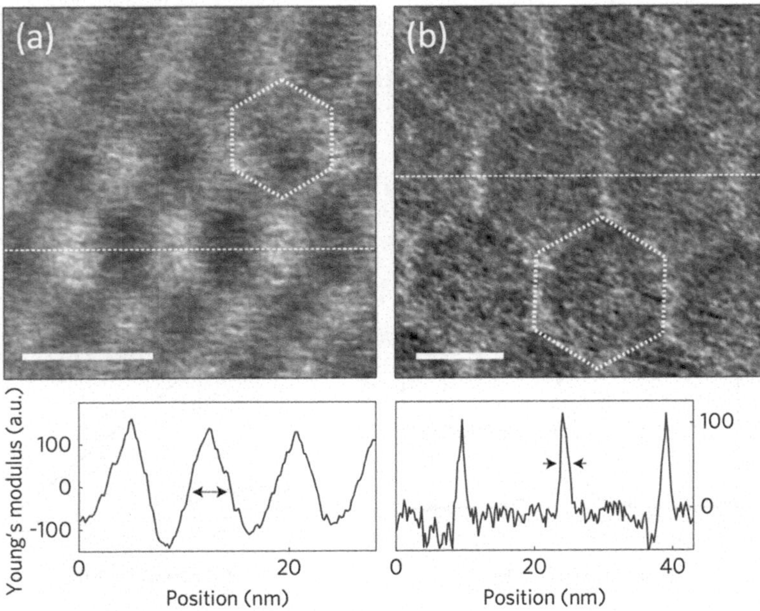

Fig. 4.11 (**a**–**b**) The top panels are the Young's modulus distribution, measured by atomic force microscopy, for graphene/BN heterostructures with 8 and 14 nm moiré patterns, respectively. The dashed hexagons highlight the moiré unit. The scale bars are 10 nm. The bottom panels show the cross sections of Young's modulus taken along the dashed lines in the corresponding top panels. The figure is adapted from Ref. [36].

incommensurable graphene and BN lattices is expected to vary gradually over the space. In contrast, the sample with $\lambda = 14$ nm ($\phi = 0^{\circ}$, perfectly aligned) exhibits low and almost constant Young's modulus in the interior areas of the hexagons, which are separated by narrow regions of high modulus (the domain walls) along the edges of the moiré hexagons. The width of the domain walls, as revealed in the cross section, is only ~2 nm. The results have been argued to signify an incommensurate-commensurate transition in the graphene/BN superlattice at small rotational angle. That is, at small enough rotational angle, part of graphene's lattice is expanded to match the BN lattice with the accumulated strain relaxed at the domain walls. As a result, graphene will have inhomogeneous strain distribution, which broadens the 2D Raman mode from the intralayer optical phonons.

4.6 Summary

Van der Waals heterostructures formed from 2D crystals can exhibit interlayer charge transfer, interlayer phonon modes and intralayer lattice distortion due to the vertical coupling between the 2D crystals. Ultralow-frequency Raman spectroscopy has revealed the formation of layer-breathing phonon modes in heterobilayers of transition metal dichalcogenides. The breathing-mode Raman intensity correlates strongly with the suppression of photoluminescence that arises from interlayer charge transfer. The layer-breathing mode is generated only in heterobilayers with atomically clean interface. Its frequency also evolves systematically with the relative orientation between the layers. These results establish the interlayer Raman mode as a sensitive probe to the interfacial environment and interlayer interactions in van der Waals materials. The observation of interlayer phonons in van der Waals structures also points to the possibility of fabricating novel phononic crystals in a bottom-up approach. By stacking or epitaxially growing different 2D crystals on one another, one may potentially fabricate artificial materials with unprecedented phonon structures and desirable vibrational properties that cannot be realized in natural crystals. On the other hand, the interlayer coupling can induce lattice distortion in the van der Waals heterostructures. In particular, when the 2D crystals of the heterostructure share the same lattice structure but with slightly different lattice constant, such as in the case of graphene/BN heterostructure, the lattice of the 2D crystals can be distorted to reduce the interlayer coupling energy. The resultant inhomogeneous distribution of strain can be monitored by Raman spectroscopy. These studies have demonstrated Raman spectroscopy to be an effective tool to characterize both the interlayer and intralayer properties of van der Waals heterostructures.

References

1. A.K. Geim, Science **324**, 1530 (2009)
2. A.K. Geim, K.S. Novoselov, Nat. Mater. **6**, 183 (2007)
3. A.K. Geim, I.V. Grigorieva, Nature **499**, 419 (2013)
4. K.S. Novoselov, A. Mishchenko, A. Carvalho, A.H. Castro Neto, Science **353**, 9439 (2016)
5. C.R. Dean, L. Wang, P. Maher, C. Forsythe, F. Ghahari, Y. Gao, J. Katoch, M. Ishigami, P. Moon, M. Koshino, T. Taniguchi, K. Watanabe, K.L. Shepard, J. Hone, P. Kim, Nature **497**, 598 (2013)
6. L.A. Ponomarenko, R.V. Gorbachev, G.L. Yu, D.C. Elias, R. Jalil, A.A. Patel, A. Mishchenko, A.S. Mayorov, C.R. Woods, J.R. Wallbank, M. Mucha-Kruczynski, B.A. Piot, M. Potemski, I.V. Grigorieva, K.S. Novoselov, F. Guinea, V.I. Fal'ko, A.K. Geim, Nature **497**, 594 (2013)
7. B. Hunt, J.D. Sanchez-Yamagishi, A.F. Young, M. Yankowitz, B.J. LeRoy, K. Watanabe, T. Taniguchi, P. Moon, M. Koshino, P. Jarillo-Herrero, R.C. Ashoori, Science **340**, 1427 (2013)
8. P. Rivera, K.L. Seyler, H. Yu, J.R. Schaibley, J. Yan, D.G. Mandrus, W. Yao, X. Xu, Science **351**, 688 (2016)
9. L. Britnell, R.M. Ribeiro, A. Eckmann, R. Jalil, B.D. Belle, A. Mishchenko, Y.-J. Kim, R.V. Gorbachev, T. Georgiou, S.V. Morozov, A.N. Grigorenko, A.K. Geim, C. Casiraghi, A.H.C. Neto, K.S. Novoselov, Science **340**, 1311 (2013)
10. L. Britnell, R.V. Gorbachev, R. Jalil, B.D. Belle, F. Schedin, A. Mishchenko, T. Georgiou, M.I. Katsnelson, L. Eaves, S.V. Morozov, N.M.R. Peres, J. Leist, A.K. Geim, K.S. Novoselov, L.A. Ponomarenko, Science **335**, 947 (2012)
11. C.-H. Lee, G.-H. Lee, A.M. van der Zande, W. Chen, Y. Li, M. Han, X. Cui, G. Arefe, C. Nuckolls, T.F. Heinz, J. Guo, J. Hone, P. Kim, Nat. Nanotechnol. **9**, 676 (2014)

12. C.R. Dean, A.F. Young, I. Meric, C. Lee, L. Wang, S. Sorgenfrei, K. Watanabe, T. Taniguchi, P. Kim, K.L. Shepard, J. Hone, Nat. Nanotechnol. **5**, 722 (2010)
13. A.V. Kretinin, Y. Cao, J.S. Tu, G.L. Yu, R. Jalil, K.S. Novoselov, S.J. Haigh, A. Gholinia, A. Mishchenko, M. Lozada, T. Georgiou, C.R. Woods, F. Withers, P. Blake, G. Eda, A. Wirsig, C. Hucho, K. Watanabe, T. Taniguchi, A.K. Geim, R.V. Gorbachev, Nano Lett. **14**, 3270 (2014)
14. K.F. Mak, C. Lee, J. Hone, J. Shan, T.F. Heinz, Phys. Rev. Lett. **105**, 136805 (2010)
15. W. Zhao, Z. Ghorannevis, L. Chu, M. Toh, C. Kloc, P.-H. Tan, G. Eda, ACS Nano **7**, 791 (2012)
16. Y. Liang, S. Huang, R. Soklaski, L. Yang, Appl. Phys. Lett. **103**, 042106 (2013)
17. W. Qianwen, W. Ping, C. Gengyu, H. Min, J. Phys. D: Appl. Phys. **46**, 505308 (2013)
18. H.-P. Komsa, A.V. Krasheninnikov, Phys. Rev. B **88**, 085318 (2013)
19. X. Hong, J. Kim, S.-F. Shi, Y. Zhang, C. Jin, Y. Sun, S. Tongay, J. Wu, Y. Zhang, F. Wang, Nat. Nanotechnol. **9**, 682 (2014)
20. C.H. Lui, Z. Ye, C. Keiser, X. Xiao, R. He, Nano Lett. **14**, 4615–4621 (2014)
21. C.H. Lui, Z. Ye, C. Keiser, E.B. Barros, R. He, Appl. Phys. Lett. **106**, 041904 (2015)
22. C.H. Lui, T.F. Heinz, Phys. Rev. B **87**, 121404(R) (2013)
23. P.H. Tan, W.P. Han, W.J. Zhao, Z.H. Wu, K. Chang, H. Wang, Y.F. Wang, N. Bonini, N. Marzari, N. Pugno, G. Savini, A. Lombardo, A.C. Ferrari, Nat. Mater. **11**, 294 (2012)
24. Y. Zhao, X. Luo, H. Li, J. Zhang, P.T. Araujo, C.K. Gan, J. Wu, H. Zhang, S.Y. Quek, M.S. Dresselhaus, Q. Xiong, Nano Lett. **13**, 1007 (2013)
25. H. Rui, B. Jeremiah Van, Y. Jia-An, X. Xiaoxiang, Y. Zhipeng, Y. Gaihua, I.H. Lu, S.M. Leong, C.H. Lui, 2D Materials **3**, 031008 (2016)
26. R. He, J.-A. Yan, Z. Yin, Z. Ye, G. Ye, J. Cheng, J. Li, C.H. Lui, Nano Lett. **16**, 1404 (2016)
27. C.H. Lui, Z. Ye, C. Ji, K.-C. Chiu, C.-T. Chou, T.I. Andersen, C. Means-Shively, H. Anderson, J.-M. Wu, T. Kidd, Y.-H. Lee, R. He, Phys. Rev. B **91**, 165403 (2015)
28. R. He, T.-F. Chung, C. Delaney, C. Keiser, L.A. Jauregui, P.M. Shand, C.C. Chancey, Y. Wang, J. Bao, Y.P. Chen, Nano Lett. **13**, 3594 (2013)
29. A.M. van der Zande, J. Kunstmann, A. Chernikov, D.A. Chenet, Y. You, X. Zhang, P.Y. Huang, T.C. Berkelbach, L. Wang, F. Zhang, M.S. Hybertsen, D.A. Muller, D.R. Reichman, T.F. Heinz, J.C. Hone, Nano Lett. **14**, 3869 (2014)
30. L.Z. Kaihui Liu, T. Cao, C. Jin, D. Qiu, Q. Zhou, A. Zettl, P. Yang, S.G. Louie, F. Wang, Nat. Commun. **5**, 4966 (2014)
31. A.M. van der Zande, P.Y. Huang, D.A. Chenet, T.C. Berkelbach, Y. You, G.-H. Lee, T.F. Heinz, D.R. Reichman, D.A. Muller, J.C. Hone, Nat. Mater. **12**, 554 (2013)
32. J.V. Lauritsen, J. Kibsgaard, S. Helveg, H. Topsoe, B.S. Clausen, E. Laegsgaard, F. Besenbacher, Nat. Nanotechnol. **2**, 53 (2007)
33. J. Kang, J. Li, S.-S. Li, J.-B. Xia, L.-W. Wang, Nano Lett. **13**, 5485 (2013)
34. L. Wang, I. Meric, P.Y. Huang, Q. Gao, Y. Gao, H. Tran, T. Taniguchi, K. Watanabe, L.M. Campos, D.A. Muller, J. Guo, P. Kim, J. Hone, K.L. Shepard, C.R. Dean, Science **342**, 614 (2013)
35. M. Yankowitz, J. Xue, D. Cormode, J.D. Sanchez-Yamagishi, K. Watanabe, T. Taniguchi, P. Jarillo-Herrero, P. Jacquod, B.J. LeRoy, Nat. Phys. **8**, 382 (2012)
36. C.R. Woods, L. Britnell, A. Eckmann, R.S. Ma, J.C. Lu, H.M. Guo, X. Lin, G.L. Yu, Y. Cao, R.V. Gorbachev, A.V. Kretinin, J. Park, L.A. Ponomarenko, M.I. Katsnelson, Y.N. Gornostyrev, K. Watanabe, T. Taniguchi, C. Casiraghi, H.J. Gao, A.K. Geim, K.S. Novoselov, Nat. Phys. **10**, 451 (2014)
37. A. Eckmann, J. Park, H. Yang, D. Elias, A.S. Mayorov, G. Yu, R. Jalil, K.S. Novoselov, R.V. Gorbachev, M. Lazzeri, A.K. Geim, C. Casiraghi, Nano Lett. **13**, 5242 (2013)
38. H.R. Gutiérrez, N. Perea-López, A.L. Elías, A. Berkdemir, B. Wang, R. Lv, F. López-Urías, V.H. Crespi, H. Terrones, M. Terrones, Nano Lett. **13**, 3447 (2012)

Chapter 5
Disorder and Defects in Two-Dimensional Materials Probed by Raman Spectroscopy

Ado Jorio and Luiz Gustavo Cançado

Abstract This paper describes the fundamentals of using Raman spectroscopy to characterize disorder in two-dimensional (2D) systems caused by the presence of defects. From the dimensionality point of view, in 2D crystalline structures disorder can be described as addition of point-like (zero-dimensional, 0D) or line-like (one-dimensional, 1D) defects. To characterize the amount of 0D and 1D defects separately, two spectral parameters are needed. The two basic parameters are related to defect-induced activation of forbidden Raman modes and to defect-induced confinement of phonons. A two-dimensional Raman phase diagram can be built based on geometrical considerations, and the geometrical parameters are governed by fundamental aspects such as phonon and electron coherence lengths and Raman cross sections. We apply the general picture to the well-studied case of graphene amorphization, which has been studied since the 70'ies, with the two basic parameters being represented by the peak linewidths (Γ) and by the integrated intensity ratio (A_D/A_G) between the defect-induced (D) mode and the Raman allowed graphene (G) mode. The amorphization of graphene has been fully described in the terms presented here thanks to the development of standard materials with well-controlled amount of either point-like or line-like defects.

5.1 Introduction

A given material can be fully ordered, exhibiting a crystalline structure, fully disordered, exhibiting an amorphous structure, or anything in between, with a mixture of crystalline and amorphous regions. An amorphization trajectory can be described by the addition of defects to the periodic structure, in different quantities and locations. For a two-dimensional (2D) system, like graphene [1, 2], phosphorene [3–5], hexagonal boron-nitride [6] and transition metal dichalcogenides [7, 8],

A. Jorio (✉) · L. G. Cançado
Departamento de Física, Universidade Federal de Minas Gerais, Belo Horizonte, MG, Brazil
e-mail: adojorio@fisica.ufmg.br

P.-H. Tan (ed.), *Raman Spectroscopy of Two-Dimensional Materials*, Springer Series in Materials Science 276, https://doi.org/10.1007/978-981-13-1828-3_5

defects can be punctual, i.e. zero-dimensional (0D), such as atomic vacancies, or line-extended, i.e. one-dimensional (1D), such as dislocations. The material's mechanical, optical, thermal and transport properties change significantly with the addition of defects, such changes being highly dependent on the dimensionality of the defect.

Raman spectroscopy, the inelastic scattering of light [9], can be used to generate an amorphization phase diagram for two-dimensional systems, based on two spectral characteristics, which can be used to quantify the amount of point-like and line-like defects separately [10]. This is the subject of this chapter, which is organized in two main sections, the first presenting the model, the second showing its application to describe the amorphization trajectories in graphene. The final section presents a summary and perspectives of this work.

5.2 Raman Spectroscopy of Defects in Two-Dimensional Crystals

5.2.1 General Concepts About the Raman Signature of Defects in Crystalline Materials

A Raman spectrum is characterized by the presence of peaks related to Raman active modes, these peaks having specific frequencies (ω, Raman shift), linewidths (Γ) and integrated intensities (i.e. peak area A) [9]. The Raman shift is given by the phonon vibrational frequency, which is governed by the mass of the atoms and the strength of the chemical bonds. The linewidth is governed by the phonon frequency uncertainty (Δω). The integrated intensity A is related to the differential Raman cross-section (C), which is bound to the phonon-induced variation of material's polarizability. Integrated intensity (peak area A), rather than intensity (peak hight I), is considered because the area under each peak represents the probability of the whole phonon-related scattering process.

Symmetry dictates the phonons which can scatter light (Raman active modes). The most general selection rule comes from the periodic structure of crystals, which imposes momentum conservation. Therefore, any Raman scattering phenomenon in crystals has to obey momentum conservation, i.e. $k_s = k_i \pm q$, where k_s, k_i and q stands for momentum of the scattered photon, incident photon and phonon, respectively. Considering light has a very large wavelength when compared to the distance between neighboring atoms in a crystal (500 nm as compared to 1 Å), this momentum conservation selection rule basically implies that only $q \approx 0$ phonons are Raman active. Point-symmetries also impose selection rules that depend on the specific material.

There are two main aspects that change the Raman signature of a crystalline material, when defects are added. The first is the breakdown of momentum conservation, since the translational symmetry is broken. This symmetry selection rule relaxation can give rise to defect-induced modes, which are new peaks in the Raman spectrum of the material. The intensity of these defect-induced peaks will depend on the differential Raman cross-section (C) and the amount of defects in the crystalline network and, therefore, it can be used as a measure of the amount of disorder, if properly calibrated. The second aspect is the increase in the Raman peak linewidths ($\Delta\Gamma$) due to phonon confinement. In a perfectly ordered system, the frequency uncertainty $\Delta\omega$ is governed by the phonon lifetime in the crystal lattice (uncertainty principle). Defects, however, generate boundaries that can confine the phonons within a region of length Δx, thus generating an uncertainty Δq in the phonon momentum value. Since the phonon frequency is a function of its momentum, as represented by the phonon dispersion relations $\omega(q)$, an increase in uncertainty Δq generates an increase in uncertainty $\Delta\omega$(q), i.e. an increase in the Raman peak linewidh Γ. This effect is significant in the Raman spectrum of a material when the confinement within a region Δx generates an effect in the frequency $\Delta\omega(q)$ that overcomes the natural linewidth given by the phonon coherence length ℓ_{ph}, i.e. when $\Delta x \approx \ell_{\text{ph}}$ or smaller.

5.2.2 *Defect Raman Signatures in Two-Dimensional Crystals*

As discussed in the previous sections, defects are expected to change significantly two characteristics of the Raman spectrum of crystalline structures, which are the broadening of the Raman allowed peaks, and the appearance of new defect-induced peaks. In 2D crystalline structures, disorder can be added as point-like (zero-dimensional, 0D) defects or as line-like (one-dimensional, 1D) defects. This is a favorable situation: to characterize the amount of two types of defects (0D and 1D) separately, two spectral parameters are available, the defect-induced activation of forbidden Raman modes and the defect-induced confinement of phonons. How 0D and 1D defects change those two spectral properties is the key factor to be explored.

The disorder is characterized by: (i) the structurally-damaged area (S); (ii) the Raman cross-section for the defect-induced modes in S (C_S); (iii) the crystalline region around the structurally-damaged S area where, despite the local ordered structure, the defect-induced modes are still activated due to momentum selection rule breakdown of a nearby defect. This activated-area (A) is dictated by the electron coherence-length ℓ_e of the electrons involved in the Raman scattering process (note here A is used for activated area, differently from A, which is used for Raman peak integrated intensity); (iv) the Raman cross-section for the defect-induced modes in A (C_A); (v) the geometrical distribution of the line and point defects, and how they confine phonons with coherence length ℓ_{ph}.

The contribution of the defects to the integrated intensity of new Raman peaks related to defect-induced active modes (A_D) are a direct measure of the A and S regions, which are directly related to the amount of defects. The total area $T = S + A$ will depend on defect distribution, giving a measure of the amount of point-like and line-like defects. Besides, the width of the observed Raman peaks (Γ for either Raman active or defect-induced modes) will provide a measure of the phonon confinement, which is also related to defect distribution. Therefore, a study of a pair of parameters (A_D, Γ) should provide a measure of point-like and line-like defects.

5.2.3 Geometrical Model for the Defect Raman Signature in Two-Dimensional Crystals

Consider a point defect (0D) as a small disk of disordered atomic arrangement in the crystalline structure (see dots in Fig. 5.1, including inset). Small here means the disk is smaller than the probe, e.g. the light wavelength in a far-field scattering or the tip apex in a tip-enhanced Raman spectroscopy (TERS) experiment. The disorder is characterized by: (i) the radius r_S of the disk defining the structurally-damaged area (S^{0D}); (ii) the normalized differential Raman cross-section for the defect-induced modes in S ($C_S{}^{0D}$). Measuring absolute intensities in Raman spectroscopy is not trivial, being much simpler to work with Raman mode intensities normalized by a standard reference with differential Raman cross section (C^{norm}); (iii) the crystalline region around the structurally-damaged S^{0D} area where, despite the local ordered structure, the defect-induced modes are still activated due to momentum selection rule breakdown by a nearby defect. This activated-area (A^{0D}) is dictated by the electron coherence-length ℓ_e of the electrons involved in the Raman scattering

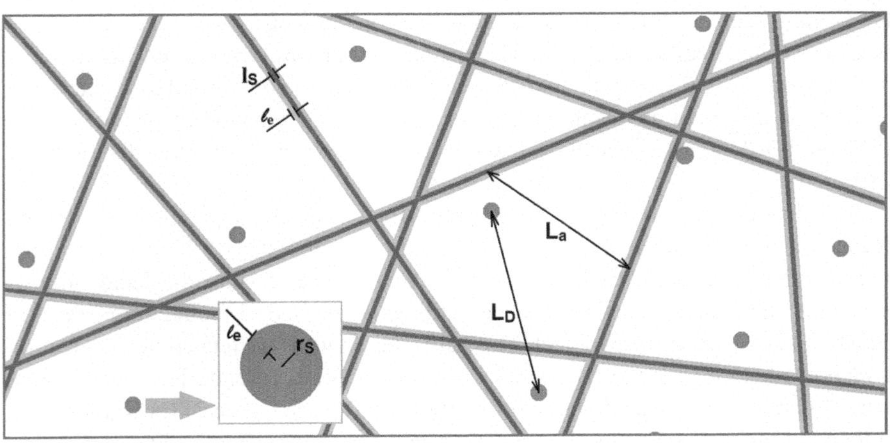

Fig. 5.1 Schematics of point-like and line-like defects in a two-dimensional (2D) system. The white area indicates the two-dimensional crystalline region. The one-dimensional (1D) and the zero-dimensional (0D) structurally disordered areas are indicated by line traces (S^{1D}) and small circles (S^{0D}), respectively. Around the S areas are the activated areas, indicated by dark and light shadow for 0D (A^{0D}) and 1D (A^{1D}) defects, respectively. Also shown are the geometrical definitions of r_S, l_S, ℓ_e, L_D and L_a (see inset and text for definitions)

process; (iv) the normalized differential Raman cross-section for the defect-induced modes in A ($C_A{}^{0D}$); (v) the average distance between two disk-defects L_D. The defect density is given by $\sigma = L_D{}^{-2}$.

Consider a line-defect (1D) as a thin ribbon of a disordered atomic arrangement in the crystalline structure (see lines in Fig. 5.1). Again, here thin means the thickness of the ribbon is smaller than the probe. The disorder is characterized by: (i) the width of the ribbon l_S defining the structurally-damaged area (S^{1D}); (ii) the normalized differential Raman cross-section for the defect-induced modes in S ($C_S{}^{1D}$); (iii) the activated-area (A^{1D}) crystalline region around the S^{1D} ribbon where the defect-induced modes are activated, as determined by the electron coherence-length ℓ_e; (iv) the normalized differential Raman cross-section for the defect-induced modes in A^{1D} ($C_A{}^{1D}$); (v) the average distance between two line-defects L_a. Since many line-defects enclosure a crystalline area, L_a is also the average size of a monocrystalline region within a crystal cut by line-defects, and it can be measured as the distance between the centers of neighboring crystallites as well.

The integrated intensity of any defect induced mode appearing in the Raman spectra of a 2D defective structure is given by:

$$A_D\left(L_a, L_D\right)/A^{norm} = C_S^{0D} S^{0D} + C_A^{0D} A^{0D} + C_S^{1D} S^{1D} + C_A^{1D} A^{1D}, \tag{5.1}$$

normalized here by the integrated intensity of a Raman-active mode (A^{norm}). Considering the disk and ribbon structures for 0D and 1D defects, respectively, and considering the scattering cross-section at the activated-area decreases exponentially with the distance from the structurally disordered area, having the electron coherence length ℓ_e parameterizing the decay rate, the intensity of a defect-induced mode can be found analytically [10], given by:

$$\begin{aligned}\frac{A_D}{A^{norm}} &= C_S^{0D}\left(1 - e^{-\frac{\pi r_S^2}{L_D^2}}\right) + \frac{2\pi}{L_D^2} C_A^{0D} \ell_e\left(\ell_e + r_S\right)\left[1 - 4l_s\frac{\left(L_a - l_S\right)}{L_a^2}\right] e^{-\frac{\pi r_S^2}{L_D^2}}\\ &+ \frac{4}{L_a^2} C_S^{1D} l_S\left(L_a - l_S\right) e^{-\frac{\pi r_S^2}{L_D^2}} + \frac{2}{L_a^2} C_A^{1D} \ell_e\left(L_a - 2l_S\right)\left(1 - e^{-\frac{L_a - l_S}{\ell_e}}\right) e^{-\frac{\pi r_S^2}{L_D^2}}\end{aligned} \tag{5.2}$$

The phonon linewidths Γ increases exponentially as the phonon localization length Δx decreases with respect to the phonon coherence length ℓ_{ph}, and it can be given by [10]:

$$\Gamma\left(L_a, L_D\right) = \Gamma_\infty + C_\Gamma e^{-\Delta x/\ell_{ph}}. \tag{5.3}$$

Γ_∞ stands for the peak full-width at half maxima (FHWM) for the pristine material. For samples with only line defects, $\Delta x = L_a$ since phonons are confined within a crystallite of size L_a. For samples with pure point defects, $\Delta x = \alpha L_D$ with

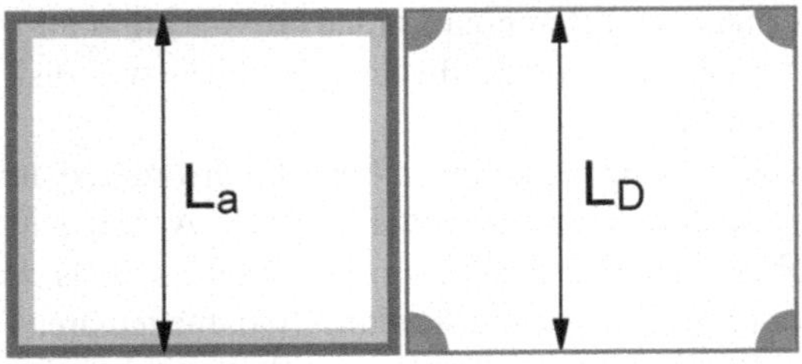

Fig. 5.2 A square network of zero-dimensional (0D, right) and one-dimensional (1D, left) defects enclosing a crystalline area (white) with similar structural parameters $L_a = L_D$, where L_a is the distance between line-defects (or the crystallite size) and L_D is the distance between point-defects

$\alpha > 1$, because point defects are less effective in localizing phonons when compared to line defects [10].

As evidence in Fig. 5.2, for similar $L_a = L_D$, both the area contribution $T = S + A$ and the effective phonon confinement are expected to be different, i.e. in general, both $A_D(L_a,L_D)$ and $\Gamma(L_a,L_D)$ will vary differently with respect to changes on L_a or L_D. Therefore, samples with varying values of L_a and L_D follow different trajectories in a two-dimensional plot of A_D/A^{norm} vs. Γ, which allows building a two-dimensional Raman phase diagram based on the spectral pair [A_D/A^{norm}, Γ].

5.3 Raman Spectroscopy of Defects in graphene

5.3.1 Raman Spectra of Disordered sp² Carbons

Figure 5.3 shows the Raman spectra of different graphene-related carbon nanostructures [11]. There are several similarities and differences among the spectra, which enables to identify the different carbon structures and to study their properties in great details. Especially interesting for the subject of this chapter is the presence of the so-called D band at $\sim$1350 cm^{-1} in the four top spectra, which comes from different types of disordered two-dimensional graphene-like materials. The D band is absent in the "graphene" and "HOPG" (highly ordered pyrolytic graphite) spectra, which are perfect crystalline structures. The two lowest spectra also exhibit the D band but they come from one-dimensional carbon nanotube structures and will not be treated here. Also important is the observation of the so-called G band, which is the graphene Raman active mode related to the stretching of the C–C bonds, appearing at $\sim$1584 cm^{-1}. This peak can be used for normalization purposes, providing $A_G = A^{norm}$. Finally, notice that the peak area ratios (A_D/A_G) and FHWM (both Γ_G and Γ_D) are different for the distinct materials, and this fact will be explored here.

There is a vast literature addressing the problem of using Raman spectroscopy to quantify defects in graphene related materials [12–32], and they have actually

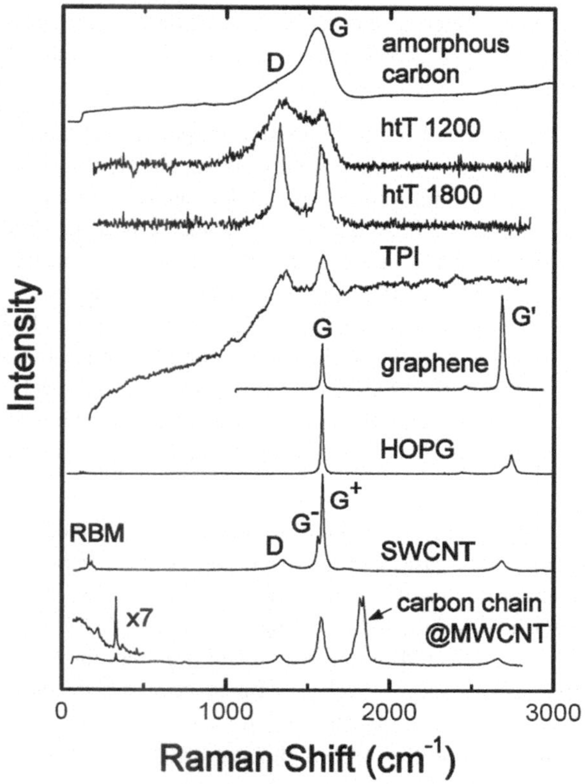

Fig. 5.3 Raman spectra of different graphene-based systems. From top to bottom: as produced amorphous carbon, amorphous carbon heat treated at 1200 °C (htT 1200), and at 1800 °C (htT 1800), charcoal grains from the "Terra Preta do índio" (TPI), single layer graphene, highly ordered pyrolytic graphite (HOPG), single wall carbon nanotube (SWCNT) bundles, and multi-wall carbon nanotubes bundles (MWCNT) filled with linear carbon chains. The different Raman peaks discussed in the text are assigned. (Reproduced from Ref. [11])

been behind the development of the geometrical model introduced in this chapter. The activation of new peaks in the Raman spectra is related to a multiple-internal-resonance mechanism that happens in the Raman scattering of graphene and it is responsible for a peak dispersive behavior (frequency change with changing excitation laser energy) [33–36]. This aspect is explored in Chap. 4 of this book. However, the focus here is on the variation of intensities and linewidths, which can be used to quantify defects using the model introduced in Sect. 5.2.

5.3.2 *Standard Reference Materials for Calibration of the Geometrical Model*

Structural characterization of defects in graphene by Raman spectroscopy has already produced well-established protocols for the quantification of both point-defect concentrations [23–26] and crystallite sizes [12, 15–19, 21, 31] separately. Based on this work and to properly parameterize the spectral signatures of 0D and 1D defects in the Raman spectrum of graphitic materials, two sets of standard reference materials varying from pristine 2D hexagonal sp^2 carbon lattices to highly disordered structures, following two different routes, were prepared and measured [10]:

- Sample set 1 starts as pristine graphene prepared by the mechanical exfoliation method of HOPG, and it is ion-bombarded with different ion doses, generating an increasing number of 0D defects, down to a fully disordered structure [23]. Pristine graphene has $L_D \rightarrow \infty$ and fully disordered graphene has $L_D \rightarrow 0$;
- Sample set 2 starts as a fully amorphous carbon material prepared by laser ablation of HOPG, and it is heat treated at different temperatures, thus generating sp^2 crystalline structures of increasingly larger crystallite sizes, up to a highly crystalline turbostratic graphitic structure for the highest heat treatment temperature [31]. Geometrically, this sample can be thought as a graphene layer "cut" by several 1D defects, or an ensemble of nanographite crystallites delimited by their borders. Pristine graphene has $L_a \rightarrow \infty$ and fully disordered graphene has $L_a \rightarrow 0$;

Figure 5.4a shows the plot of A_D/A_G as a function Γ_G, with A_D/A_G normalized by the fourth power of the excitation laser energy ($E_L{}^4$) in order to compare results obtained using different laser energies [10, 18]. Figures 5.4b, c show exemplary spectra with the corresponding $[(A_D/A_G)E_L{}^4, \Gamma_G]$-values placed in the plot of Fig. 5.4a, as indicated by the letters and numbers inside circles. Dashed and solid lines give $[(A_D/A_G)E_L{}^4, \Gamma_G]$ as a function of the structural parameters that define the sample degree of disorder, L_D for the sample with point-defects (dashed line), and L_a for the sample with line-defects (solid line). These lines were obtained by fitting the spectra from the reference standard materials set 1 and set 2, using Eqs. (5.2) and (5.3), finding the parameters shown in Table 5.1 [10]. $[(A_D/A_G)E_L{}^4, \Gamma_G]$ follow different functions of L_D and L_a, depending on the defect dimensionality, except for the two extremes (L_a,$L_D \rightarrow 0,\infty$), where the values converge. Pristine graphene appears in the plot at the smallest Γ_G value and $(A_D/A_G)E_L{}^4 = 0$, while amorphous samples appear in the plot at the higher values of Γ_G and $(A_D/A_G)E_L{}^4 \neq 0$. Samples with mixed 0D and 1D defects fall inside the area limited by the solid and dashed lines, except for the highly defective region ($\Gamma_G > 40$ cm^{-1}), where the measured values can overcome the dashed line (for more details, see Ref. [10]).

5.4 Summary and Perspectives

A geometrical model to quantify defects in two-dimensional systems, differentiating 0D and 1D defects, was presented here. The model was originally developed for graphene and applied in the analyses of the structure of two-dimensional nanocarbons generated from renewable gas, used to produce inks and conducting coatings [10].

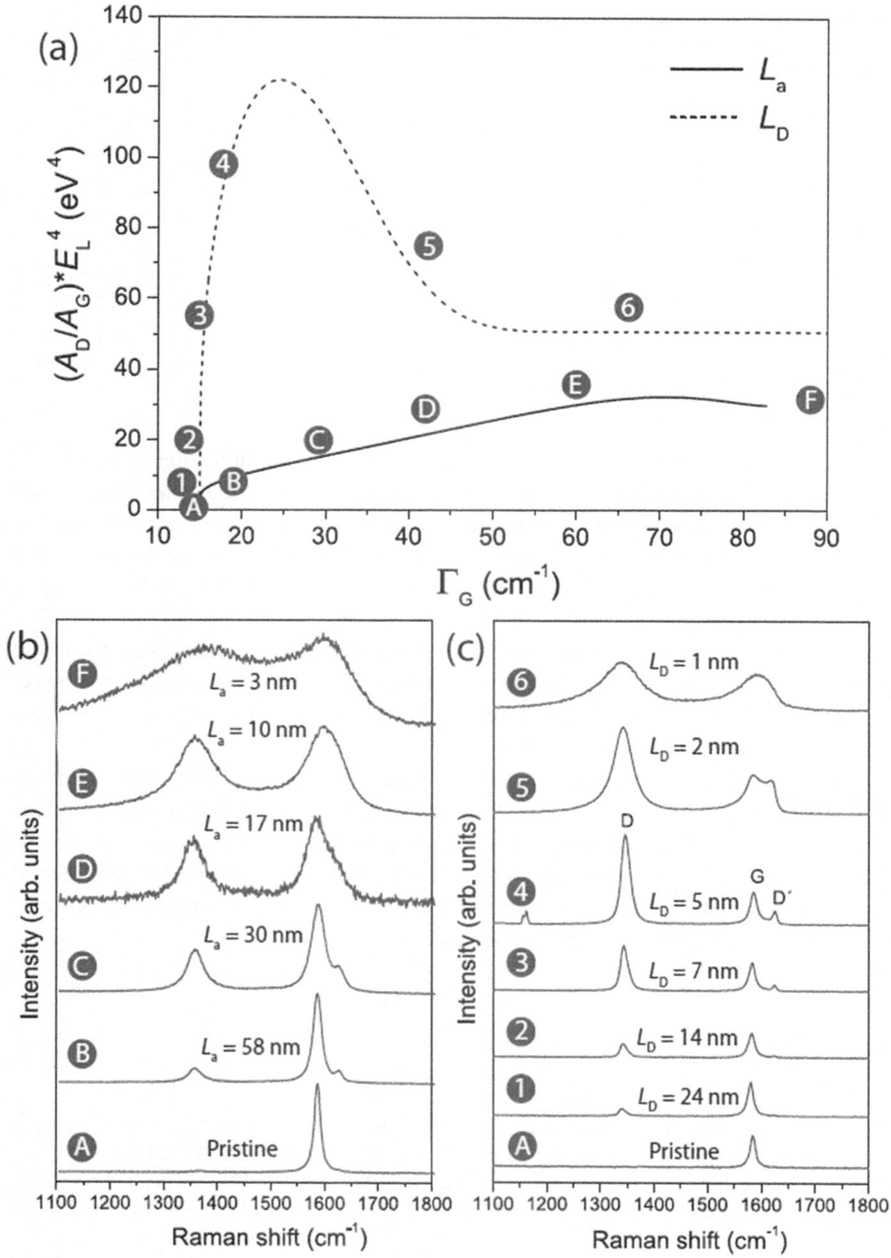

Fig. 5.4 (**a**) The Raman amorphization diagram for graphene, plotting $(A_D/A_G)E_L^4$ as a function of Γ_G. Solid and dashed lines stand for the theoretical $[(A_D/A_G)E_L^4, \Gamma_G]$ values for pure 1D and 0D defects, respectively. (**b**) Exemplary Raman spectra for samples following the amorphization route of pure line defects. (**c**) Exemplary Raman spectra for samples following the amorphization route of pure point defects. The letters/numbers correlate the specific spectrum in (**b**, **c**) with the spectral location at the Raman phase diagram in (**a**). (Reproduced from Ref. [10])

Table 5.1 Numerical values for the amorphization phase diagram of graphene [10], to be used with Eqs. (5.2) and (5.3)

Parameter	Value (unit)
C_Γ	87 cm^{-1}
$\Gamma_\infty{}^G$	15 cm^{-1}
ℓ_{ph}	16 nm
ℓ_e	4 nm
l_S	2 nm
r_S	2.2 nm
$C_S{}^{1D}$	30.3 eV^4
$C_A{}^{1D}$	30.4 eV^4
$C_S{}^{0D}$	26.5 eV^4
$C_A{}^{0D}$	30.4 eV^4
α	10

The methodology can be broadly applied to other two-dimensional systems [37, 38]. However, it is necessary to develop standard reference materials for 0D and 1D defects, and to develop a parameterization protocols to generate the specific amorphization diagram. At this point, it is not clear whether special resonance Raman conditions, which plays a role in the defect-induced Raman scattering in graphene (see Chap. 4), may be needed for the effectiveness of the model. This aspect has to be clarified for the extension of the model to other 2D systems. Other important aspects, such as the influence of crystallographic orientations with respect to light polarization and electro-defect scattering also play a role [10, 39], and should be explored in other 2D materials.

In conclusion, the rather simple geometrical model discussed here works wonderfully to quantify point-like and line-like defects in graphene, and its generality points towards a broader application in other 2D systems and in 1D systems. Proving the generality of the model is a research direction to be pursued. A similar geometrical model can also be developed for 1D systems, such as carbon nanotubes, but also requiring the development of a parameterization protocol.

References

1. A.C. Ferrari, J.C. Meyer, V. Scardaci, C. Casiraghi, M. Lazzeri, F. Mauri, et al., Raman spectrum of graphene and graphene layers. Phys. Rev. Lett. **97**(18), 187401 (2006)
2. L.M. Malard, M.H.D. Guimarães, D.L. Mafra, M.S.C. Mazzoni, A. Jorio, Group-theory analysis of electrons and phonons in N-layer graphene systems. Phys. Rev. B **79**, 125426 (2009)
3. J. Ribeiro-Soares, R.M. Almeida, L.G. Cançado, M.S. Dresselhaus, A. Jorio, Group theory for structural analysis and lattice vibrations in phosphorene systems. Phys. Rev. B **91**, 205421 (2015)
4. X. Ling, H. Wang, S. Huang, F. Xia, M.S. Dresselhaus, The renaissance of black phosphorus. Proc. Natl. Acad. Sci. **112**(15), 4523–4530 (2015)
5. L. Xi, S. Huang, E.H. Hasdeo, L. Liang, W.M. Parkin, Y. Tatsumi, A.R.T. Nugraha, A.A. Puretzky, P.M. Das, B.G. Sumpter, D.B. Geohegan, J. Kong, R. Saito, M. Drndic, V. Meunier,

M.S. Dresselhaus, Anisotropic electron-photon and electron-phonon interactions in black phosphorus. Nano Lett. **16**(4), 2260–2267 (2016)
6. R.V. Gorbachev, I. Riaz, R.R. Nair, R. Jalil, L. Britnell, B.D. Belle, et al., Hunting for monolayer boron nitride: optical and Raman signatures. Small **7**(4), 465–468 (2011)
7. J. Ribeiro-Soares, R.M. Almeida, E.B. Barros, P.T. Araujo, M.S. Dresselhaus, L.G. Cançado, A. Jorio, Group theory analysis of phonons in two-dimensional transition metal dichalcogenides. Phys. Rev. B **90**, 115438 (2014)
8. Y. Zhao, X. Luo, H. Li, J. Zhang, P.T. Araujo, C.K. Gan, J. Wu, H. Zhang, S.Y. Quek, S. Mildred, Dresselhaus, and Qihua Xiong, interlayer breathing and shear modes in few-trilayer MoS2 and WSe2. Nano Lett. **13**(3), 1007–1015 (2013)
9. A. Jorio, M.S. Dresselhaus, R. Saito, *Raman Spectroscopy in Graphene Related Systems* (Wiley-VCH, Weinheim, 2011)
10. L.G. Cançado, M.G. Silva, E.H.M. Ferreira, F. Hof, K. Kampioti, K. Huang, A. Pénicaud, C.A. Achete, B.R. Capaz, A. Jorio, 2D Materials. **4**, 025039 (2017)
11. A. Jorio, A.G. Souza Filho, Raman studies of carbon nanostructures. Annu. Rev. Mater. Res. **46**, 357–382 (2016)
12. F. Tuinstra, J.L. Koenig, J. Chem. Phys. **53**, 1126 (1970)
13. M.S. Dresselhaus, G. Dresselhaus, K. Sugihara, I.L. Spain, H.A. Goldberg, *Graphite Fibers and Filaments*, vol 5 (Springer, Dordrecht, 2013)
14. M.S. Dresselhaus, R. Kalish, *Ion Implantation in Diamond, Graphite and Related Materials*, vol 22 (Springer, Dordrecht, 2013)
15. A. Ferrari, J. Robertson, Phys. Rev. B **61**, 14095 (2000)
16. K. Takai, M. Oga, H. Sato, T. Enoki, Y. Ohki, A. Taomoto, K. Suenaga, S. Iijima, Phys. Rev. B **67**(21), 214202 (2003)
17. A. Ferrari, J. Robertson, Phil. Trans. R. Soc. Lond. A **362**, 2477 (2004)
18. L.G. Cançado, K. Takai, T. Enoki, M. Endo, Y.A. Kim, H. Mizusaki, A. Jorio, L.N. Coelho, R. Magalhes-Paniago, M.A. Pimenta, Appl. Phys. Lett. **88**, 163106 (2006)
19. A. Ferrari, Solid State Comm. **143**, 47 (2007)
20. C. Casiraghi, A. Hartschuh, H. Qian, S. Piscanec, C. Georgi, A. Fasoli, K. Novoselov, D. Basko, A. Ferrari, Nano Lett. **9**, 1433 (2009)
21. M.A. Pimenta, G. Dresselhaus, M.S. Dresselhaus, L.G. Cancado, A. Jorio, R. Saito, Phys. Chem. Chem. Phys. **9**(11), 1276 (2007)
22. M.S. Dresselhaus, A. Jorio, M. Hofmann, G. Dresselhaus, R. Saito, Nano Lett. **10**, 751 (2010)
23. M.M. Lucchese, F. Stavale, E.H.M. Ferreira, C. Vilani, M.V.O. Moutinho, R.B. Capaz, C.A. Achete, A. Jorio, Carbon **48**, 1592 (2010)
24. E.M. Ferreira, M.V. Moutinho, F. Stavale, M. Lucchese, R.B. Capaz, C. Achete, A. Jorio, Phys. Rev. B **82**, 125429 (2010)
25. A. Jorio, M.M. Lucchese, F. Stavale, E.H.M. Ferreira, M.V. Moutinho, R.B. Capaz, C.A. Achete, J. Phys. Condens. Matter **22**, 334204 (2010)
26. L. Cançado, A. Jorio, E.M. Ferreira, F. Stavale, C. Achete, R. Capaz, M. Moutinho, A. Lombardo, T. Kulmala, A. Ferrari, Nano Lett. **11**, 3190 (2011)
27. R. Beams, L.G. Cançado, L. Novotny, Nano Lett. **11**, 1177 (2011)
28. A. Eckmann, A. Felten, A. Mishchenko, L. Britnell, R. Krupke, K.S. Novoselov, C. Casiraghi, Nano Lett. **12**, 3925 (2012)
29. A. Jorio, ISRN Nanotech. (2012)
30. A. Jorio, L.G. Cançado, Phys. Chem. Chem. Phys. **14**, 15246 (2012)
31. J. Ribeiro-Soares, M. Oliveros, C. Garin, M. David, L. Martins, C. Almeida, E. Martins-Ferreira, K. Takai, T. Enoki, R. Magalhães Paniago, A. Malachias, A. Jorio, B. Archanjo, C. Achete, L. Cançado, Carbon **95**, 646 (2015)
32. R. Beams, L.G. Cançado, L. Novotny, J. Phys. Condes. Matter **27**, 83002 (2015)
33. C. Thomsen, S. Reich, Double resonant Raman scattering in graphite. Phys. Rev. Lett. **85**, 5214 (2000)

34. R. Saito, A. Jorio, A.G. Souza Filho, G. Dresselhaus, M.S. Dresselhaus, M.A. Pimenta, Probing phonon dispersion relations of graphite by double resonance Raman scattering. Phys. Rev. Lett. **88**, 027401 (2001)
35. D.M. Basko, Theory of resonant multiphonon Raman scattering in graphene. Phys. Rev. B **78**(12), 125418 (2008)
36. P. Venezuela, M. Lazzeri, F. Mauri, Theory of double-resonant Raman spectra in graphene: intensity and line shape of defect-induced and two-phonon bands. Phys. Rev. B **84**(3), 035433 (2011)
37. S. Mignuzzi, A.J. Pollard, N. Bonini, B. Brennan, I.S. Gilmore, M.A. Pimenta, D. Richards, D. Roy, Effect of disorder on Raman scattering of single-layer MoS2. Phys. Rev. B **91**, 195411 (2015)
38. Z. Lin, A. McCreary, N. Briggs, S. Subramanian, K. Zhang, Y. Sun, X. Li, N.J. Borys, H. Yuan, S.K. Fullerton-Shirey, 2D materials advances: from large scale synthesis and controlled heterostructures to improved characterization techniques, defects and applications. 2D Mater. **3**, 4 (2016)
39. L.G. Cançado, M.A. Pimenta, B.R.A. Neves, M.S. Dantas, A. Jorio, Influence of the atomic structure on the Raman spectra of graphite edges. Phys. Rev. Lett. **93**, 247401 (2004)

Chapter 6
Raman Spectroscopy Study of Two-Dimensional Materials Under Strain

Chunxiao Cong, Yanlong Wang, and Ting Yu

The exceptionally high stretchability of atomically thin materials enables extensive manipulation of their properties and exploration of rich physics through the application of external strain. Therefore, it is important to understand strain effects on two-dimensional materials both for fundamental studies and developing various applications, especially in flexible and wearable devices. In this chapter, we will give several examples of how Raman spectroscopy can be utilized to investigate the strain effects on fundamental properties of atomically thin materials.

6.1 Strain Effects on Graphene

Atomically thin materials, so called two-dimensional (2D) materials become the hub of nanomaterials after the rediscovery and unveiling exceptional properties of graphene, a single atomic carbon layer since 2004 [1]. Raman spectroscopy has been widely adopted as a key diagnostic tool for probing the properties of nanocarbon materials, such as graphite, carbon nanotube, fullerene, and graphene

---

C. Cong
School of Information Science and Technology, Fudan University, Shanghai, China

Y. Wang
Division of Physics and Applied Physics, School of Physical and Mathematical Sciences, Nanyang Technological University, Singapore, Singapore

Key Laboratory of Chemical Lasers, Dalian Institute of Chemical Physics, Chinese Academy of Sciences, Dalian, China

T. Yu (✉)
Division of Physics and Applied Physics, School of Physical and Mathematical Sciences, Nanyang Technological University, Singapore, Singapore
e-mail: yuting@ntu.edu.sg

P.-H. Tan (ed.), *Raman Spectroscopy of Two-Dimensional Materials*, Springer Series in Materials Science 276, https://doi.org/10.1007/978-981-13-1828-3_6

[2–4]. During last decade, Raman spectroscopy has already been used as a versatile tool for studying the properties of graphene, such as electronic structure, electron-phonon interaction, life times of phonons, stacking order, doping level, defects, edge chirality, and so on [5–13]. As the lattice vibrational motions of graphene, also quantized as phonon modes, are extremely sensitive to the strain applied to the graphene layers, the in-situ strain Raman spectroscopy should and has emerged great potential in investigating properties of graphene under strain.

6.1.1 Phonon Softening

At the early stage of graphene studies, Yu's group firstly demonstrated how to successfully apply uniaxial strain to graphene and conducted Raman mapping under strain [14, 15]. Instead of the typical 300 nm SiO_2/Si substrate, flexible and foldable substrate – polyethylene terephthalate (PET) was employed. Graphene layers with different thickness were mechanically exfoliated from a highly ordered pyrolytic graphite (HOPG) bulk and transferred onto pre-cleaned PET substrate. It is worth to note that the pre-cleaning process is important because usually PET surface covered with some compounds that could influence the adhesion and quality of Raman spectra of graphene negatively. The careful cleaning process could effectively remove such compounds. After the graphene layers being identified under an optical microscope by the same procedure as using the SiO_2/Si substrate, the Raman spectra and mapping measurements are conducted.

Reading the width of 2D mode peak and examining its line shape offer a quite trustable way to further determine the exact thickness of graphene layers. The uniaxial strain is applied by bending the PET thin film as shown in Fig. 6.1a. With the increase of the bending angles, the strain applied right at the very top of the arc, where the graphene locates and the laser beam focuses, also increases, and are obtained by a geometric analysis. The single Raman spectrum and the Raman mapping constructed by plotting the 2D mode positions are recorded under different mounts of controlled strains. It is obvious that the tensile strain causes significant softening of graphene phonons, *i.e.* 2D phonon in this work, as reflected by the dramatic red-shift of 2D peak (see Fig. 6.1b). The unique advantage of strained Raman mapping comparing to single Raman spectrum is to clearly picture out the local response of graphene to the strain or the strain distribution inside graphene layers. It can be clearly seen that the strain distribution is not perfectly uniform. The nonuniformity might be due to the contact, and consequently, the van der Waals force between the graphene and PET differs from region to region. Such nonuniformity in fact calls for caution of developing graphene based flexible electronics considering the sensitivity of graphene properties to the strains. The further carefully analysis of spectral features of 2D peak, such as fitting the 2D modes and plotting its position as a function of strain, then reading the slope, reveals the strain sensitivity of graphene is in the range of 5 cm^{-1} to 15 cm^{-1}/% for this work. This is much larger than that of C_{60} (-0.13 cm^{-1}/%) or multiwalled

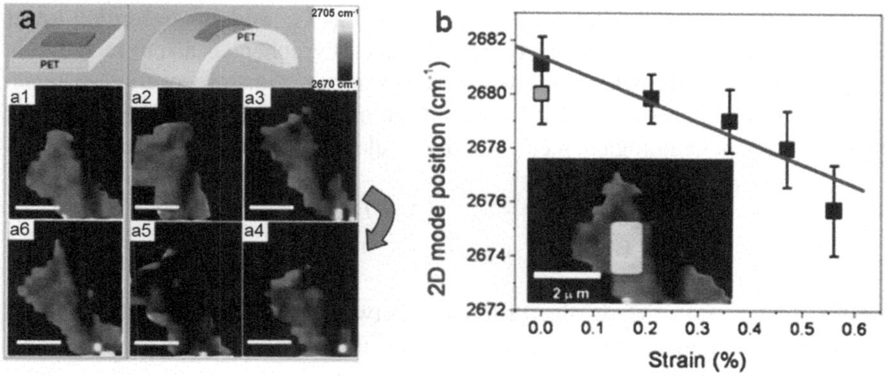

Fig. 6.1 (**a**) Schematic diagram showing the flat and bent PET substrates. Raman images of (*a*1) unstrained graphene, (*a*2–*a*5) strained graphene, (*a*6) relaxed graphene by extracting 2D mode frequency. (**b**) Mean value of the 2D mode frequency from the entire interested area subjected to the statistical analysis as a function of strain. The error bars are the standard deviations. The data point in green/gray is from the relaxed graphene. The inset shows the interested area subjected to the statistical analysis [14]. (Copyright 2008, American Chemical Society)

carbon nanotubes (MWNTs) (-0.48 cm^{-1}/%) [16], indicating the great potential of graphene as the ultrasensitive strain sensor. It has also been demonstrated that the phonon softening process in graphene under tensile strain is reversible before any slipping occurs. The happening of slipping could be signed by a sudden blue-shift of 2D peak.

6.1.2 G Mode Splitting and Crystal Orientation Determination

For applying strain to graphene and further perform strain study, the selection of flexible substrates and extra processes are developed. For example, Ferrari's team spun coats 400 nm SU8 on top of PET thin film before peeling off graphene layers to make an enhanced contrast for identifying graphene flakes [17]. Heinz's team transferred graphene layers from typical SiO_2/Si substrate onto PDMS and then clamped the graphene flakes by several narrow Ti strips to further anchoring graphene on the flexible substrate [18]. With such well-designed and prepared samples, both groups conducted systematic Raman spectral study of graphene G mode under uniaxial strain.

Inside graphene, the carbon atoms pattern themselves into closely packed honeycomb structure, represented as 2D point group D_6 and displays C_6 and m symmetry. Once strain is applied, only C_2 symmetry could generally survive as well m symmetry if the strain happens to be along either zigzag or armchair direction. Therefore, the degeneracy of the doubly degenerated optical G phonon mode of E_{2g} symmetry will be lifted or broken, leading to a splitting into two singlet peaks of different vibrational energies, usually assigned as G^{-} and G^{+} according lower and higher frequencies, respectively. The G mode of graphene is a zero momentum optical phonon mode, where the two sublattice move with respect to each other. Analysis of this G phonon mode of uniaxially strained graphene could be carried out by using a phenomenological method and a secular equation could be developed as

$$\begin{vmatrix} A\varepsilon_{xx} + B\varepsilon_{yy} - \lambda & (A - B)\,\varepsilon_{xy} \\ (A - B)\,\varepsilon_{xy} & B\varepsilon_{xx} + A\varepsilon_{yy} - \lambda \end{vmatrix} = 0 \tag{6.1}$$

where $\lambda \equiv \omega^2 - \omega_0^2$ is the difference between the square of the strained and unstrained phonon energies, A and B are the phonon deformation potential coefficients, and $\varepsilon_{ij}$is the strain tensor. By introducing the Poisson ratio v (=0.16 for graphene) [19], the strain tensor could be further expressed as $\varepsilon_{xx} = \varepsilon$; $\varepsilon_{yy} = -v\varepsilon$, and $\varepsilon_{xy} = 0$, where x is the direction of the applied stress, ε is the magnitude of the resulting strain. Therefore, the energies of the G^{-} and G^{+} modes are:

$$\begin{cases} \omega_{G^+} = \omega_0 + \frac{B - \upsilon A}{2\omega_0}\varepsilon \\ \omega_{G^-} = \omega_0 + \frac{A - \upsilon B}{2\omega_0}\varepsilon \end{cases} \tag{6.2}$$

Obtaining the G peak (including both G^{-} and G^{+}) positions at different stress/strain (see Fig. 6.2) and fitting the curves according to the equation above, the G phonon deformation potential coefficients A and B are found to be

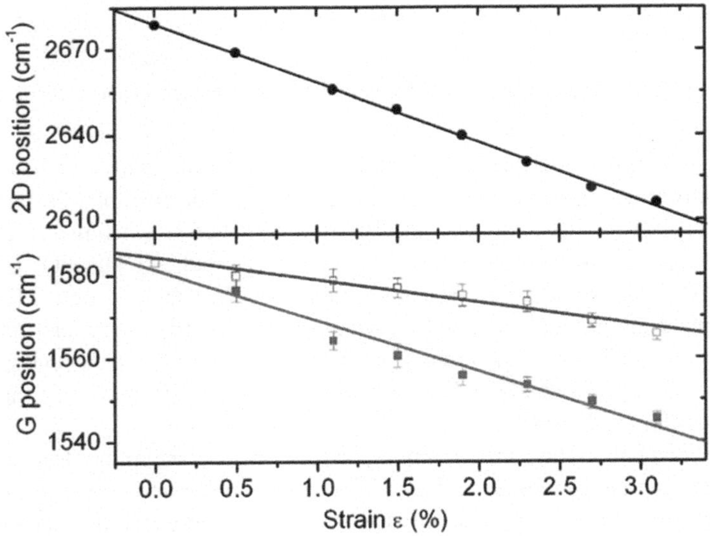

Fig. 6.2 The variation of the phonon frequencies of 2D, G^{+} and G^{-} bands as a function of strain. The solid lines are linear fits [18]. (Copyright 2009 National Academy of Sciences)

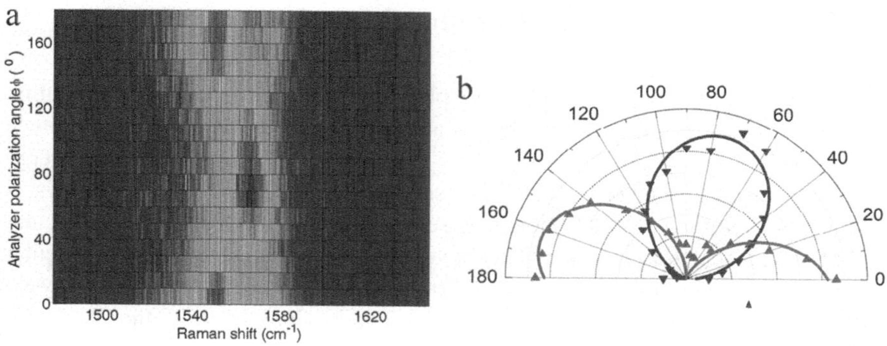

Fig. 6.3 (**a**) False-color image of the intensity of the Raman scattered light as a function of the Raman shift and angle of the analyzer φ for detection of the Raman signal. The angle φ was measured with respect to the incident light polarization, which was aligned along the strain axis. The data were obtained by measuring Raman spectra every 10°. (**b**) Raman scattering intensity for the G^+ (inverted triangle) and G^- (triangle) bands as a function of φ. The solid lines are fits to the form of $sin^2(\varphi - \varphi_0)$ [18]. (Copyright 2009 National Academy of Sciences)

$-4.4 \pm 0.8 \times 10^6 cm^{-2}$ and $-2.5 \pm 0.5 \times 10^6 cm^{-2}$, respectively [18]. With these two coefficients, the G mode Grüneisen constant and shear phonon deformation potential (SDP) could also be obtained as 0.69 ± 0.14 and 0.38 ± 0.08, respectively [18].

Applying sufficient uniaxial strain would deform the original graphene lattice and break the isotropic symmetry. The strain direction with respect to the crystal orientation, such as armchair or zigzag could cause the different lattice evolution or symmetry changes, which could be reflected by the intensity changes of the strain induced split G peaks. Polarized Raman spectroscopy is the best technique to probe such crystal structures. Figure 6.3a shows the Raman intensity of G mode as a function of Raman shift and angle between the incident and scattered light polarization (E-field direction). Remarkable polarization behavior (color changes in Fig. 6.3a) of both G^- and G^+ presents. Such polar properties are further quantified by plotting the intensities of G^- and G^+ modes as a function of analyzer angles (Fig. 6.3b) and further well fit by $sin^2(\phi - \phi_0)$ with $\phi_0 = 165.3^o, 75.8^o$ for the G^+ and G^- bands, respectively [18], implying the truth of the linear polarization of both two modes and the orthogonal polarization between each other. Bringing the Raman tensors for both G^- and G^+, which carry the crystal structural information under strain and working them on both incident and scattering light could properly describe intensities of these modes under different polar figurations and eventually tell the strain direction to the either zigzag or armchair. An angle of 25.2° is obtained for the sample shown in Fig. 6.3 [18].

6.1.3 2D Mode Splitting and Inner-Scattering Process

Comparing to the zero momentum G phonon mode, the nonzero momentum, so called double resonant intervalley scattering 2D mode of graphene is much more complicated. The 2D phonons are required to conserve both the energy and momentum of electrons, which are scattered between two energy valleys in graphene. Therefore, the 2D phonon is sensitive to the electronic band structure and phonon dispersion. As discussed above, applying sufficient strain, in particular along some certain direction such as zigzag or armchair could deform the crystal lattice or change the crystal symmetry very much. Consequently, the electronic bands and phonon dispersion would evolve significantly. These evolutions certainly lead to the changes of 2D mode or alternatively could be probed by the 2D mode. Cheong's team performs polarized micro-Raman measurement and first-principles calculations on such project [20].

Figure 6.4 shows the Raman spectra of the graphene layers under the different amount of strains along near armchair (A-strain) (see Fig. 6.4a) and zigzag (Z-strain) (see Fig. 6.4b) direction, respectively [20]. Similar as the G mode, but with very different mechanism, the 2D peak splits into two peaks under strain and the separation between two split peaks increases with the increase of the strain. Substantially different to the G mode, the frequency shift rates (the slopes) of the $2D^{-}$ and $2D^{+}$ show remarkable differences when the strain is along the armchair or the zigzag directions (Fig. 6.4c, d).

As illustrated in Fig. 6.5a, two possible scattering paths or processes could be responsible for the 2D mode. One is the smallest momentum transfer path (inner process) and the other is largest momentum transfer path (outer process). For pristine (unstrained) graphene, the scattering processes involving the three K' points around a give K point are completely equivalent. Therefore, the 2D band appears as a single peak. Once uniaxial strain is applied along or nearly along the armchair direction, the real and the reciprocal lattice are distorted (Fig. 6.5b): the horizontal K' point moves away, whereas the other two K' points move closer to the K point. Hence, the momenta of the phonons are different and the corresponding phonon energies/frequencies are different, reflected as the splitting of the 2D mode. The same anisotropic situation and Raman peak splitting happen to the case, where the zigzag strain is applied (Fig. 6.5c).

To further understand such strain-direction depended 2D mode splitting and nonuniversal phonon softening rates, the electronic band structure of strained graphene is studied based on the first-principles self-consistent pseudopotential method using the generalized gradient approximation for exchange-correlation function and the phonon dispersion is explored by using the density-functional perturbation theory [20]. The calculation results are plotted together with the experimental data in Fig. 6.6. A few points could be concluded: (i) under strain along either zigzag or armchair direction, for both inner- and outer-scattering processes, the required resonant phonon momenta are linearly changes with the increase of the strain; (ii) along different directions, the phonon dispersion are quite different

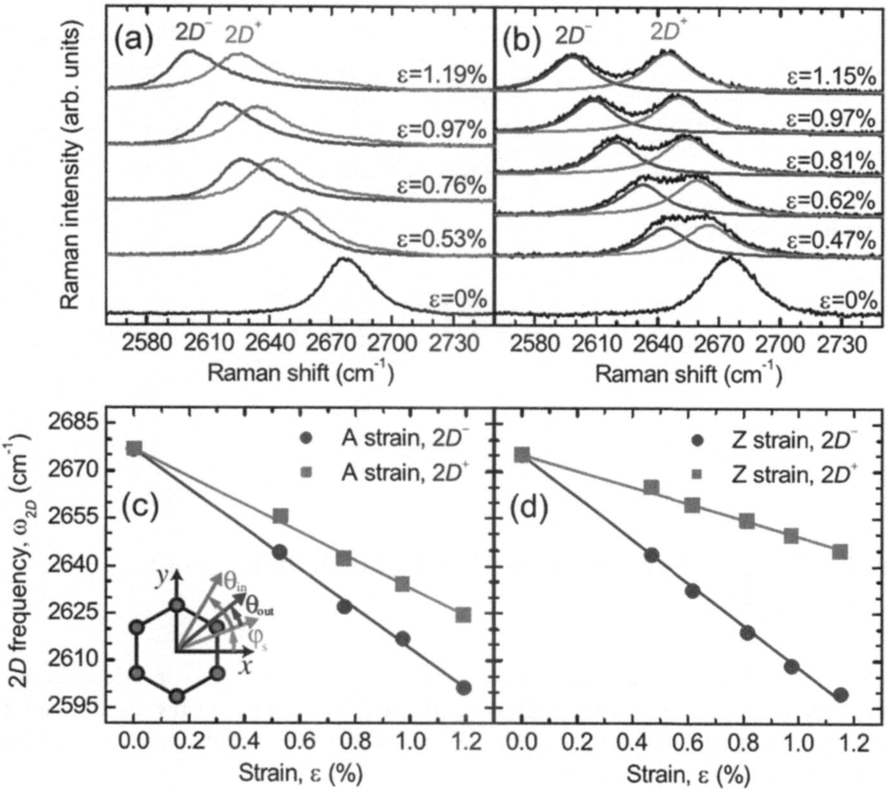

Fig. 6.4 Evolutions of the 2D bands of (**a**) A-strain and (**b**) Z-strain samples as a function of uniaxial strain. In (**a**), the $2D^-$ and $2D^+$ peaks are obtained with $\theta_{in} = \theta_{out} = 0^\circ$ and $\theta_{in} = \theta_{out} = 90^\circ$, respectively. In (**b**), the spectra are measured with $\theta_{in} = \theta_{out} = 50^\circ$ and deconvoluted into two Lorentzian peaks. The positions of the $2D^-$ and $2D^+$ peaks of (**c**) the A-strain and (**d**) Z-strain samples as a function of strain. The solid lines are linear fits to the data. The inset shows the polarization geometry, where θ_{in}, θ_{out}, and φ_s are the angles that the incident laser polarization, the analyzer axis, and the zigzag direction make with respect to the strain axis, respectively [20]. (Copyright 2011 American Physical Society)

under certain strain (i.e. 2%); and (iii) most importantly, the inner-scattering process shows good agreement to the experimental observations with regards of the different red shift rates for differently oriented strain. Thus, combining the strain Raman spectroscopy study and the first-principles calculations, Cheong and his coauthors show that the anisotropic modifications of the 2D phonon dispersion together with the changes in electronic structures are the origins of the strain-induced 2D mode splitting and strain-direction sensitive redshifts.

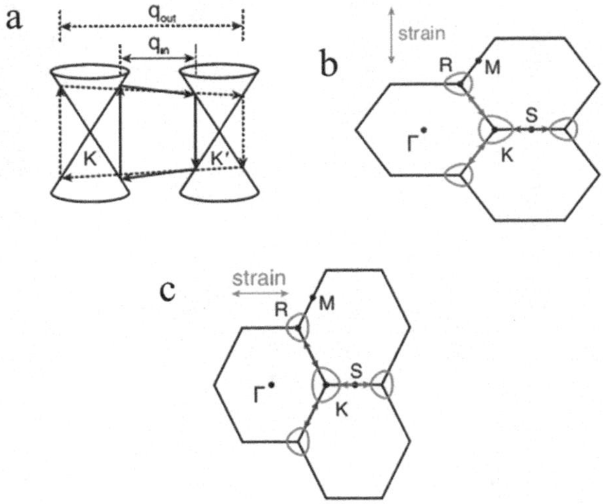

Fig. 6.5 (**a**) Double-resonance Raman scattering process. Inner (q_{in}) and outer (q_{out}) processes are indicated. Reciprocal lattice diagram for (**b**) A-strained, and (**c**) Z-strained graphene, showing strain-induced distortions [20]. (Copyright 2011 American Physical Society)

6.2 Strain Effects on Transition Metal Dichalcogenides (TMDs)

As with graphene, Raman spectroscopy is also a powerful tool to probe the strain-dependent properties of other 2D systems. In this section, we review the strain effects on the fundamental properties of transition metal dichalcogenides (TMDs) revealed by Raman spectroscopy.

6.2.1 Strain-Induced Phonon Modification in MoS_2

Due to the native bandgap and the indirect-direct bandgap transition in monolayer limit [21–23], group-VIB semiconducting TMDs with the general form of MX_2 (M = Mo or W; X = S or Se) are deemed as the potential alternatives to graphene. Moreover, monolayer MX_2 demonstrates a high on/off ratio [24, 25] and superior mechanical properties [26, 27], which make them hold promise for flexible optoelectronic applications. Because semiconducting 2H-MX_2 share similar strain dependent phonon behaviors as a result of the same crystal structure [28–30], monolayer MoS_2 is adopted as an example to demonstrate the strain effects on MX_2.

Figure 6.7a shows the Raman spectra of monolayer MoS_2 under increasing uniaxial strain induced by stretching the flexible PET substrate [28]. Though no obvious shift occurs to the A_{1g} mode, the frequencies of the E^1_{2g} mode redshift with increasing strain. This initially doubly degenerate optical E^1_{2g} mode further splits into two singlet sub-modes, termed E^{1+}_{2g} and E^{1-}_{2g}, under the strain larger than 1% (see Fig. 6.7c), which lowers the original symmetry of monolayer MoS_2 [28]. As

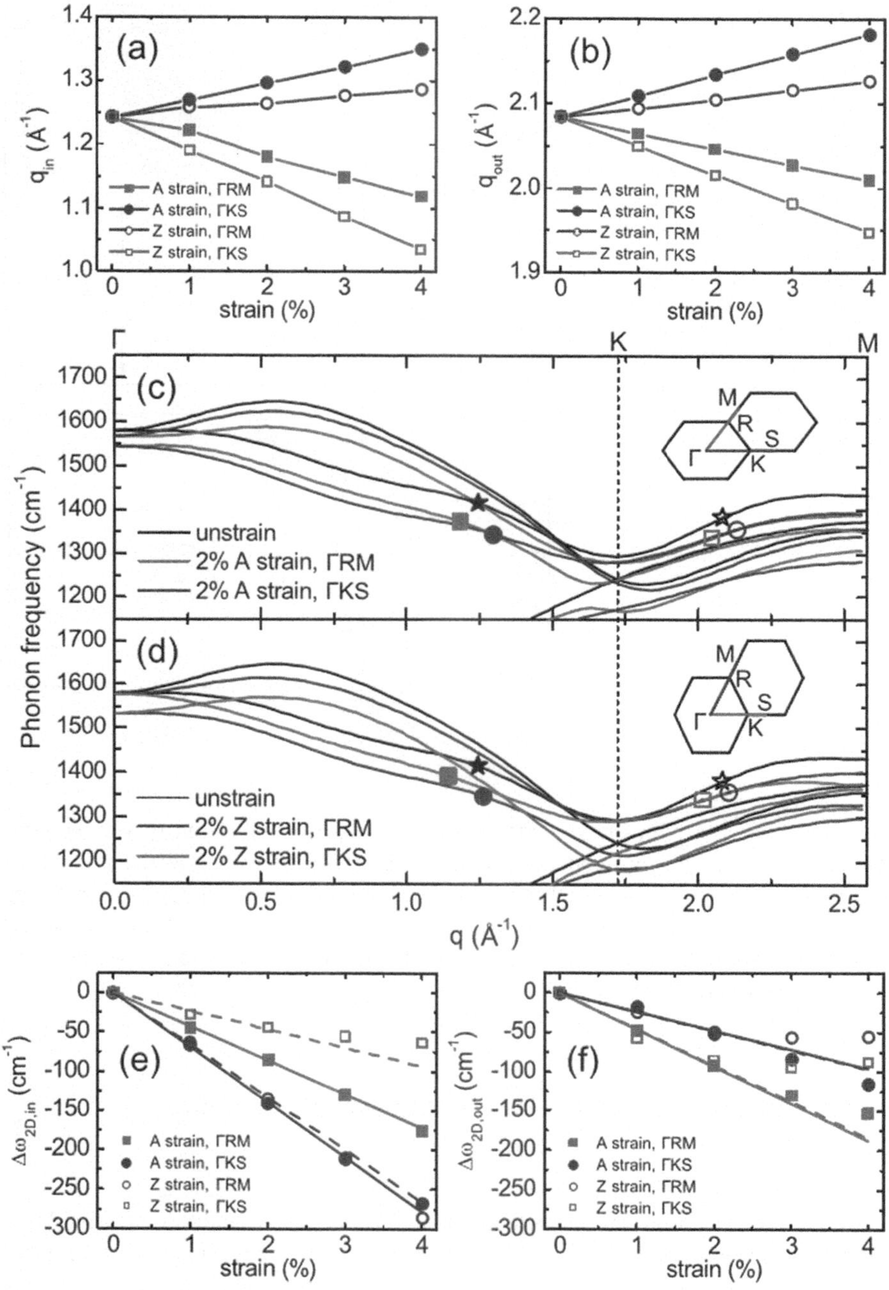

Fig. 6.6 Calculated phonon momenta involved in the (**a**) inner (q_{in}) and (**b**) outer (q_{out}) processes for the scattering in the ΓKS and ΓRM directions and A and Z strains. Phonon dispersions near the K or R points for 2% strain applied in the (**c**) armchair and (**d**) zigzag direction. Filled and open stars are for the 2D band in unstrained graphene, filled and open squares for the $2D^+$ peak, and filled and open circles for the $2D^-$ peak. Calculated strain dependences of the 2D peaks for (**e**) inner and (**f**) outer processes [20]. (Copyright 2011 American Physical Society)

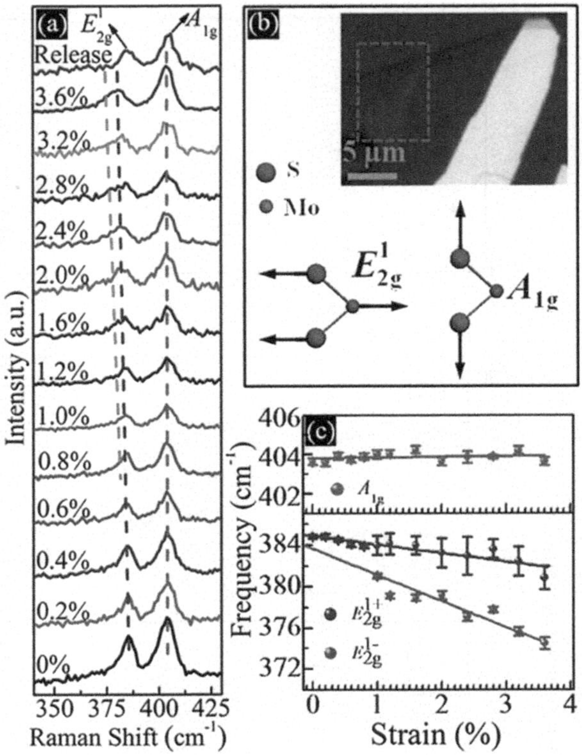

Fig. 6.7 (**a**) Raman spectra of monolayer MoS_2 as a function of uniaxial strain. The peak positions are guided by the dash lines. (**b**) The schematic diagram of the atomic displacements of the E^1_{2g} and A_{1g} modes. The inset shows the optical image of the single-layer MoS_2 (highlighted by the dashed rectangle) on PET. (**c**) Fitted vibrational frequencies of E^1_{2g} and A_{1g} modes under increasing uniaxial strain. The solid lines denote the linear fitting. (Adapted from Ref. [28]. Copyright 2013, Wiley-VCH Verlag GmbH)

displayed in Fig. 6.7b, the E^1_{2g} mode involves opposite vibration of two sulphur atoms relative to the molybdenum atom in the basal plane while the A_{1g} mode arises from the out-of-plane vibration of mere sulphur atoms in reverse directions [28]. The experimental observations undoubtedly demonstrate that the dominated covalent bonding between sulphur and molybdenum atoms is rather susceptible to the in-plane uniaxial strain.

6.2.2 *Strain-Assisted Crystal Orientation Determination in MoS_2*

This section describes the identification of the crystal orientation of monolayer MoS_2 by the polar behaviour of the splitted E^1_{2g} modes under uniaxial strain [28]. Figure 6.8a demonstrates the configuration of the polarization dependent Raman measurements. The polarizations of the incoming and scattered lights have ψ and φ angles relative to the strain direction, and θ denotes the angle between the strain axis and zigzag direction. By rotating an analyzer, Wang *et al.* obtained the Raman spectra of the monolayer MoS_2 under 3.6% uniaxial strain as a function of angle

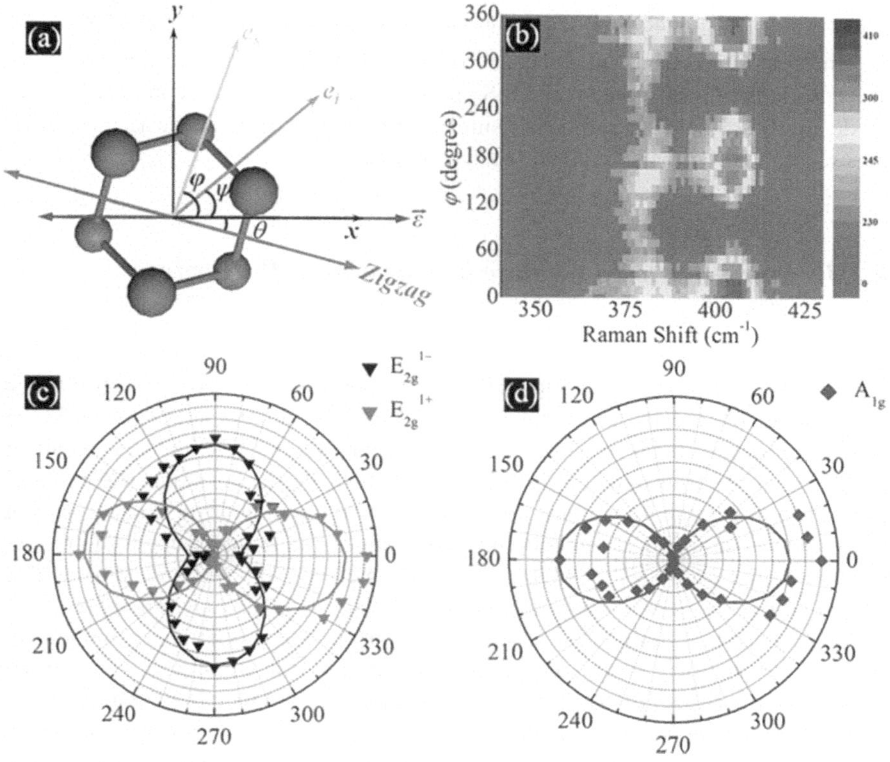

Fig. 6.8 (**a**) Schematic illustration of the configuration for the polarization dependent Raman measurements. The strain direction ($\rightarrow \varepsilon$) is fixed along the x-axis, with a θ angle relative to the zigzag direction. e_i and e_s mark the polarization of the incoming and scattered lights, which deviate from the strain direction by angles of ψ and φ, respectively. (**b**) False-color image of the Raman spectrum intensity as a function of the angle φ and the Raman shift under uniaxial strain of 3.6%. (**c**) and (**d**) show polar plots of the intensities of E_{2g}^{1+}, E_{2g}^{1-}, and A_{1g} modes. (Adapted from Ref. [28]. Copyright 2013, Wiley-VCH Verlag GmbH)

φ and plotted the color map of the spectrum intensity versus both polar angles and peak positions (Fig. 6.8b) [28]. It can be seen that the strongest E_{2g}^{1+} and E_{2g}^{1-} peaks turn up at different polarization angles, revealed as the alternation of the intensity maxima within the E_{2g}^{1} mode frequency range while the maxima and minima of the A_{1g} band appear in the same position of the Raman shift. As can be seen from the fitted polarization dependent phonon intensities in Fig. 6.8c, the splitted E_{2g}^{1+} and E_{2g}^{1-} modes induced by the applied strain are very sensitive to the polarization angle and can be fitted well by the two equations shown below:

$$I_{E_{2g}^{1+}} \propto d^2 \cos^2(\varphi + \psi + 3\theta) \tag{6.3}$$

$$I_{E_{2g}^{1-}} \propto d^2 \sin^2(\varphi + \psi + 3\theta) \tag{6.4}$$

This demonstrates that the linear polarizations of the scattered light from E_{2g}^{1+} and E_{2g}^{1-} modes are perpendicular to each other and could be used to determine the crystal orientation of monolayer MoS_2 by extracting θ value. In contrast, the A_{1g} mode follows a $\cos^2\varphi$ dependence similar to that in pristine state, as it is rather unsusceptible to in-plane uniaxial strain (Fig. 6.8d).

6.2.3 Strain-Induced Change in Vibrational Properties of $ReSe_2$

$ReSe_2$, a representative of Group-VIIB TMDs, possesses one extra electron in the d-orbitals of the metal atom compared to Group-VIB counterparts such as MoS_2 and has attracted much research interest in recent years [31–33]. As a result of the different electron numbers in the outer shell of the metal atom, $ReSe_2$ lacks indirect-to-direct gap transition from thick layers to the monolayer counterpart but possesses vanishing interlayer coupling and strong in-plane anisotropy [32], in stark contrast to the group-VIB TMDs. The weak interlayer interaction indicates that strain engineering could be applied onto $ReSe_2$ flakes with any layer number [33].

Yang *et al.* has studied the strain effects on phonon properties of monolayer $ReSe_2$ [33]. In their work, the application of uniaxial strain was realized by generation of wrinkles on the $ReSe_2$ flake, as illustrated in Fig. 6.9. First, thin $ReSe_2$ sheets were mechanically exfoliated onto the elastomeric gel-film substrate, which was prestretched to some degree. Following that, the prestrained substrate was released and local strain in $ReSe_2$ sheets was successfully achieved by generation of wrinkles on the samples.

As shown in Fig. 6.10, two peaks at 125 and 155 cm^{-1} dominate the Raman spectrum of the pristine monolayer $ReSe_2$ and are predicted by the calculated phonon dispersion relation based on Density Functional Theory (DFT) [33]. According to the symmetry analysis, these two peaks result from a combination of several fundamental Raman modes and become associated with Eg-like and Ag-like modes, respectively, unlike the well-defined E_{2g}^{1} or A_{1g} peaks as observed in other TMDs, such as MoS_2 [33]. The modification of phonon modes by local strain could be revealed by a comparison between the Raman spectra of pristine and strained monolayer $ReSe_2$ (Fig. 6.10b). It can be clearly seen that the E_g-like peak splits into two sub-peaks, while the Ag-like mode only slightly blueshifts under strain. The distinct phonon response to uniaxial strain is similar to the observations for strained monolayer MoS_2 [28] and attributed to the different atomic displacements involved for the two phonon modes [33].

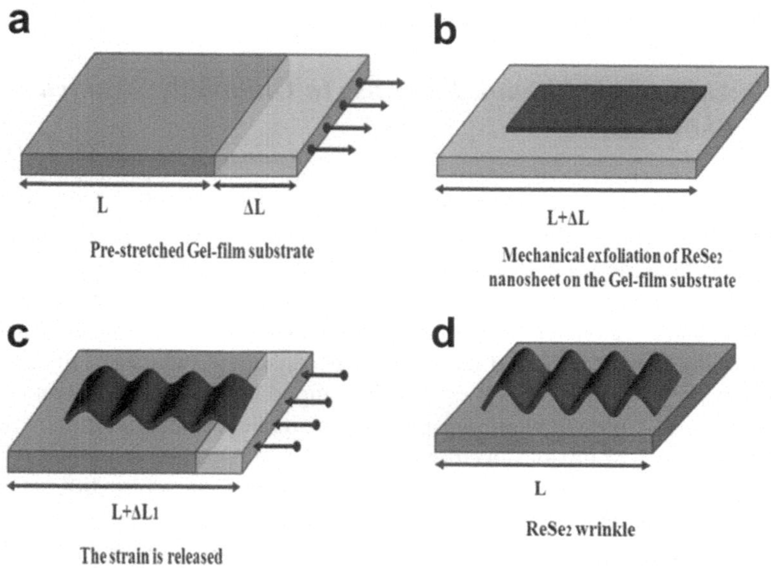

Fig. 6.9 Schmatic illustration of the process of wrinkle generation in the $ReSe_2$ sample. (**a**) First, the gel-film substrate needs to be prestretched; (**b**) $ReSe_2$ samples are then mechanically exfoliated onto the prestretched gel-film substrate; (**c**) following (**b**), the prestretched gel-film substrate is released; (**d**) uniaxial strain is successfully realized by generation of wrinkles on the $ReSe_2$ flake. (Adapted from Ref. [33]. Copyright 2015, American Chemical Society)

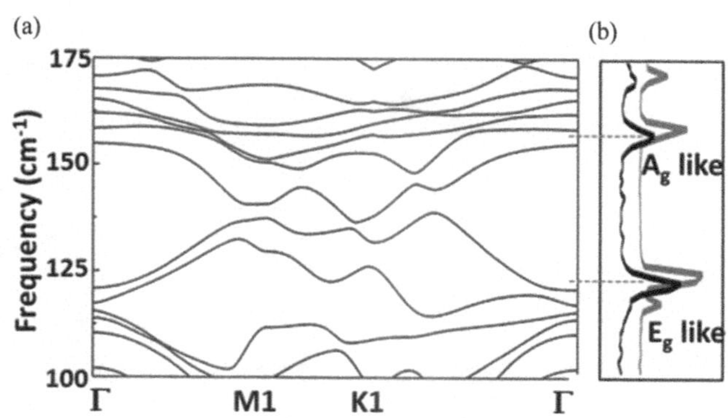

Fig. 6.10 (**a**) Calculated phonon dispersion relation of monolayer $ReSe_2$ from DFT calculations. (**b**) Raman spectra of pristine (black) and strained (red/gray) $ReSe_2$. (Adapted from [33]. Copyright 2015, American Chemical Society)

6.3 Strain Effects on Other 2D Systems

6.3.1 Anisotropic Phonon Response in Uniaxially Strained Phosphorene

Phosphorene possesses a direct bandgap for any layer number [34], with the thickness dependent bandgaps bridging the gap between TMDs and graphene [35], making phosphorene perfectly suitable for mid- and near- infrared optoelectronics. Moreover, the on/off ratio and mobility combination of phosphorene happens to fill the blank region unreachable by TMDs and graphene [35]. Of particular interest is the unique puckered structure of phosphorene, which results in many anisotropic properties [36–41]. Therefore, it is important to understand strain effects on the anisotropic properties of phosphorene for its better integration into various technologies, such as flexible electronics.

Wang *et al.* have applied uniaxial strain onto phosphorene along different directions and observed the remarkable anisotropic phonon response by *in situ* strained Raman measurements [42]. As shown in Fig. 6.11a, b, the A_g^1 mode which vibrates

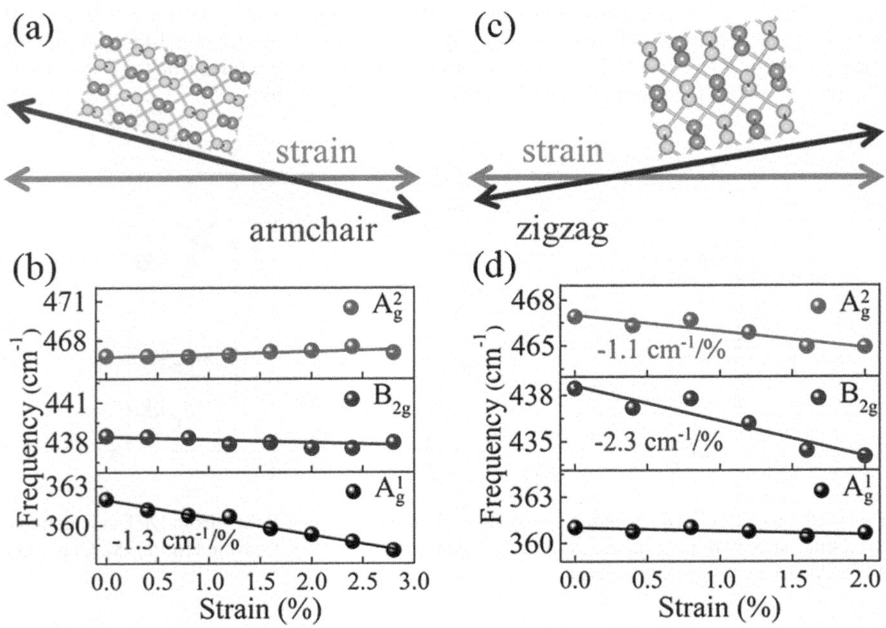

Fig. 6.11 (**a**) Schematic diagram of the relationship between the uniaxial strain direction and armchair orientation of phosphorene for the *in situ* strained Raman measurement. (**b**) Measured phonon frequencies of A_g^1, B_{2g}, and A_g^2 modes of phosphorene under uniaxial strain along the near-armchair direction, as shown in (**a**). (**c**) and (**d**) are same as (**a**) and (**b**), but for tensile strain along the near-zigzag directions. (Adapted from Ref. [42]. Copyright 2015, Tsinghua University Press and Springer-Verlag Berlin Heidelberg)

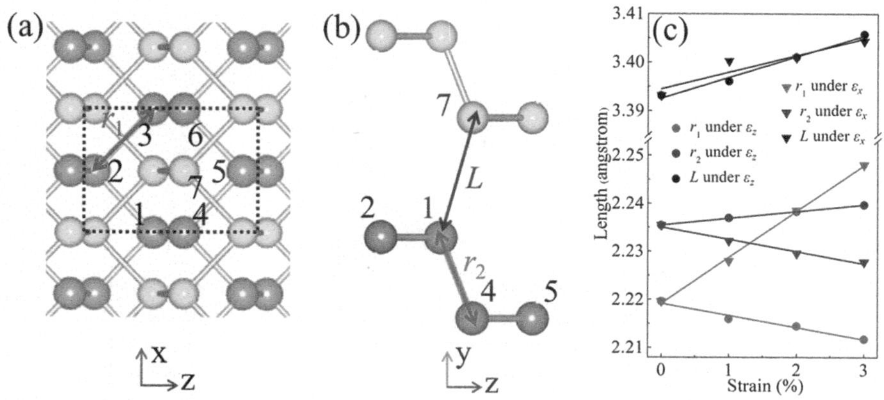

Fig. 6.12 (**a**) Top and (**b**) side view of bilayer phosphorene. r_1 and r_2 are in-plane and out-of-plane phosphorus-phosphorus bond lengths, respectively. L denotes the inter-distance. (**c**) Calculated bond lengths (r_1, r_2) and inter-distance (L) as a function of uniaxial tensile strain along armchair and zigzag directions for quinary-layer phosphorene. (Adapted from Ref. [42]. Copyright 2015, Tsinghua University Press and Springer-Verlag Berlin Heidelberg)

along the out-of-plane direction linearly redshifts at a rate of $-1.3\ \text{cm}^{-1}/\%$ while the shift rates of B_{2g} and A_g^2 modes involving in-plane vibrations of phosphorus atoms are negligible with the mechanical tension applied near the armchair direction. In contrast, B_{2g} and A_g^2 modes linearly soften with the strain (with the shift rates of $-2.3\ \text{cm}^{-1}/\%$ and $-1.1\ \text{cm}^{-1}/\%$, respectively) while the A_g^1 mode is inert to it under uniaxial tension along the near-zigzag direction (Fig. 6.11c, d).

To clarify the observed prominent dependence of phonon behaviors on the crystal orientation in uniaxially strained phosphorene, Wang *et al.* calculated the evolution of bond lengths and the inter-distance under both armchair and zigzag strain (0–3%) [42]. As a result of the unique puckered crystal structure, phosphorene shows the anisotropic mechanical response to uniaxial strain (Fig. 6.12), which could well explain the pronounced dependence of strained phonon frequencies on crystallographic orientations in phosphorene based on simple Harmonic spring oscillator model [42].

6.3.2 Strain-Induced Position Variations of E_{2g} Mode in Hexagonal Boron Nitride (hBN)

Thin layers of hexagonal boron nitride (hBN), a typical 2D insulator, have a structure similar to graphene where nitrogen and boron atoms are bound by covalent force within the layers [43]. Because of its chemical inertness, atomic smoothness, and lack of dangling bonds, hBN serves as an excellent insulating substrate for electronic devices based on various 2D materials [43, 44].

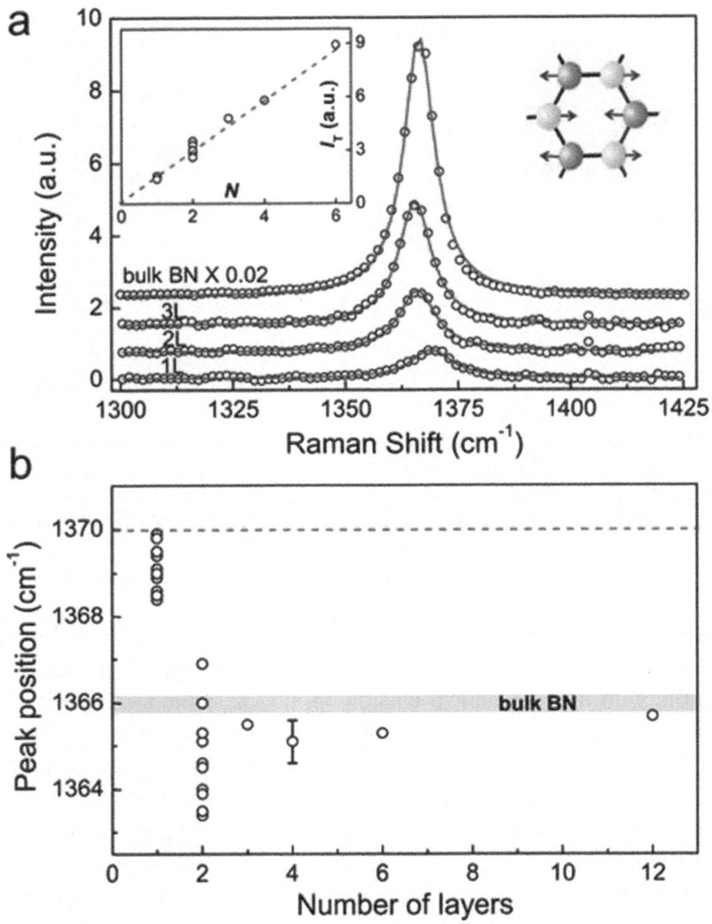

Fig. 6.13 (**a**) Raman spectra of hBN with different thickness. The left inset plots evolutions of the peak intensity for the E_{2g} mode as a function of layer numbers and the right inset shows the schematic diagram of atomic displacements for the E_{2g} mode. (**b**) Peak positions of the E_{2g} mode for hBN flakes with different layer numbers. The dash line denotes the peak position of the E_{2g} mode predicted for single-layer hBN and the gray bar marks the measured position of the E_{2g} mode for bulk hBN. (Adapted from Ref. [45]. Copyright 2011, Wiley-VCH Verlag GmbH)

Gorbachev *et al.* have reported the variations of peak positions for E_{2g} mode in ultrathin hBN flakes caused by unintentional strain generated during the exfoliation process [45]. As shown in Fig. 6.13a, the characteristic Raman peak for hBN, E_{2g} mode, involves in-plane oscillations with boron and nitrogen atoms vibrating out of phase. The intensity of the E_{2g} mode becomes stronger with increasing sample thickness and could be used to accurately determine layer numbers for hBN samples thinner than 7 layers. When it comes to the peak position, it is observed that the E_{2g} mode blueshifts in monolayers and mostly redshifts in bilayers with respect to that in bulk hBN (Fig. 6.13b). In addition, monolayer and bilayer hBN flakes show significantly larger variations in peak positions of the E_{2g} mode compared with those of thicker layers. Unintentional strain, possibly generated during the cleavage process, is proposed as the main mechanism to account for the observed large variations [45]. The maximum hardening of the E_{2g} mode in single-layer hBN samples compared with the bulk counterpart is around 4 cm^{-1}, in agreement with the calculated position difference between them, caused by the shorter length of B-N bond in exfoliated single-layer hBN samples [46]. The observed sample-dependent softening in single-layer hBN flakes relative to the inherent position of around 1370 cm^{-1} demonstrates that local tensile strain exists in monolayer hBN and its strength is not constant. This argument can also explain the large variations in Raman shifts of the E_{2g} mode for bilayer hBN sheets and a slight intrinsic hardening of around 1 cm^{-1} can be deduced by the observed maximum blueshift for bilayer hBN samples compared with the bulk [45].

6.4 Summary and Outlook

In this chapter, we have studied Raman spectroscopy of two-dimensional layered materials under strain including graphene, transition metal dichalcogenides, and other 2D systems like black phosphorene. Two-dimensional layered materials are widely considered to have significant prospects for future flexible electronics systems due to its novel electronic properties and mechanical properties. So it is imperative and necessary to thoroughly understand how strain effects affect both the electronic and mechanical properties of two-dimensional layered materials. We have shown that Raman spectroscopy can be used to probe the strain effects on the 2D layers. By applying strain on the 2D layers, their electronic band structures, phonon properties, electron-phonon interactions, crystal orientations, anisotropic behaviors, and so on, can be revealed through Raman spectroscopy.

In conclusion, the nature with high flexibility and bendability of such 2D layered crystals enables them as most promising candidates for future flexible smart and wearable systems. However, there are still many issues that need to be solved before the practical applications such as how to apply uniform strain, how to obtain the exact strain magnitude applied on the 2D layers, and how to control the distribution of the strain, and so on. Raman spectroscopy as a non-destructive, easily-operated tool will play a great role in stain studies of 2D layered materials.

References

1. K.S. Novoselov, A.K. Geim, S.V. Morozov, D. Jiang, Y. Zhang, S.V. Dubonos, I.V. Grigorieva, A.A. Firsov, Electric field effect in atomically thin carbon films. Science **306**, 666–669 (2004)
2. M. Dresselhaus, A. Jorio, R. Saito, Characterizing graphene, graphite, and carbon nanotubes by Raman spectroscopy. Annu. Rev. Condens. Matter Phys. **1**, 89–108 (2010)

3. A.C. Ferrari, D.M. Basko, Raman spectroscopy as a versatile tool for studying the properties of graphene. Nat. Nanotechnol. **8**, 235–246 (2013)
4. R. Saito, M. Hofmann, G. Dresselhaus, A. Jorio, M. Dresselhaus, Raman spectroscopy of graphene and carbon nanotubes. Adv. Phys. **60**, 413–550 (2011)
5. C. Cong, T. Yu, K. Sato, J. Shang, R. Saito, G.F. Dresselhaus, M.S. Dresselhaus, Raman characterization of ABA- and ABC-stacked trilayer graphene. ACS Nano **5**, 8760–8768 (2011)
6. L.M. Malard, D.L. Mafra, S.K. Doorn, M.A. Pimenta, Resonance Raman scattering in graphene: probing phonons and electrons. Solid State Commun. **149**, 1136–1139 (2009)
7. L.M. Malard, M.A. Pimenta, G. Dresselhaus, M.S. Dresselhaus, Raman spectroscopy in graphene. Phys. Rep. **473**, 51–87 (2009)
8. J. Yan, Y. Zhang, P. Kim, A. Pinczuk, Electric field effect tuning of electron-phonon coupling in graphene. Phys. Rev. Lett. **98**, 166802-1–166802-4 (2007)
9. K. Kang, D. Abdula, D.G. Cahill, M. Shim, Lifetimes of optical phonons in graphene and graphite by time-resolved incoherent anti-Stokes Raman scattering. Phys. Rev. B **81**, 165405-1–165405-6 (2010)
10. A. Das, S. Pisana, B. Chakraborty, S. Piscanec, S.K. Saha, U.V. Waghmare, K.S. Novoselov, H.R. Krishnamurthy, A.K. Geim, A.C. Ferrari, A.K. Sood, Monitoring dopants by Raman scattering in an electrochemically top-gated graphene transistor. Nat. Nanotechnol. **3**, 210–215 (2008)
11. L. Cançado, M. Pimenta, R. Saito, A. Jorio, L. Ladeira, A. Grueneis, A. Souza-Filho, G. Dresselhaus, M. Dresselhaus, Stokes and anti-stokes double resonance Raman scattering in two-dimensional graphite. Phys. Rev. B **66**, 035415 (2002)
12. Y.M. You, Z.H. Ni, T. Yu, Z.X. Shen, Edge chirality determination of graphene by Raman spectroscopy. Appl. Phys. Lett. **93**, 163112-1–163112-3 (2008)
13. C. Cong, T. Yu, H. Wang, Raman study on the G mode of graphene for determination of edge orientation. ACS Nano **4**, 3175–3180 (2010)
14. T. Yu, Z. Ni, C. Du, Y. You, Y. Wang, Z. Shen, Raman mapping investigation of graphene on transparent flexible substrate: the strain effect. J. Phys. Chem. C **112**, 12602–12605 (2008)
15. Z.H. Ni, T. Yu, Y.H. Lu, Y.Y. Wang, Y.P. Feng, Z.X. Shen, Uniaxial strain on graphene: Raman spectroscopy study and band-gap opening. ACS Nano **2**, 2301–2305 (2008)
16. C.A. Cooper, R.J. Young, Investigation of structure/property relationships in particulate composites through the use of Raman spectroscopy. J. Raman Spectrosc. **30**, 929–938 (1999)
17. T. Mohiuddin, A. Lombardo, R. Nair, A. Bonetti, G. Savini, R. Jalil, N. Bonini, D. Basko, C. Galiotis, N. Marzari, Uniaxial strain in graphene by Raman spectroscopy: G peak splitting, Grüneisen parameters, and sample orientation. Phys. Rev. B **79**, 205433-1–205433-8 (2009)
18. M. Huang, H. Yan, C. Chen, D. Song, T.F. Heinz, J. Hone, Phonon softening and crystallographic orientation of strained graphene studied by Raman spectroscopy. Proc. Natl. Acad. Sci. U. S. A **106**, 7304–7308 (2009)
19. B. Kelly, *Physics of graphite* (Applied Science, London, 1981), p. 477
20. D. Yoon, Y.-W. Son, H. Cheong, Strain-dependent splitting of the double-resonance Raman scattering band in graphene. Phys. Rev. Lett. **106**, 155502-1–155502-4 (2011)
21. K. Mak, C. Lee, J. Hone, J. Shan, T. Heinz, Atomically thin MoS_2: a new direct-gap semiconductor. Phys. Rev. Lett. **105**, 136805-1–136805-4 (2010)
22. A. Splendiani, L. Sun, Y. Zhang, T. Li, J. Kim, C.-Y. Chim, G. Galli, F. Wang, Emerging photoluminescence in monolayer MoS_2. Nano Lett. **10**, 1271–1275 (2010)
23. W. Zhao, Z. Ghorannevis, L. Chu, M. Toh, C. Kloc, P.-H. Tan, G. Eda, Evolution of electronic structure in atomically thin sheets of WS_2 and WSe_2. ACS Nano **7**, 791–797 (2012)
24. B. Radisavljevic, M.B. Whitwick, A. Kis, Integrated circuits and logic operations based on single-layer MoS_2. ACS Nano **5**, 9934–9938 (2011)
25. B. Radisavljevic, J. Brivio, V. Giacometti, A. Kis, A. Radenovic, Single-layer MoS_2 transistors. Nat. Nanotechnol. **6**, 147–150 (2011)
26. R.C. Cooper, C. Lee, C.A. Marianetti, X. Wei, J. Hone, J.W. Kysar, Nonlinear elastic behavior of two-dimensional molybdenum disulfide. Phys. Rev. B **87**, 035423-1–035423-11 (2013)

27. S. Bertolazzi, J. Brivio, A. Kis, Stretching and breaking of ultrathin MoS_2. ACS Nano **5**, 9703–9709 (2011)
28. Y. Wang, C. Cong, C. Qiu, T. Yu, Raman spectroscopy study of lattice vibration and crystallographic orientation of monolayer MoS_2 under uniaxial strain. Small **9**, 2857–2861 (2013)
29. Y. Wang, C. Cong, W. Yang, J. Shang, N. Peimyoo, Y. Chen, J. Kang, J. Wang, W. Huang, T. Yu, Strain-induced direct–indirect bandgap transition and phonon modulation in monolayer WS_2. Nano Res. **8**, 2562–2572 (2015)
30. H.J. Conley, B. Wang, J.I. Ziegler, R.F. Haglund Jr., S.T. Pantelides, K.I. Bolotin, Bandgap engineering of strained monolayer and bilayer MoS_2. Nano Lett. **13**, 3626–3630 (2013)
31. H.-X. Zhong, S. Gao, J.-J. Shi, L. Yang, Quasiparticle band gaps, excitonic effects, and anisotropic optical properties of the monolayer distorted 1T diamond-chain structures ReS_2 and $ReSe_2$. Phys. Rev. B **92**, 115438-1–115438-7 (2015)
32. H. Zhao, J. Wu, H. Zhong, Q. Guo, X. Wang, F. Xia, L. Yang, P. Tan, H. Wang, Interlayer interactions in anisotropic atomically thin rhenium diselenide. Nano Res. **8**, 3651–3661 (2015)
33. S. Yang, C. Wang, H. Sahin, H. Chen, Y. Li, S.S. Li, A. Suslu, F.M. Peeters, Q. Liu, J. Li, S. Tongay, Tuning the optical, magnetic, and electrical properties of $ReSe_2$ by nanoscale strain engineering. Nano Lett. **15**, 1660–1666 (2015)
34. A. Castellanos-Gomez, L. Vicarelli, E. Prada, J.O. Island, K.L. Narasimha-Acharya, S.I. Blanter, D.J. Groenendijk, M. Buscema, G.A. Steele, J.V. Alvarez, H.W. Zandbergen, J.J. Palacios, H.S.J. van der Zant, Isolation and characterization of few-layer black phosphorus. 2D Mater **1**, 025001 (2014)
35. X. Ling, H. Wang, S. Huang, F. Xia, M.S. Dresselhaus, The renaissance of black phosphorus. Proc. Natl. Acad. Sci. **112**, 4523–4530 (2015)
36. G. Qin, Q.-B. Yan, Z. Qin, S.-Y. Yue, M. Hu, G. Su, Anisotropic intrinsic lattice thermal conductivity of phosphorene from first principles. Phys. Chem. Chem. Phys. **17**, 4854–4858 (2015)
37. J. Qiao, X. Kong, Z.-X. Hu, F. Yang, W. Ji, High-mobility transport anisotropy and linear dichroism in few-layer black phosphorus. Nat. Commun. **5**, 4475 (2014)
38. J. Wu, N. Mao, L. Xie, H. Xu, J. Zhang, Identifying the crystalline orientation of black phosphorus using angle-resolved polarized Raman spectroscopy. Angew. Chem. Int. Ed. **54**, 2366–2369 (2015)
39. F. Xia, H. Wang, Y. Jia, Rediscovering black phosphorus as an anisotropic layered material for optoelectronics and electronics. Nat. Commun. **5**, 4458 (2014)
40. A. Jain, A.J. McGaughey, Strongly anisotropic in-plane thermal transport in single-layer black phosphorene. Sci. Rep. **5**, 8501 (2015)
41. H.B. Ribeiro, M.A. Pimenta, C.J.S. de Matos, R.L. Moreira, A.S. Rodin, J.D. Zapata, E.A.T. de Souza, A.H. Castro Neto, Unusual angular dependence of the Raman response in black phosphorus. ACS Nano **9**, 4270–4276 (2015)
42. Y. Wang, C. Cong, R. Fei, W. Yang, Y. Chen, B. Cao, L. Yang, T. Yu, Remarkable anisotropic phonon response in uniaxially strained few-layer black phosphorus. Nano Res. **8**, 3944–3953 (2015)
43. I. Stenger, L. Schué, M. Boukhicha, B. Berini, B. Plaçais, A. Loiseau, J. Barjon, Low frequency Raman spectroscopy of few-atomic-layer thick hBN crystals. 2D Mater **4**, 031003 (2017)
44. A. Falin, Q. Cai, E.J.G. Santos, D. Scullion, D. Qian, R. Zhang, Z. Yang, S. Huang, K. Watanabe, T. Taniguchi, M.R. Barnett, Y. Chen, R.S. Ruoff, L.H. Li, Mechanical properties of atomically thin boron nitride and the role of interlayer interactions. Nat. Commun. **8**, 15815 (2017)
45. R.V. Gorbachev, I. Riaz, R.R. Nair, R. Jalil, L. Britnell, B.D. Belle, E.W. Hill, K.S. Novoselov, K. Watanabe, T. Taniguchi, A.K. Geim, P. Blake, Hunting for monolayer boron nitride: optical and Raman signatures. Small **7**, 465–468 (2011)
46. R. Arenal, A.C. Ferrari, S. Reich, L. Wirtz, J.Y. Mevellec, S. Lefrant, A. Rubio, A. Loiseau, Raman spectroscopy of single-wall boron nitride nanotubes. Nano Lett. **6**, 1812–1816 (2006)

Chapter 7
Double Resonance Raman Spectroscopy of Two-Dimensional Materials

R. Saito, Y. Tatsumi, T. Yang, H. Guo, S. Huang, L. Zhou, and M. S. Dresselhaus

Abstract In this chapter, we overview double resonance Raman spectra of two dimensional materials. Many weak Raman spectral peaks are observed in the two dimensional materials which can be attributed to second order, double resonance Raman spectra. It is useful for material characterization to understand not only first order Raman spectra but also second order Raman spectra since the second order Raman spectra has more information on electronic structure of the materials than the first order Raman spectra. Combined with the conventional first order resonance Raman theory, we will explain why the double resonance condition can be strong in the two dimensional materials. Since the double resonance Raman spectra give the information of phonon with non-zero wavevectors in the Brillouin zone, both the resonant wavevector and corresponding Raman spectra can shift with changing the incident laser energy. Here we will discuss the physics of double resonance Raman spectra of graphene, transition metal dichalcogenides by theoretical analysis using the first principles calculation.

M. S. Dresselhaus (deceased, February, 2017)

R. Saito (✉) · Y. Tatsumi
Department of Physics, Tohoku University, Sendai, Japan
e-mail: rsaito@flex.phys.tohoku.ac.jp

T. Yang
Shenyang National Laboratory for Materials Science, Institute of Metal Research, Chinese Academy of Sciences, Shenyang, China

H. Guo
College of Sciences, Liaoning Shihua University, Fushun, China

S. Huang
Electrical Engineering Department, Pennsylvania State University, University Park, PA, USA

L. Zhou
School of Chemistry and Chemical Engineering, Shanghai Jiao Tong University, Shanghai, China

Department of Electrical Engineering and Computer Science, Massachusetts Institute of Technology, Cambridge, MA, USA

P.-H. Tan (ed.), *Raman Spectroscopy of Two-Dimensional Materials*, Springer Series in Materials Science 276, https://doi.org/10.1007/978-981-13-1828-3_7

7.1 Raman Intensity Formula

7.1.1 First Order Raman Scattering

When we shed the light on a material, we can see the material by the scattered light from the material. Then we can know the color or texture of the material. Among the scattered light whose energy does not generally change after scattering (that is elastic scattering of light or Rayleigh scattering), we know in physics that there is inelastic scattering, too, though it is very weak compared with the inelastic scattering (or Raman scattering). In Raman spectroscopy, we measure the inelastic scattered light in which the energy loss (or gain) of the light occurs in the scattered light. The energy loss corresponds to the elementary excitation in the materials such as phonon. The phenomena of light scattering of a solid material can be expressed by time dependent perturbation theory of physics in which Raman scattering consists of three elemental processes (see Fig. 7.1a); (1) light absorption of an electron, (2) a phonon emission (or absorption) from the photo excited electron, and (3) light emission by recombining the scattered photo excited electron with a hole. The Raman scattering with emitting (or absorbing) a single phonon is called the first order Stokes (anti-Stokes) Raman, while the Raman scattering with emitting two phonons (or emitting a phonon + one elastic scattering) is called the second order Raman. The intensity of Raman spectra in the first order Raman process is expressed by square of the sum of the scattered amplitudes as a function of Raman shift E_{RS} (the difference of energy between the incident and scattered light, $E_{\text{RS}} = \hbar\omega_\nu$) and the incident laser energy E_{L} as follows [1–5],

$$I(E_{\text{L}}, E_{\text{RS}}) = \sum_{\nu} \left| \sum_{\boldsymbol{k}} \sum_{i,m,m'} \frac{M_{\text{op}}^{fm'}(\boldsymbol{k}) M_{\text{ep}}^{m'm\nu}(\boldsymbol{k}) M_{\text{op}}^{mi}(\boldsymbol{k})}{(E_{\text{L}} - \Delta E_{mi}(\boldsymbol{k}))(E_{\text{L}} - \hbar\omega_\nu - \Delta E_{m'i}(\boldsymbol{k}))} \right|^2, \tag{7.1}$$

where $M_{\text{op}}(\boldsymbol{k})$ and $M_{\text{ep}}(\boldsymbol{k})$ are, respectively, matrix elements of electron-photon (or optical) and electron-phonon interactions. The electron-photon matrix element $M_{\text{op}}^{mi}(\boldsymbol{k}) = \langle m|\nabla|i\rangle$ depends on the wavefunctions of the initial states $|i\rangle$ and the intermediate states $\langle m|$ for a given wavevector $\boldsymbol{k}$ [6]. The electron-phonon matrix element $M_{\text{ep}}(\boldsymbol{k})$ which is the matrix element of so-called deformation potential depends on the two intermediate states, $\langle m'|, |m\rangle$, and the phonon modes ν [7, 8].

In Eq. (7.1), each scattered amplitude has two energy denominators for a given photon energy E_{L} and phonon energy $\hbar\omega_\nu$ in which ΔE_{mi} denotes an energy separation between the initial states E_i and the intermediate states E_m of electron, which is expressed by

$$\Delta E_{mi}(\boldsymbol{k}) = E_m(\boldsymbol{k}) - E_i(\boldsymbol{k}) + i\gamma, \tag{7.2}$$

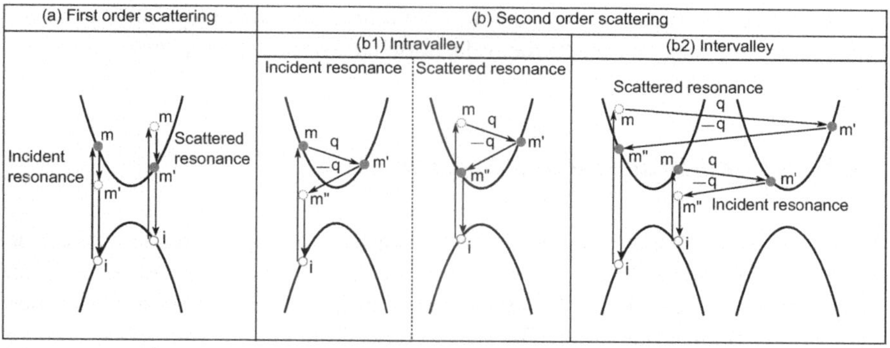

Fig. 7.1 Raman process of (**a**) first order scattering and (**b**) second order scattering. There are two kinds of resonance conditions for each case; incident resonance and scattered resonance conditions in which solid and open circles at intermediate states (m or m') denote, respectively, real and virtual electronic states. In the case of the real electronic structure, the corresponding energy denominator of the Raman intensity formula ((7.1) and (7.3)) becomes zero which makes the Raman intensity large (resonance Raman effect). In the case of double resonance conditions, two of three intermediate states are real electronic states. If there are two energy bands in the Brillouin zone, there are two possibility of second order scattering, that is, (b1) intravalley scattering and (b2) intervalley scattering (see detail in the text)

where $i\gamma$ represents a spectral width of the Raman spectra that is related to the uncertainty relation of energy for the photo excited electron with a finite life time. Here we usually take the value of $\gamma = 0.1$ eV. When one of the two energy denominators becomes zero for specific E_L's such as $E_L = \Delta E_{mi}$ or $E_L = \Delta E_{m'i} + \hbar\omega_\nu$ in Eq. (7.1), the Raman scattering amplitude becomes significantly large, which is known as the resonance Raman effect. The corresponding E_L's are called incident and scattered resonance (See Fig. 7.1a) energies, respectively. On the other hand, when the energy denominator is large, we observe relatively weak Raman signal as is called non-resonant Raman spectra whose intensity does not depend on E_L.

It is important to note that in the case of non-resonant condition, we do not need to have a real unoccupied electronic state at $E_m = E_i(\boldsymbol{k}) + E_L$. In the perturbation theory, such a non-existing state at $E_m = E_i(\boldsymbol{k}) + E_L$ is called "virtual state" which is expressed by a linear combination of eigenstates of unoccupied states $|u\rangle$; ($|m\rangle = \sum_u C_u|u\rangle$) and the coefficient C_u becomes large when $E_u - (E_i(\boldsymbol{k}) + E_L)$ is small. In some cases, we can approximate $|m\rangle$ by the energetically closest state $|u\rangle$.

Usually a photo excited electron in the solid material with the wavevector $\boldsymbol{k}$ can emit a phonon with the wavevector $\boldsymbol{q}$. In this case the scattered electron has the wavevector of $\boldsymbol{k} - \boldsymbol{q}$. However, in order for the scattered electron to recombine with a hole at the $\boldsymbol{k}$ in the valence band, the phonon wavevector $\boldsymbol{q}$ should be zero. It means that we can see only $\boldsymbol{q} = 0$ (or zone-centered) phonon in first order Raman spectroscopy (Fig. 7.1a) even though there are so many other $\boldsymbol{q} \neq 0$ inelastic scattering of photo excited electrons. The scattered photo excited electrons with $\boldsymbol{k} - \boldsymbol{q}$ ($\boldsymbol{q} \neq 0$) can not recombine with a hole and thus they can stay in the conduction band and wait for further relaxation by emitting the second phonon.

7.1.2 Second Order Raman Spectra

When the second phonon is emitted with the wavevector $-\boldsymbol{q}$, the scattered electron can return to the original electronic state at $\boldsymbol{k}$ and thus the photo-excited electron can recombine with a hole (Fig. 7.1b). Such Raman process with two phonons is called second order Raman scattering. The second order Raman scattering has a freedom for selecting any values of $\boldsymbol{q}$ and thus the sum of phonon energy $\hbar\omega_1(\boldsymbol{q}) + \hbar\omega_2(-\boldsymbol{q})$ that generally gives a broad Raman spectra as a function of E_{RS}. If the two phonon modes are identical ($\omega_1(\boldsymbol{q}) = \omega_2(-\boldsymbol{q})$), it is called an overtone mode, while if not ($\omega_1(\boldsymbol{q}) \neq \omega_2(-\boldsymbol{q})$), it is called a combination mode. If the two phonons do not have a pair of wavevectors $\boldsymbol{q}$ and $-\boldsymbol{q}$, the electron will emit the third phonon before recombination, which corresponds to the higher order Raman process that generally gives the background Raman spectra without any peaks. The intensity of two-phonon Raman scattering is given by

$$
\begin{aligned}
& I(E_{\mathrm{L}}, E_{\mathrm{RS}}) \\
&= \sum_{\nu,\nu'} \left| \sum_{\boldsymbol{k}} \sum_{f=i,m,m',m''} \right. \\
&\times \left. \frac{M_{\mathrm{op}}^{fm''}(\boldsymbol{k}) M_{\mathrm{ep}}^{m''m'\nu'}(\boldsymbol{k}-\boldsymbol{q},\boldsymbol{k}) M_{\mathrm{ep}}^{m'm\nu}(\boldsymbol{k},\boldsymbol{k}-\boldsymbol{q}) M_{\mathrm{op}}^{mi}(\boldsymbol{k})}{(E_{\mathrm{L}} - \hbar\omega_\nu - \hbar\omega_{\nu'} - \Delta E_{m''i}(\boldsymbol{k}))(E_{\mathrm{L}} - \hbar\omega_\nu - \Delta E_{m'i}(\boldsymbol{k}))(E_{\mathrm{L}} - \Delta E_{mi}(\boldsymbol{k}))} \right|^2
\end{aligned}
\tag{7.3}
$$

When we compare Eq. (7.3) with Eq. (7.1), we have two electron-phonon matrix elements and three energy denominators appear, which correspond to, respectively, two-phonon scattering and the fourth order time-dependent perturbation theory. Generally the second order Raman intensity is much weaker than the first order process. However, the second order Raman intensity can become comparable with (or even larger than) the first order process when the two of the three resonance conditions in the denominator of Eq. (7.3) ($E_{\mathrm{L}} = \hbar\omega_\nu + \hbar\omega_{\nu'} + \Delta E_{m''i}(\boldsymbol{k})$, $E_{\mathrm{L}} = \hbar\omega_\nu + \Delta E_{m'i}(\boldsymbol{k})$, and $E_{\mathrm{L}} = \Delta E_{mi}(\boldsymbol{k})$) are satisfied at the same time. A simple example is that the G′ band (second order) of graphene is stronger than the G band (first order).

When the photo excited electron emits a phonon with the wavevector of $\boldsymbol{q}$, if hole emits a phonon with the same $\boldsymbol{q}$, then another second order Raman processes occurs at $\boldsymbol{k} - \boldsymbol{q}$ in the Brillouin zone though this process is not considered in Eq. (7.3) for simplicity. Since electron-photon matrix elements at $\boldsymbol{k}$ and $\boldsymbol{k} - \boldsymbol{q}$ (or energy difference between valence and conduction bands at $\boldsymbol{k}$ and $\boldsymbol{k} - \boldsymbol{q}$) are generally different from each other on the symmetry point of view, we can distinguish the two cases for the Raman process.

It is noted that photoluminescence occurs when the photo excited electron and the hole relax, respectively, to the energy bottom of the conduction band and to the energy top of the valence band by emitting many phonons at which they recombine as spontaneous emission of a photon. Photoluminescence is a different phenomena from Raman process. In fact photoluminescence process (ns) is much slower than Raman processes (sub ps) and emitted photon represents an energy gap of semiconductors.

7.1.3 Double Resonance Raman Spectra

When two of the three energy denominators in Eq. (7.3) become zero at the same time, the two-phonon Raman spectra are enhanced significantly, which should be comparable to the first order resonance Raman spectra. Such a condition is called double resonance condition. There are two possibilities of the double resonance condition, that is, incident light resonance and scattered light resonance as shown in Fig. 7.1b. In the case of the incident resonance condition, both the intermediate states m and m' are real electronic (not virtual) states, while in the scattered resonance condition, both m' and m'' states are real electronic states.

In the case of the double resonance conditions, since the phonon wavevector is restricted so as to satisfy the double resonance conditions, double resonance Raman signal becomes sharp which is comparable to that of the first order resonance Raman spectra. Further, when the electronic density of states (DOS) is singular as is known as van Hove singularity (VHS) [9] for one-dimensional (1D) or two-dimensional (2D) materials, Raman intensities for both first order and second order scattering are enhanced. This enhancement by VHS is independent of the resonance Raman enhancement. In particular, for the 2D materials, when the electronic energy dispersion $E(\boldsymbol{k})$ shows so called a saddle point structure at a given $\boldsymbol{k}_{\mathrm{VHS}}$, the DOS becomes logarithmically singular such as $\log|E - E(\boldsymbol{k}_{\mathrm{VHS}})|$. In the case of 2D materials with the hexagonal lattice such as graphene, transition metal dichalcogenides (TMDs), electronic energy dispersion shows a saddle point structure at the M point (the center of hexagonal edge of the Brillouin zone). Thus when E_{L} coincides to the optical transition at the M point, we expect a strong Raman signal, which will be seen in the case of TMDs (see Sect. 7.3) [10, 11].

What is special for double resonance Raman spectra is that the phonon energy that satisfies the double resonance condition can change by changing E_{L}, since the corresponding intermediate states m and m' are determined for given E_{L}. When

E_L increases in the case of Fig. 7.1b, the optical absorption occurs at $\boldsymbol{k}$ that are away from the energy minimum of the conduction bands. Then the $\boldsymbol{q}$ that satisfies the double resonance condition becomes large and corresponding phonon energies $\hbar\omega_1(\boldsymbol{q})+\hbar\omega_2(-\boldsymbol{q})$ changes along the phonon dispersion relations [12]. Thus we can check if the unknown spectra are related to double resonance Raman spectra or not by using more than two laser energies to see so-called "dispersive" Raman spectra. If the Raman peak position changes (or does not change) by changing the E_L, we can say that the peak is double resonance Raman (or first order Raman) peak.

In order to assign the phonon modes for experimentally observed Raman peaks, we need theoretical analysis based on first principles calculations. In Sect. 7.1.4, we will explain what are needed for theoretical analysis and how to assign Raman spectra in particular for double resonance Raman spectra.

7.1.4 Analysis by First Principles Calculation

Now let us describe how to assign the phonon modes in the double resonance Raman spectra by first principles calculation. We use the following four steps; (1) electronic and phonon calculation for obtaining $E_i(\boldsymbol{k})$ and $\hbar\omega_j(\boldsymbol{q})$, (2) by calculating DOS, we obtain the van Hove singular $\boldsymbol{k}_{\mathrm{VHS}}$ point, (3) connecting two different $\boldsymbol{k}_{\mathrm{VHS}}$ points that have a similar energy each other, we get the double resonance Raman wavevector $\boldsymbol{q}$, and (4) comparing calculated phonon frequencies at the $\boldsymbol{q}$ with the experimentally observed Raman spectra, we assign the phonon modes. Hereafter we explain each step one by one.

7.1.4.1 Electronic and Phonon Calculation

Both electronic energy dispersion $E_i(\boldsymbol{k})$ and phonon dispersion relation $\hbar\omega_j(\boldsymbol{q})$ are essential to determine electron-photon resonance and electron-phonon resonance conditions for the double resonance Raman spectra. In order to get the possible electron-photon resonance condition, we calculate the electronic band structure of 2D materials by using first principles density functional theory (DFT) within the local density approximation (LDA) or generalized gradient approximation (GGA) as implemented, for example, in the Quantum-Espresso code [13]. But it is well-known that the single-particle picture from DFT calculations always underestimate electronic energy band gap and also neglect the excitonic interaction between an electron and a hole. Many-body picture based on GW approximation [14] or Bethe-Salpeter equation (BSE) [15–17] is sometimes employed to improve the conduction band calculations but is very intensive and time-consuming computationally. Though we notice that optical transition matrix elements that are common in the resonant Raman process are enhanced by the exciton effect, but the *relative* Raman intensity thus obtained does not change much even though we do not consider the exciton effect [17, 18]. Thus in the following theoretical analysis, we upshift the conduction bands from the GGA/LDA calculation to fit the experimental band gap. Phonon energy dispersion relations of 2D materials are calculated based on density functional perturbation theory [19].

When we have $E_i(\boldsymbol{k})$ and $\hbar\omega_\nu(\boldsymbol{q})$, we can find double resonance condition for a given $E_{\rm L}$. In fact, we need to find $\boldsymbol{k}$ and $\boldsymbol{q}$ which satisfy two resonance conditions at the same time from the three energy denominators in Eq. (7.3), that is

$$E_{\rm L} = \begin{cases} \Delta E_{mi}(\boldsymbol{k}) \\ \hbar\omega_\nu(\boldsymbol{q}) + \Delta E_{m'i}(\boldsymbol{k}) \end{cases}, \tag{7.4}$$

for incident resonance condition or

$$E_{\rm L} = \begin{cases} \hbar\omega_\nu(\boldsymbol{q}) + \Delta E_{m'i}(\boldsymbol{k}) \\ \hbar\omega_\nu(\boldsymbol{q}) + \hbar\omega_{\nu'}(-\boldsymbol{q}) + \Delta E_{m''i}(\boldsymbol{k}) \end{cases}, \tag{7.5}$$

for scattered resonance condition. In Sect. 7.1.4.2, we will go into detail how to obtain such $\boldsymbol{k}$ and $\boldsymbol{q}$.

7.1.4.2 Electron-Photon Resonance Condition, the Corresponding Wave Vector *k*

A photon of the laser light excites an electron vertically at $\boldsymbol{k}$ in the electronic band structure from occupied valence band $\epsilon_i(\boldsymbol{k})$ to unoccupied conduction band $\epsilon_m(\boldsymbol{k})$, which gives rise to electron-photon resonance subject to several conditions including (1) the energy of the photon $E_{\rm L}$ matches the energy difference between the valence and conduction bands, $\epsilon_m(\boldsymbol{k}) - \epsilon_i(\boldsymbol{k})$ (2) $M_{\rm op}^{mi}(\boldsymbol{k})$ in Eq. (7.3) should not be zero by symmetry. Optical selection rule for $M_{\rm op}^{mi}(\boldsymbol{k})$ which decides whether photo excitation of electrons occurs from the symmetry of the product of electronic states and electron-photon interaction, $M_{\rm op}^{mi}(\boldsymbol{k}) = \langle m|\nabla|i\rangle$; (3) Joint density of states (JDOS) should be large. Around a VHS point, there are many $\boldsymbol{k}$ states that satisfy the resonance condition. Here JDOS is defined by

$$\text{JDOS}(E_{\rm L}) = \sum_{m,i,\boldsymbol{k}} \delta(\epsilon_m(\boldsymbol{k}) - \epsilon_i(\boldsymbol{k}) - E_{\rm L}), \tag{7.6}$$

in which the summation is taken over $\boldsymbol{k}$ points in the Brillouin zone and for all the occupied valence bands $\epsilon_i(\boldsymbol{k})$ and the unoccupied conduction bands $\epsilon_m(\boldsymbol{k})$ involved in the vertical excitation at each $\boldsymbol{k}$ point.

The electron-photon matrix element $M_{\rm op}^{mi}(\boldsymbol{k})$ is given by [6],

$$M_{\rm op}^{mi}(\boldsymbol{k}) = \frac{e\hbar^2}{m_e E_{\rm L}}\sqrt{\frac{I}{c\epsilon_0}}e^{i(\omega_m - \omega_i \pm E_{\rm L}/\hbar)t}\boldsymbol{D}^{mi}(\boldsymbol{k})\cdot\boldsymbol{P}, \tag{7.7}$$

in which m_e is the electron mass, I is the intensity of the incident laser in units of W/m^2, ϵ_0 is the dielectric constant for vacuum, $\boldsymbol{D}^{mi}(\boldsymbol{k}) \equiv \langle \psi_m^c(\boldsymbol{k})|\nabla|\psi_i^v(\boldsymbol{k})\rangle$ is the dipole vector [6] and $\boldsymbol{P}$ is the unit vector of the polarization of the laser light. Incorporating the electron-photon matrix element into JDOS in Eq. (7.6), we get the optical absorption probability, α, as a function of $\boldsymbol{k}$ per unit area and per unit time,

$$\alpha(E_L, \boldsymbol{k}) = \frac{2\pi}{\hbar} \sum_{m,i} | M_{op}^{mi}(\boldsymbol{k}) |^2 \delta(\epsilon_m(\boldsymbol{k}) - \epsilon_i(\boldsymbol{k}) - E_L). \tag{7.8}$$

For a given polarization of light that is expressed by $\boldsymbol{P}$, we can calculate $\boldsymbol{D} \cdot \boldsymbol{P}$ in Eq. (7.7) for obtaining the polarized Raman spectra [20]. In the case of polarized Raman spectra, since we can change the polarization directions of both incident and scattered (or in-going or out-going) light, $\boldsymbol{P}_{in}$ and $\boldsymbol{P}_{out}$, $\boldsymbol{D} \cdot \boldsymbol{P}_{in}$ and $\boldsymbol{D} \cdot \boldsymbol{P}_{out}$ are respectively calculated for $M_{op}^{mi}(\boldsymbol{k})$ and $M_{op}^{fm'}(\boldsymbol{k})$ to obtain the polarized Raman intensity as a function of the polarization direction. For unpolarized light of laser, $| M_{op}^{mi}(\boldsymbol{k}) |^2$ is calculated based on $\sum_{\beta=x,y,z} \frac{|D_\beta^{mi}(\boldsymbol{k})|^2}{3}$. The optical absorption probability α in unit of 1/s is evaluated through this manuscript by the units of $\alpha_0 = \frac{2\pi e^2 \hbar^3 I}{m^2 c \epsilon_0 E_L^3 a_0^2}$ with a_0 as the lattice constant,

$$\alpha(E_L, \boldsymbol{k}) = \alpha_0 a_0^2 E_L \sum_{m,i} \left| \boldsymbol{D}^{mi}(\boldsymbol{k}) \cdot \boldsymbol{P} \right|^2 \delta(\epsilon_m(\boldsymbol{k}) - \epsilon_i(\boldsymbol{k}) - E_L). \tag{7.9}$$

From first principles DFT calculations, we get not only the electronic band dispersion $\epsilon_{m,i}$(k) but also the wave function coefficients expanded by the plane-wave basis as follows,

$$\psi^i(\boldsymbol{k}, \boldsymbol{r}) = \sum_{\boldsymbol{G}} C_{\boldsymbol{G}}^i e^{i(\boldsymbol{k}+\boldsymbol{G})\cdot \boldsymbol{r}}. \tag{7.10}$$

The $C_{\boldsymbol{G}}$ can be used to calculate the dipole vector [21],

$$\boldsymbol{D}^{mi}(\boldsymbol{k}) \equiv \langle \psi_m^c(\boldsymbol{k})|\nabla|\psi_i^v(\boldsymbol{k})\rangle = \sum_{\boldsymbol{G}} i(\boldsymbol{k}+\boldsymbol{G}) C_{\boldsymbol{G}}^{m*} C_{\boldsymbol{G}}^i. \tag{7.11}$$

Thereby, we can draw contour lines of equal optical absorption $\alpha(E_L, k_x, k_y)$ over the Brillouin zone of 2D materials, or the wave vectors $\boldsymbol{k}$ where electron-photon resonance occurs.

7.1.4.3 Electron-Phonon Resonance, the Double Resonance Raman Wavevector

Once we draw the equi-optical-absorption contour lines of the wave vectors $\boldsymbol{k}$ for electron-photon resonance $E_{\mathrm{L}} = \Delta E_{mi}(\boldsymbol{k})$ in Eq. (7.4), we can go further to analyze the second resonance condition $E_{\mathrm{L}} = \hbar\omega_{\nu}(\boldsymbol{q}) + \Delta E_{m'i}(\boldsymbol{k})$ in which the phonon dispersion relation is needed. The photo excited electron can be scattered by emitting a phonon with a phonon wave vector $\boldsymbol{q}$ from the wave vector $\boldsymbol{k}$ to $\boldsymbol{k} - \boldsymbol{q}$, either within one contour line (called intra-valley scattering) as shown in Fig. 7.1(b1) or between different contour lines in the Brillouin zone (inter-valley scattering) as shown in Fig. 7.1(b2).

7.1.4.4 Phonon Mode Assignment

With all possible phonon wave vectors $\boldsymbol{q}$ for double resonance Raman spectra and the calculated phonon dispersion relation available, we are ready for the assignment of phonon modes to the experimental Raman peaks. By adding or subtracting the frequencies of two selected calculated phonon modes (or the combinational or difference mode, respectively) or doubling the frequency of one mode (or the overtone mode) at the specific wave vector $\boldsymbol{q}$, we match the combinatory frequency values to the experimental Raman shifts and assign the phonon modes. To confirm or to rule out some possible phonon mode assignments, the dependence of Raman shifts on laser energy can be used and a similar trend between the experiment and theoretical assignment should be observed. To test the validity of different modes at $\hbar\omega_1(\boldsymbol{q}) - \hbar\omega_2(-\boldsymbol{q})$ which usually arise from one Stokes and another anti-Stokes mode, Raman spectra at very low temperature can be measured since the corresponding Raman peak should disappear due to the fact that phonon for the anti-Stokes Raman process can not be thermally activated.

It should be mentioned that we do not yet calculate the electron-phonon matrix elements for general $\boldsymbol{q}$ by first principles calculation since we do not have the computational program to calculate it. Once electron-phonon matrix element can be calculated, we can calculate double resonance Raman spectra of any materials which should be a goal of this project. Recently, so called EPW (electron-phonon Wannier function) package is open for public. Using the EPW package [22], we now make a program to take out the electron-phonon matrix element at $\boldsymbol{q} = 0$ for calculating the first order resonant Raman spectra.

In the case of graphene and carbon nanotubes, since we have developed the computer program to calculate electron-phonon matrix element within tight binding calculations [7], we can calculate the double resonance Raman spectra of graphene and carbon nanotubes. In the next section, we overview the double resonance Raman spectra of graphene.

7.2 Double Resonance Raman Spectra of Graphene

7.2.1 G′ Band

In this section we overview double resonance Raman spectra of graphene. In Fig. 7.2, we show the Raman spectra of graphene and graphite. In the case of single layer graphene (SLG), there are two strong Raman spectra around 1600 and 2700 cm^{-1}, which are intrinsic spectra for graphene related materials [23], that is free from the effect of any defects and they are called G and G′ bands, respectively. In the recent papers, they also call 2D band for G′ band [24]. The G band spectra is first order Raman spectra of in-plane optical phonon modes (longitudinal and in-plane tangential optical modes, LO and iTO) that are degenerate the E_{2g} mode in the D_{6h} symmetry. The G′ band is second order Raman spectra in which the K point phonon of the iTO phonon dispersion in the Brillouin zone are relevant as an overtone mode [25]. The reasons why we have a strong G′ band in graphene are that (1) the double resonance condition is satisfied between two Dirac cones near the K and K' points and thus the photo-excited electron can be scattered from the K to K' valleys by emitting the K point phonon (intervalley scattering, Fig. 7.1(b2)) and that (2) electron-phonon interaction for the iTO phonon mode is relatively strong at the K point since the vibrational mode shows "breathing mode of the hexagonal ring" that makes the deformation potential large [4].

The G′ band of graphene is even stronger than the G band in the case of monolayer graphene, while the intensity ratio of the G′ to G bands monotonically decreases with increasing the number of graphene layer [26]. When the number of layers increases, the energy bands of Dirac cone are split into energy subbands by interlayer interactions between the graphene layers. In such cases, there are many different the resonance $\boldsymbol{q}$ that connect the two Dirac cone. Since the corresponding

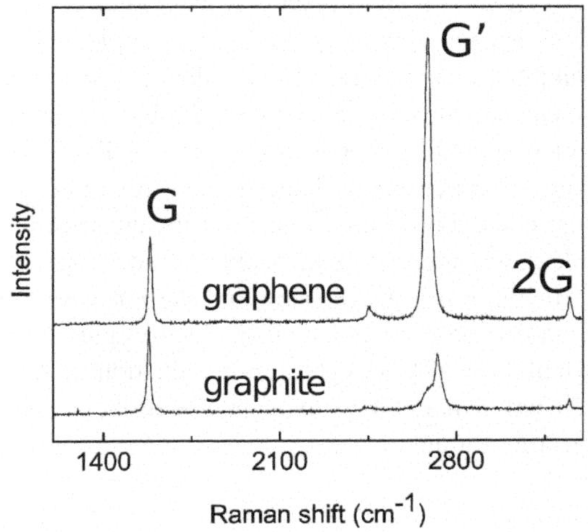

Fig. 7.2 The Raman spectra of graphene and graphite [23]. The laser excitation energy for the measurement is 2.41 eV (514 nm)

phonon frequencies for the resonance $\boldsymbol{q}$ spread compared with those in the case of monolayer graphene, the Raman spectra becomes broad and the peak value of the G′ band becomes small. On the other hand, for the case of G band, since the resonance $\boldsymbol{q}$ is always zero, the spectra of G band do not change with increasing the number of layer. Thus the G band intensity is proportional to the number of layers. The layer number dependence of G′/G peak intensity ratio and the spectral width is frequently used for characterizing layer number of graphene [24].

An important feature of G′ band is that the phonon frequency of the G′ band depends on E_L. For $E_L = 2.54$ eV, the G′ band frequency appears around 2700 cm^{-1}. When E_L increases energy by 1 eV, the Raman frequency of the G′ band increases 106 cm^{-1} from which the dispersion of the G′ band is given by $d\omega/dE_L = 106\,\text{cm}^{-1}/\text{eV}$. This dependence is almost linear to E_L because the both electronic energy dispersion and the phonon dispersion of iTO phonon modes around the K point is the linear to $\boldsymbol{k}$ and $\boldsymbol{q}$ measured from the K point. Thus the wave vectors $\boldsymbol{k}$ and $\boldsymbol{q}$ that satisfy the double resonance condition linearly increase with increasing E_L.

When graphene has structural (short range) defects such as a point defect, we expect elastic and intervalley scattering of the photo excited electron. In such case, one of two scattering processes in the double resonance Raman scattering as shown in Fig. 7.1b can be the elastic scattering. Since there is no energy loss in the elastic scattering, the Raman shift becomes half of G' band, around 1350 cm^{-1} for $E_L = 2.54$ eV and $d\omega/dE_L = 53\ \text{cm}^{-1}/\text{eV}$ [4, 27], which is called D band.

For the other phonon modes, since the electron-phonon interaction is relatively small compared with that for the K point iTO phonon modes (G′ band), we can observe the double resonance Raman peaks as weak peaks if we carefully measured by taking a sufficient long duration time of the measurement. P. Tan et al. have shown many double resonance Raman spectra in graphite whisker that depend on E_L [28]. All the Raman spectra for given phonon $\boldsymbol{q}$ vectors and for many E_L appear on the phonon dispersion relations [12, 29], if we decomposed the two-phonon Raman spectra into two-phonon frequencies. Thus if we have many laser energies for Raman measurement, we can probe the phonon dispersion relation by double resonance Raman spectroscopy.

7.2.2 M Band

Let us show some example of weak Raman peaks of graphene and graphene related materials such as carbon nanotubes. Around 1750 cm^{-1}, we can see a weak Raman peak that is called M band [30]. The reason why we call the peak the M band is that the M band always show two split (by $\sim$20 cm^{-1}) peaks denoted by M^+ and M^- peaks like a character of M [31]. The M band is overtone mode of oTO mode around the Γ point (intravalley scattering). Since the vibration of the oTO mode is an odd function of z in the direction perpendicular to the graphene surface, the oTO mode is not Raman active for SLG or incommensurate bilayer graphene (IBLG).

We can measure the M band only for AB stacked bilayer graphene (BLG) or few layer graphene (FLG) in which the out-of-plane modes are split into in-phase and out-of-phase motions [32]. It is the reason why the M band shows two spectra. The existence of M band is a reliable way to distinguish BLG from SLG or IBLG. Further a useful information of the M band is that the frequency of the M band monotonically increases with increasing the number of layers up to 8 which is a direct way to measure the number of layers of FLG by Raman spectra [30].

7.2.3 *iTOLA Band and Other Bands*

Around 1950 cm^{-1}, there is a dispersive Raman peak that is combination mode of iTO and LA phonon mode at the Γ point and thus we call iTOLA band [31]. Since the LA phonon modes have a large phonon dispersion around the Γ point, the Raman spectra of the iTOLA band are very dispersive, $d\omega/dE_{\rm L} = 230\,{\rm cm}^{-1}$/eV. Further, the Raman shift increases with increasing number of layers of FLG [30], which is similar to the M band.

In Table 7.1 we show Raman spectra that can be observed in graphene related materials for not only double resonance Raman spectra with one phonon + one elastic scattering (DR1), with two phonons (DR2) or single resonance (SR). At the low frequency region below 100 cm^{-1}, we can observe so called C band which is

Table 7.1 Raman mode frequencies of graphene related materials, given from Ref. [3] with the modification. The frequency ω is observed at $E_{\rm L} = 2.41$ eV. d_t denotes diameter of a carbon nanotube [35]. DR1, DR2 and SR denote, respectively, double resonance (1 phonon + 1 elastic scattering), double resonance (2 phonons) and single resonance. The phonon dispersion $d\omega/dE_{\rm L}$ in units of cm^{-1}/eV denotes how the phonon frequency in cm^{-1} changes by changing $E_{\rm L}$ by 1 eV

Name	ω[cm^{-1}]	Resonance	$d\omega/dE_{\rm L}$	Origin
C	42	SR	0	Shear E_{2g} modes
iTA	288	DR1	129	TA mode, intravalley scattering
LA	453	DR1	216	LA mode, intravalley scattering
RBM	248/d_t	SR	0	Nanotube only, vibration of radius
oTO(1)	860	DR1	0	oTO mode, intravalley scattering, $\boldsymbol{q}$=0
oTO(2)	865	DR1	13	oTO mode, intravalley scattering, $\boldsymbol{q}$=2$\boldsymbol{k}$
D	1350	DR1	53	LO mode, intervalley scattering, $\boldsymbol{q}$=2$\boldsymbol{k}$
LO	1450	DR1	0	LO mode, intervalley scattering, $\boldsymbol{q}$=0
BWF	1550	SR	0	Only metallic carbon
G	1590	SR	0	Raman active mode of graphite
M	1750	DR2	26	Overtone of oTO mode
iTOLA	1950	DR2	230	Combinational mode of iTO and LA
G'	2700	DR2	106	Overtone of D mode
2LO	2900	DR2	0	Overtone of LO mode
2G	3180	DR2	0	Overtone of G mode

the shear mode of graphene [33]. The shear mode has E_{2g} symmetry of D_{6h} that has the same symmetry as the G band. Because of the interlayer interaction between graphene layers, the C band frequencies depend on the number of layers [34]. Since SR Raman modes are limited among the Raman spectra, it is important to investigate the double resonance Raman spectra for obtaining more precise information of structure and electronic structure. It is noted that similar ω's with the same origin in Table 7.1 such as (oTO(1) and oTO(2)) or (D and LO) comes from double resonance Raman spectra for the same phonon modes and the same intra/inter valley scattering but different $\boldsymbol{q}$ values. In the double resonance condition, there are two strong peaks for $\boldsymbol{q} = 0$ and $\boldsymbol{q} = 2\boldsymbol{k}$ measured from either the Γ or K point [4, 12], which we will not go in detail. In the next section, we will discuss the double resonance Raman spectra of transition metal dichalcogenides (TMDs).

7.3 Double Resonance Raman Spectra of TMDs

In this section, we show experimental results of double resonance Raman spectra of transition metal dichalcogenides (TMDs), i.e., $MoTe_2$ (Sect. 7.3.1) and MoS_2 (Sect. 7.3.2). Since the phonon assignment of the double resonance Raman peaks by theory is lengthy, the theoretical assignment will be shown in Sect. 7.4. The final results of the phonon assignments are given in Table 7.2 for $MoTe_2$ and Table 7.3 for MoS_2 in Sect. 7.4.

7.3.1 Double Resonance Raman Spectra of $MoTe_2$

In Fig. 7.3, we show (a) 2H, (b) 1T and (c) 1T′ lattice structures of MX_2 (M = Mo and W, X = S, Se, Te), in which the 2H phase (hexagonal structure, Fig. 7.3a) and the 1T′ phase (monoclininic structure, Fig. 7.3c) are stable for ambient conditions [36]. Bulk 2H $MoTe_2$ is an indirect bandgap semiconductor, with a bandgap of 1.0 eV. Monolayer 2H $MoTe_2$ has a direct bandgap of 1.1 eV, which is similar to the indirect bandgap of silicon [37]. $MoTe_2$ has the smallest energy gap and the energy band width among TMDs, since both Mo and Te are heavy elements. Since the energy position of the M point van Hove singularity (VHS) is not so high compared with other TMDs, we can see the double resonance Raman peaks due to VHS in visible light energy region. The Raman spectra of the monolayer 2H $MoTe_2$ for $E_L = 2.53$ eV laser excitation has two strong first order Raman modes: A_{1g}, E^1_{2g}, while the Raman spectra of bilayer and trilayer 2H $MoTe_2$ have four first order Raman modes: E_{1g}, A_{1g}, E^1_{2g}, B^1_{2g} [38, 39].

Besides the first order Raman peaks, there are several small peaks (denoted as ω_1-ω_6) in Raman spectra of 2H $MoTe_2$ as shown in Fig. 7.4a [40]. In Fig. 7.4b, the Raman spectra of monolayer 2H $MoTe_2$ for different laser excitation energies are shown. Since the frequencies of the ω_1-ω_6 modes shift by changing E_L, the

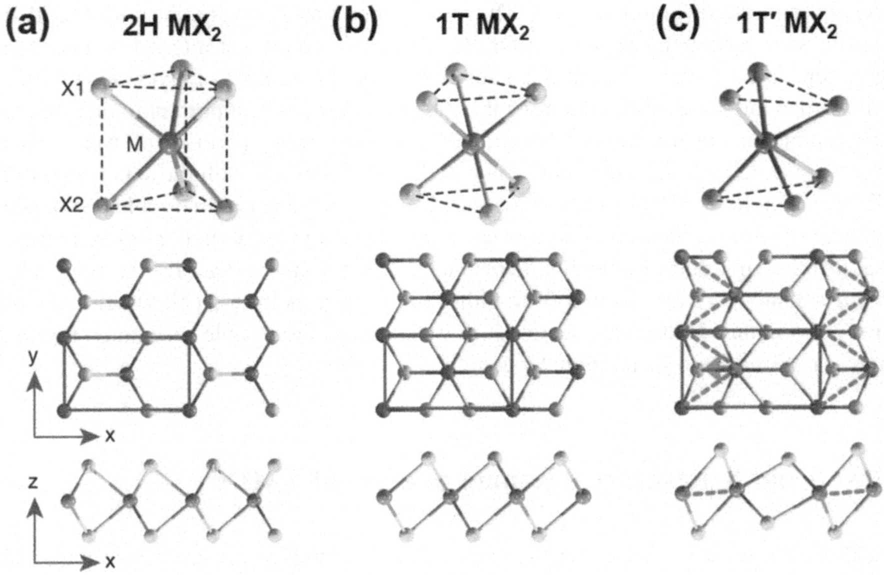

Fig. 7.3 Side and top views of crystal structures, respectively, of (**a**) 2H MX_2, (**b**) 1T MX_2, and (**c**) 1T′ MX_2. The dash lines in (**c**) indicate a one-dimensional zigzag atomic chain consisting of the distorted M atoms in the M plane of 1T′ MX_2. The arrow in (**c**) denotes the distortion for the lattice of 1T′ MX_2 from the 1T MX_2.The 1T MX_2 is not stable for ambient conditions

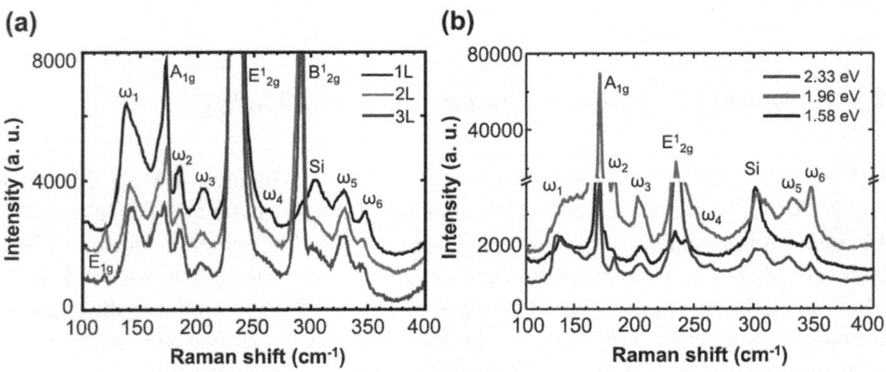

Fig. 7.4 (**a**) The Raman spectra of monolayer, bilayer and trilayer 2H $MoTe_2$ with 2.33 eV (532 nm) laser excitation. (**b**) Raman spectra of monolayer 2H $MoTe_2$ for E_L = 2.33, 1.96, and 1.58 eV [40]

ω_1-ω_6 modes can be assigned to double resonance Raman spectra [40]. As shown in Fig. 7.4b, the Raman intensity for $E_L = 1.96$ eV is larger than those for $E_L = 1.58$ eV and 2.33 eV, since $E_L = 1.96$ eV is close to the resonance condition.

In Fig. 7.5, we show the polarized Raman intensity of bulk 1T′ $MoTe_2$ for each phonon mode. Because of the in-plane anisotropic lattice structure of 1T′ $MoTe_2$ as shown in Fig. 7.3c, Raman intensity shows a dependence on the polarization angle

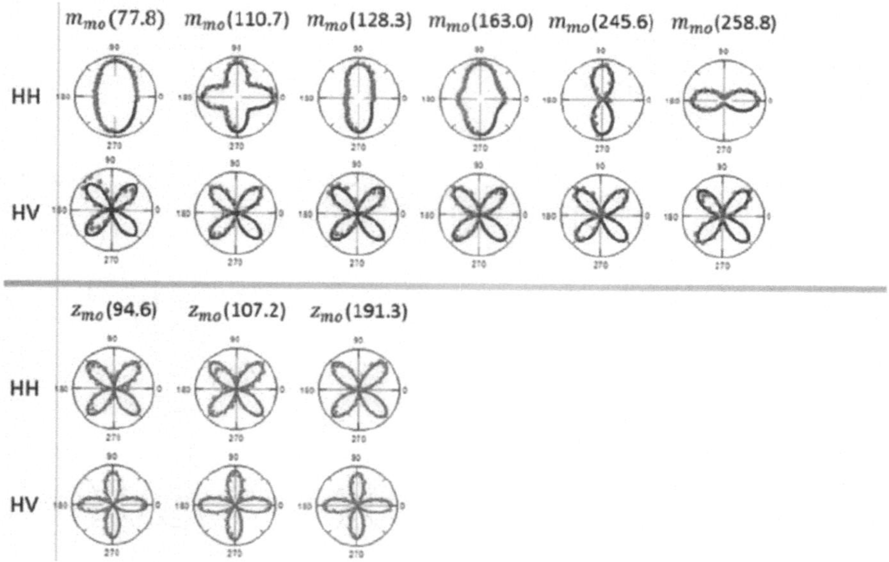

Fig. 7.5 Polarized Raman intensity of bulk 1T′ $MoTe_2$ for HH and HV in-plane polarization geometries. "m_{mo}" and "z_{mo}" denote, respectively, the modes vibrating along the mirror plane and along the zigzag Mo chain direction. The subscript "mo" means monoclinic structure [41]

of electric field of the light measured from the a-axis within the $MoTe_2$ plane. As shown in Fig. 7.5, the polarized Raman spectra show a different pattern for each mode as a function of polarization angle with keeping parallel polarization geometry (HH) and perpendicular polarization geometry (HV) for the polarization directions of incident and scattered light, reflecting the symmetry of phonon modes. In particular, for HH configurations, there are distinct Raman spectra depending on the phonon modes while for HV configuration, all the Raman spectra show 4-fold symmetry [41]. It is noted that not only first order Raman but also second order Raman show similar polarization patterns to one another, from which we can say that the polarized patters come from the polarization dependence of M_{op}.

7.3.2 Double Resonance Raman Spectra of MoS_2

Now let us discuss Raman spectra of 2H MoS_2. 2H MoS_2 has a direct energy gap at the K and K' points. With exciton binding energies involved, the lowest exciton energy is taken to be the 'band gap'. The optical band gap of 2H MoS_2 is around 1.83–1.90 eV from photoluminescence measurement [42, 43]. Unlike graphene which is always resonant to all E_L due to the linear electronic dispersion around the K point, MoS_2 has a strong resonance E_L around the energy gap. Because of 2D materials, the exciton binding energy is sufficiently large so as to exist at the

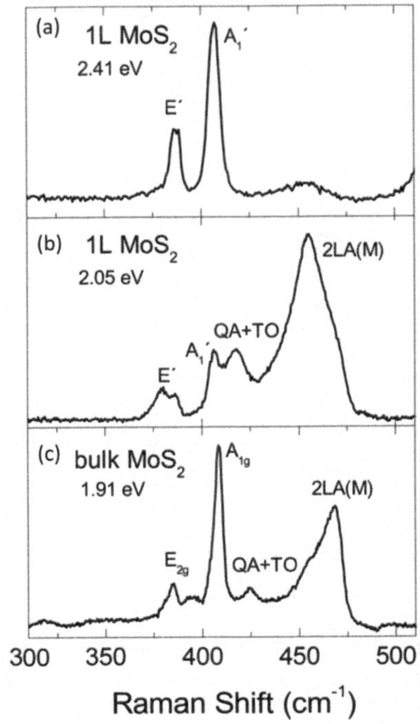

Fig. 7.6 Raman spectra of monolayer MoS_2 for E_L = (**a**) 2.41 eV and (**b**) 2.05 eV, and (**c**) Raman spectra of bulk 2H MoS_2 for E_L = 1.91 eV [44]

room temperature. Thus the strong resonance near the energy gap comes not from the JDOS singularity but the exciton effect. Further, we have the *M* point VHS in the ultra-violet region in which we can see anomalous strong double resonance peaks as shown below.

In Fig. 7.6, typical Raman spectra of monolayer MoS_2 for E_L = (a) 2.41 eV and (b) 2.05 eV, are shown [44]. As shown in Fig. 7.6a, only two first order Raman modes E′ and A_1' are observed in the case of E_L = 2.41 eV. This is because that E_L = 2.41 eV is away from the resonance energy 1.9 eV of monolayer MoS_2. In the case of E_L = 2.05 eV as shown in 7.6b, on the other hand, whose energy is close to the resonance condition, more resonant Raman spectra are observed, in which we can see double resonant modes QA(quasi-acoustic longitudinal mode)+TO and 2LA(M) [44].

The 2LA(M) mode involves two longitudinal acoustic (LA) phonons with opposite $\boldsymbol{q}$ wavevectors at the *M* point of the Brillouin zone. Therefore the double-resonant mode is called 2LA(M). In this case, the phonon wavevector connects the *K* and I (the middle point of the Γ-*K* line) points, which corresponds to the *M* point phonon [44].

For E_L = 2.05 eV, the double resonance Raman intensity is even stronger than that of the first order modes E' and A_1'. In fact, many other weak peaks observed in the Raman spectra also come from the combination of phonons at the *M* point. Since the electronic energy dispersions are flat both at the bottom of the conduction band at the *K* point and the saddle point at the *M* point, the dispersion of double

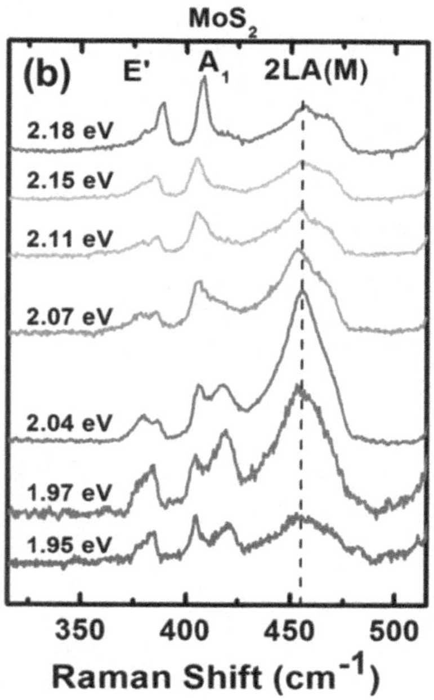

Fig. 7.7 Raman excitation profile of monolayer MoS_2 for E_L = 1.95 to 2.18 eV [44]

resonance Raman spectra as a function of E_L are not clearly seen compared with the case of graphene. Only the Raman intensity is sensitive to E_L, which we call Raman excitation profile. From the Raman excitation profile, we can know experimentally the energy position of the energy gap and VHS position which can be compared with the optical absorption spectra by Ruppert et al. [37].

In Fig. 7.7, we show the Raman excitation profile of monolayer MoS_2 measured by almost continuous laser energies [44]. As shown in Fig. 7.7, the 2LA(M) mode is not dispersive. On the other hand, the intensity of 2LA(M) changes as a function of E_L and reaches the maximum at $E_L = 2.04$ eV, which corresponds to the B exciton in monolayer MoS_2 [42]. B exciton connects to the conduction band and the lower valence band due to spin-orbit splitting. Since the enhancement of the 2LA(M) mode only occurs in a small window of laser energy, we can suggest the resonance Raman effect to the discrete exciton state.

In Fig. 7.8, Raman spectra of MoS_2 are measured for a wider region of E_L up to the deep ultra-violet light (4.66 eV) [10]. In particular, for $E_L = 3.50$ eV (354 nm), 13 Raman peaks denoted by ω_1 to ω_{13} between 200 to 900 cm^{-1} are anomalously enhanced as shown in Fig. 7.8a, b. These peaks can be assigned as the double resonance Raman spectra. The reason why so many peaks are observed with this laser energy is that the electronic joint density of states in monolayer MoS_2 is very high due to the M-point VHS. In fact, with three laser energies, 2.54, 3.50 and 4.66 eV, high joint density of states all occur, as seen in Fig. 7.12b, resulting in the appearance of many double-resonant Raman peaks, as shown in Fig. 7.8b, c. For bulk MoS_2, due to the change of energy band structure compared to monolayer, the

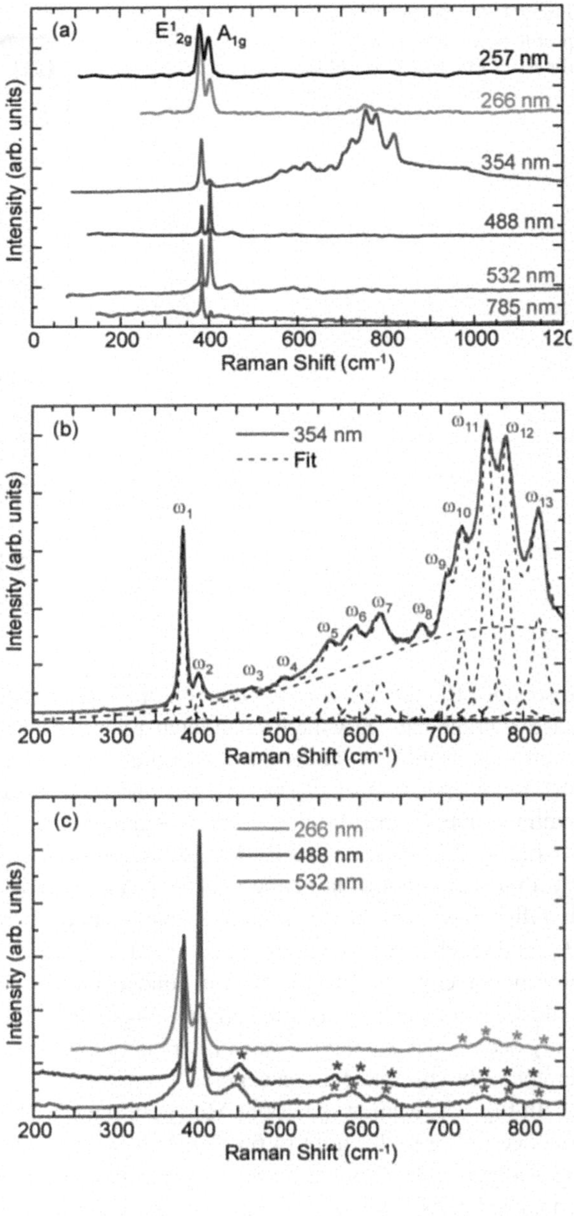

Fig. 7.8 (**a**) Raman spectra of monolayer MoS_2 for wider range of E_L from 257 to 785 nm. (**b**) The spectra at 354 nm laser excitation. Dotted lines are fitting to each peak and background. (**c**) The spectra for 266, 488 and 532 nm lasers. The label "*" denotes the double resonant Raman peaks that are observed in (**b**) for the 354 nm laser [10]

Raman spectra appear in a different way. For example, Fig. 7.6c shows the Raman spectrum of bulk MoS_2 for $E_L = 1.91$ eV. Though the positions and assignments of the Raman modes are similar to those for monolayer MoS_2, the 2LA(M) mode is less intense than the first order A_{1g} mode in the bulk MoS_2. A possible reason is that the exciton states in the bulk MoS_2 can not exists at the room temperature.

Chakraborty et al. reported the thickness dependence of MoS_2 Raman spectra measured by the 633 nm laser [45]. According to their results, the obvious thickness evolution of Raman spectra mainly lies in their relative intensities, including several double resonant spectra, which supports the dimensional crossover from 2D (monolayer to few layers) to 3D (bulk) as discussed above.

7.4 Theoretical Analysis of Double Resonance Raman Spectra of TMDs

We here give the theoretical analysis and phonon assignment of double resonance Raman spectra of TMDs.

7.4.1 MoTe$_2$

As is discussed in Sect. 7.1.2, the photo excited electron is scattered to two real electronic states or energy bands to obtain the double resonance condition. In Fig. 7.9, we show (a) the electronic energy band, the density of states (DOS) and (b) the optical absorption spectra of monolayer $MoTe_2$ [40]. The electronic energy band in Fig. 7.9a shows that both the top of the valence and bottom of the conduction bands are located at the K point, which give a direct energy band gap. The calculated energy dispersion reproduces the main observed features [37, 46, 47] such as the direct electronic band gap (1.00 eV) and a spin-orbit splitting of the valence band (~230 meV) at the K point which are close to 1.10 eV and 250 meV, respectively, from the optical absorption measurements [37]. In Fig. 7.9a right, we plot the DOS for showing the VHS. Around the K point, 2D DOS gives constant values, while at the M point, the DOS gives logarithmic $\log|E - E_0|$ divergence as is known as 2D VHS as is discussed in Sect. 7.1.3.

In Fig. 7.9b, we plot calculated real (ϵ') and imaginary (ϵ'') parts of the dielectric constant and the optical absorption coefficient, α as a function of E_L. The ϵ'' has a value from the photon energy $E_L = 1.00$ eV which corresponds to the energy gap, and a peak appears at $E_L \approx 1.80$ eV which corresponds to the VHS. It is noted that the energy band gap is underestimated by the local density functional method. In fact, the experimental value of the optical absorption in monolayer $MoTe_2$ [37] is at around 1.10 eV, which is slightly larger than 1.00 eV obtained by the present calculation. We upshift the conduction bands and the unoccupied DOS by 0.10 eV in Fig. 7.9a to fit the experimental data. By the upshift, we find that (1) a strong

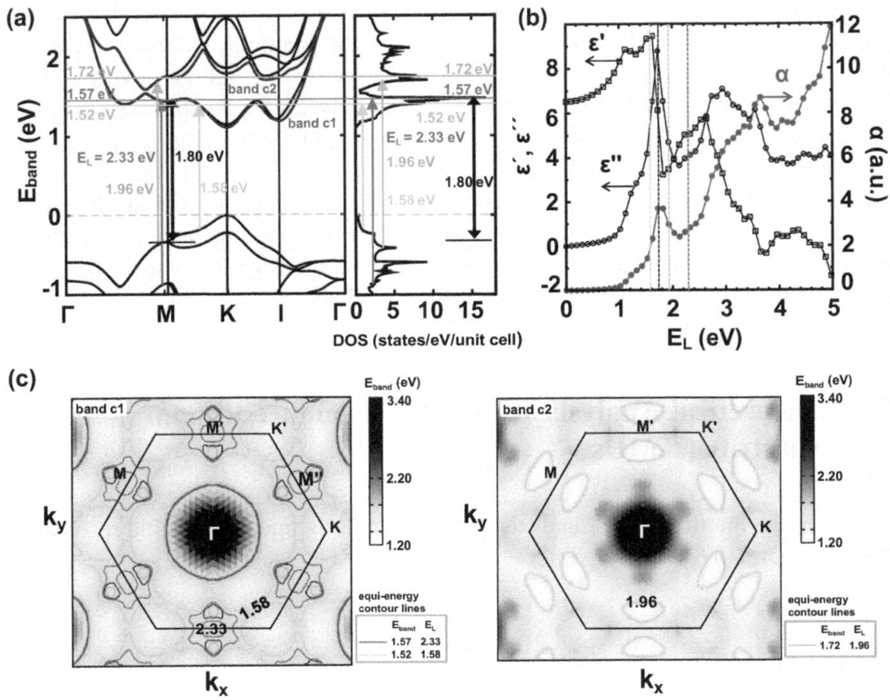

Fig. 7.9 (**a**) Electronic energy band structure and density of states (DOS), (**b**) the real (ϵ'), and imaginary (ϵ'') parts of the dielectric function and optical absorption coefficient α as a function of photon energy E_L, and (**c**) equi-energy contour plot of the two lowest conduction bands (c1 and c2) with the same spin for monolayer $MoTe_2$. When we consider energy separation between two van Hove singularities (VHSs) of the DOS, the peak at 1.80 eV in the absorption spectrum corresponds to optical transition at the M point (double headed arrow). (**c**) The equi-energy contour plot for VHS E_{band} energies E^{c1}_{band} = 1.52 and 1.57 eV (left) and E^{c2}_{band} = 1.72 eV (right). All equi-energy contours appear around the M point. The E_{band} energies of 1.52, 1.57 and 1.72 eV (horizontal lines) correspond to the experimental laser energies (2.33, 1.58 and 1.96 eV) [40]

optical absorption at 1.80 eV from our calculation agrees well with the large peak at around 1.83 eV from the optical absorption measurement on monolayer $MoTe_2$ [37]. (2) The electronic excitation energy of 2.30 eV (2.07 eV) from the second (first) valence band to the lowest (second lowest) conduction band at the M point matches well for E_L = 2.33 eV (1.96 eV) used in the experiment in Fig. 7.4, as indicated by the corresponding arrows in the band structure of Fig. 7.9a. Thus resonant optical absorption is expected at the E_L's.

In Fig. 7.9a, we highlight in gray solid lines the conduction bands c1 and c2 which have the same spin [48, 49]. The black lines near the c1 and c2 denote energy bands with opposite spin to that of c1 and c2, split by spin-orbit interaction. In Fig. 7.9c, we plot equi-energy contour for E_{band} = 1.52 and 1.57 eV for c1 and 1.72 eV for c2, which correspond to VHS energies. These three E_{band} of 1.52, 1.57

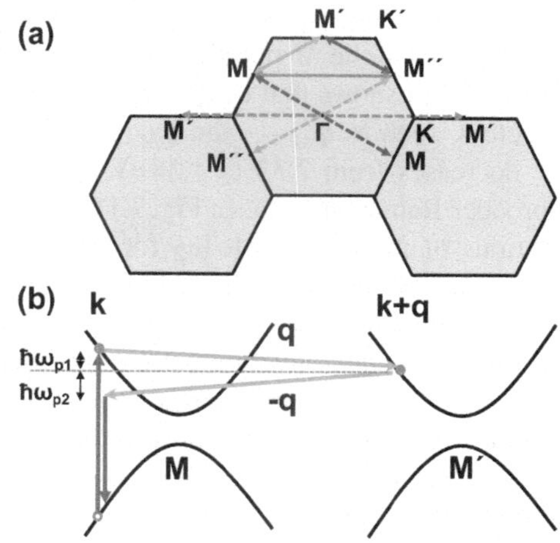

Fig. 7.10 (**a**) In the double resonance process, photo excited electrons are scattered inelastically by emitting phonons and make inter-valley (i.e., $M \to M'$, $M' \to M''$, and $M \to M''$) transitions, as shown by the solid double-end arrows in the first BZ. The related phonon vectors $\boldsymbol{q}$ measured from the Γ point are also given by the six dashed arrows. (**b**) The schematics of the double resonance process (electron-photon and electron-phonon processes). This process generates two phonons $\hbar\omega_{p1}(\boldsymbol{q})$ and $\hbar\omega_{p2}(-\boldsymbol{q})$ [40]

and 1.72 eV match $E_{\mathrm{L}} = 1.58$, 2.33 and 1.96 eV (used in Fig. 7.4), respectively, as indicated by gray arrows in Fig. 7.9a. The equi-energy contour lines E^{c1}_{band} and E^{c2}_{band} in Fig. 7.9c both show the saddle point shape of the energy dispersion in the vicinity of the M point. That is, the energy band E^{c1}_{band} (E^{c2}_{band}) starting from the M point monotonically decreases (increases) along the M-K direction, whereas this band energy increases (decreases) along the M-Γ direction. When we decrease E_{L} from 2.33 to 1.58 eV, the contour lines in Fig. 7.9c left change the shapes around the M point, suggesting that the area inside of the contour line for each M point decreases with increasing laser energy. This situation is opposite to the case of graphene in which the equi-energy contour is almost circular and the area increases with increasing energy [4, 12].

In Fig. 7.10, we illustrate double resonance Raman process around the M points [40]. If the E_{L} matches an optical transition energy at the M point, we expect that many photo excited electrons at the M point are scattered to the inequivalent M' or M'' points by an intervalley resonant electron-phonon interaction, as shown in Fig. 7.10a. Selecting one process from Fig. 7.10a, we show in Fig. 7.10b a double resonance Raman process around the M and M' valleys (intervalley scattering). As analyzed in the Fig. 7.10a, the momentum $\boldsymbol{q}$ of the phonon can only have a magnitude of $|\overrightarrow{MM'}|$ or $|\overrightarrow{MM''}|$ or $|\overrightarrow{M'M''}|$. When we measure the value of $|\boldsymbol{q}|$ from the Γ point, the $\boldsymbol{q}$ corresponds to the M point phonon [40].

In order to specify the phonons contributing to the second order Raman process, we analyzed possible intervalley phonon vectors $\boldsymbol{q}_{MM'}$ for the three laser energies and calculated phonon spectra of monolayer $MoTe_2$, as shown in Fig. 7.11. For the laser with energy of 2.33 eV, we analyze possible $\boldsymbol{q}_{MM'}$ between two elliptical equi-energy contour lines (gray lines) at the M points in the left panel of Fig. 7.11a. The dark area for $\boldsymbol{q}_{MM'}$ in Fig. 7.11a shows that the phonon vectors appear near the M point. Other phonon vectors, such as $\boldsymbol{q}_{M'M''}$ and $\boldsymbol{q}_{MM''}$, are similar to $\boldsymbol{q}_{MM'}$ (not shown). When the E_L decreases from 2.33 to 1.58 eV, the equi-energy increases and thus we expect a broader Raman spectra. In Fig. 7.11b, we show the calculated phonon dispersion relations of monolayer $MoTe_2$ for the assignment of phonon. The symmetry for each phonon band, such as E^1_{2g} and E_{1g} is in Fig. 7.11b. Here LA, TA and ZA refer to the longitudinal, transverse and out-of-plane acoustic modes, respectively. The gray arrows represent overtone or combination Raman modes of ω_i (i = 1, 2, ..., 6), respectively. Downward arrows with a "−" superscript denote different combination mode, while upward arrows with the "+" or "2" superscripts denote combination or overtone modes.

In Table 7.2, we list first order (upper) and second order (lower) Raman modes for the experiment, calculated results and the present assignment [40]. First of all, good agreement is obtained between the experiment and the calculation for the first order Raman modes. Changing the laser energy can help the assignment in Table 7.2. When E_L is changed from 2.33 to 1.58 eV, the experimental peak positions of ω_i show small dispersion about 1–3 cm^{-1}, which can be used for confirming the phonon assignment as is discussed in Sect. 7.1.4.1. By decreasing the E_L from 2.33 to 1.58 eV, we have already seen in Fig. 7.9c that the equi-energy contours become large with decreasing E_L. Thus $\boldsymbol{q}_{MM'}$ for $E_L = 1.58$ and 1.96 eV correspond to be relatively large measured from the M point. The relatively dark region in Fig. 7.11a indicates that (1) the $\boldsymbol{q}_{MM}$ for $E_L = 1.96$ eV appear both at the M point and in the vicinity of the M point (approximately by lengths of $\frac{1}{3}\overline{MK}$) along the M and K direction and (2) the $\boldsymbol{q}_{MM'}$ for $E_L = 1.58$ eV appear in the vicinity of the M point along the $M'M''$ (or MM'') direction approximately by lengths of $10\%\overline{M'M''}$ (or $6.5\%\overline{MM''}$) away from the M'' point. In Fig. 7.11b we show the second order Raman modes assigned to the M point for $E_L = 2.33$ eV. Since the phonon dispersion relation shows a small dispersion near the M point, the Raman frequency cannot be dispersive much. This agrees well with the experimental results for all three laser energies, as given in Fig. 7.4b. All the new phonon frequencies are obtained at two points away from the M point by a length of $\frac{1}{3}\overline{MK}$ and $10\%\overline{M'M''}$ for $E_L = 1.96$ eV and 1.58 eV, respectively, and the results are listed in Table 7.2 as "off M". The dependence of the second order Raman frequencies on the laser energy from our calculations agrees well with those that come from our experimental measurements.

It is noted that two Raman peaks are assigned to difference combination mode, for example, ω_1 assigned to ω_1^- (= E^1_{2g}(M) − LA(M)) and ω_2 to ω_2^- (= E^1_{2g}(M) − TA(M)). These assignments can be further tested experimentally by low-temperature anti-Stokes Raman spectroscopy.

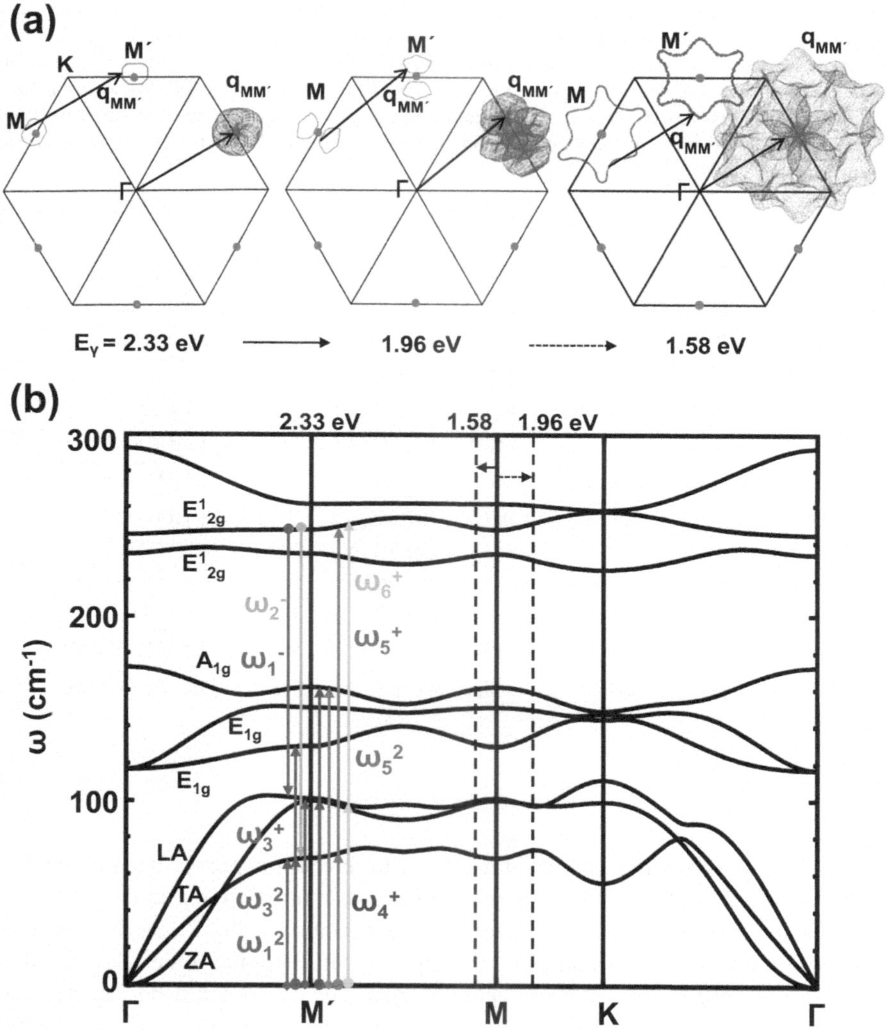

Fig. 7.11 (**a**) All possible phonon vectors $\boldsymbol{q}_{MM'}$ coming from the double resonance Raman process for E_L = 2.33 eV, 1.96 eV and 1.58 eV. The $\boldsymbol{q}_{MM'}$ is equal to either $\overline{\Gamma M}$ (for the laser energy E_L = 2.33 and 1.96 eV) or $\sim \frac{1}{3}\overline{MK}$ (for E_L = 1.96 eV) or $\sim 10\%\overline{MM'}$ (for E_L = 1.58 eV), as indicated by an arrow away from the Γ point. (**b**) The phonon dispersion relations of a monolayer $MoTe_2$. The second order phonon modes are analyzed at the M' point for the laser energy E_L = 2.33 eV, and then at two k points marked by two vertical dashed lines which are shifted leftward and rightward from the M point for the laser energy E_L = 1.58 eV and 1.96 eV, respectively. The gray vertical arrows are used for modes ω_1, ω_2, ω_3, ω_4, ω_5, and ω_6, respectively. The superscripts "+", "-", and "2" for ω refer to possible "combination modes" (i.e. ω_4^+ = A_{1g}(M) + LA(M), ω_6^+ = E^1_{2g}(M) + LA(M)), "difference modes" (i.e. ω_1^- = E^1_{2g}(M) − LA(M), ω_2^- = E^1_{2g}(M) − TA(M)) and "overtones" (i.e. ω_1^2 = 2TA(M), ω_3^2 = 2LA(M)), respectively [40]

Table 7.2 Phonon assignment of the first and second order Raman spectra of monolayer $MoTe_2$ are listed for three different laser energies E_L. A comparison for first and second order Raman spectra is given between the experimentally unassigned modes ω_i (i = 1, 2, ..., 6) and the calculated phonon frequencies both at the M point and near the M point denoted by (off M) whose $\boldsymbol{q}$ lies around $\frac{1}{3}\overline{MK}$ and 10%$\overline{MM'}$ for E_L = 1.96 eV and 1.58 eV, respectively. The inconsistent dispersion behavior for 1.96 eV between the experiment and the theory are underlined. The frequency is given in units of cm^{-1} [40]

	Experiment			Calculation			Phonon assignments
E_L	2.33	1.96	1.58 eV	2.33	1.96	1.58 eV	
E_{1g}	NA	NA	NA	116.9			First order
A_{1g}	171.1	171.3	170.0	172.2			First order
E^1_{2g}	235.8	236.2	234.1	236.9			First order
				at M	at M/off M	off M	
ω_1	138.2	140.7	139.0	137.4	137.4/146.6	140.1	2TA(M)
				132.8	132.8/133.1	133.2	E^1_{2g}(M)-LA(M)
ω_2	183.7	183.6	181.5	178.3	178.3/177.2	177.8	E^1_{2g}(M)-TA(M)
ω_3	205.5	205.3	205.9	197.9	197.9/208.2	200.5	E_{1g}(M)+TA(M)
				202.7	202.7/196.0	200.4	2LA(M)
ω_4	264.0	260~270	260~270	263.0	263.0/256.6	261.0	A_{1g}(M)+LA(M)
ω_5	329.0	333.2	328.1	323.3	323.3/317.2	321.6	$2A_{1g}$(M)
				315.7	315.7/323.8	317.9	E^1_{2g}(M)+TA(M)
ω_6	347.4	348.4	346.5	348.3	348.3/348.5	348.0	E^1_{2g}(M)+LA(M)

7.4.2 MoS_2

In Fig. 7.12, we show the calculated results of (a) electronic energy band of MoS_2, (b) JDOS and (c) optical absorption probability α as a function of $\boldsymbol{k}$ in the Brillouin zone for the E_L's [10]. From the energy band structure in Fig. 7.12a, we can see spin-orbit splitting Δ_{SO} around the K points in the Brillouin zone (BZ) for the top of the valence bands and some energy bands. The calculated band gap Δ_g and splitting Δ_{SO} occurring at the top valence band both at the K point are around 1.72 eV and 147 meV, respectively, in good agreement with the previous theoretical calculations [50] and experimental results [42, 51, 52].

In the case of MoS_2, some pairs of valence and conduction bands are parallel to each other which enhances the JDOS. E_L = 2.54, 3.50 and 4.66 eV corresponds to the JDOS peak as shown in Fig. 7.12a. In order to have a better view of JDOS as a function of $\boldsymbol{k}$ in the BZ, we calculated the JDOS as a function of $\boldsymbol{k}$ and E_L (see Eq. (7.6)). The calculated JDOS($\boldsymbol{k}$, E_L) is plotted over the BZ for E_L = 1.58 eV, 2.33 eV, 2.54 eV, 3.50 eV, 4.66 eV, and 4.82 eV in Fig. 7.12b. The first BZ is marked by a white solid hexagon with high-symmetry points Γ, $M(M')$ and K shown. For the laser line 785 nm (~1.58 eV), no electronic excitation occurs since E_L is smaller than Δ_g = 1.72 eV, as shown in Fig. 7.12b. Thus the Raman scattering spectra for 785 nm laser (Fig. 7.8a) are non-resonant and we do not expect double resonance

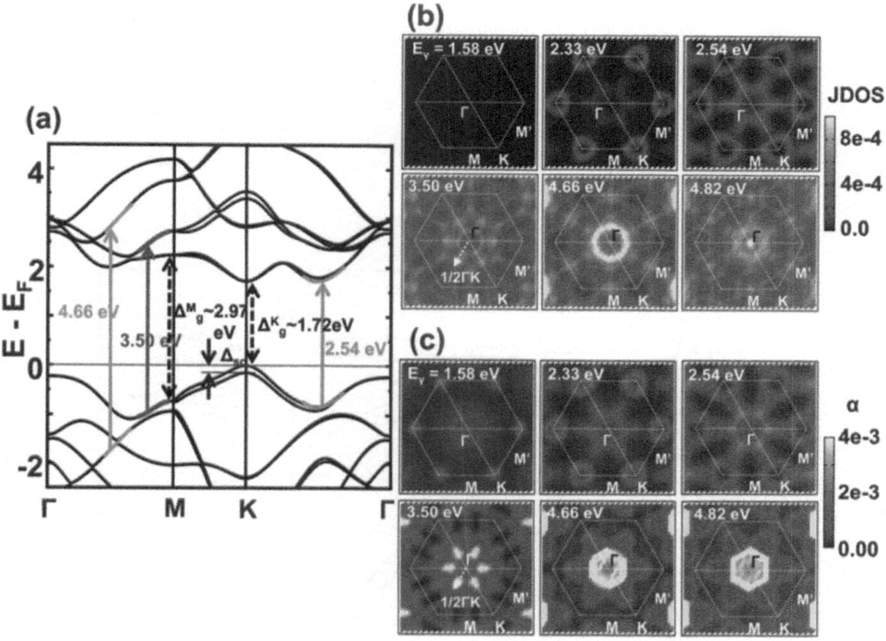

Fig. 7.12 (**a**) Energy band of monolayer MoS_2, arrows denote the $\boldsymbol{k}$ for several E_L's. (**b**) Joint density of states and (**c**) optical absorption of MoS_2 as a function of $\boldsymbol{k}$ in the Brillouin zone for the E_L's [10]

Raman peaks. When E_L is larger than Δ_g but smaller than the M point separation ($\Delta_g^M \sim 2.97$ eV), for example at 532 nm ($\sim$2.33 eV) and 488 nm ($\sim$2.54 eV), finite value of JDOS appears around the K and also the Γ point. When E_L is larger than Δ_g^M, JDOS as a function of $\boldsymbol{k}$ show clear patterns as shown in Fig. 7.12b. Significant values are distributed at a ring area around the Γ point and the M point than around the K point, suggesting Γ and M point phonons play an essential role for double resonance Raman scattering spectra in which $\boldsymbol{q} \neq 0$ phonon is relevant to the experimental data.

To better evaluate electron-photon resonance, we calculated the optical absorption probability α, as shown in Fig. 7.12c. The absorption probability α follows a trend of the JDOS except for the laser lines of 266 nm ($\sim$4.66 eV) and 257 nm ($\sim$4.82 eV) where α is not due to the M point but merely from a ring area around the Γ point as highlighted by a white circle. More importantly, there is additional large optical absorption occurring at $\frac{1}{2}\overrightarrow{\Gamma K}$ (or the Λ point [53]), which appears exclusively at $E_L = 3.50$ eV. This can account for the remarkable enhancement of the intensity as shown in Fig. 7.8b [10].

Large amplitude of optical absorption probability at six $\frac{1}{2}\overrightarrow{\Gamma K}$ vectors and on a circle centered around the Γ point usually gives rise to intervalley and intravalley electronic scattering by phonons, respectively. In Fig. 7.13a, we show possible

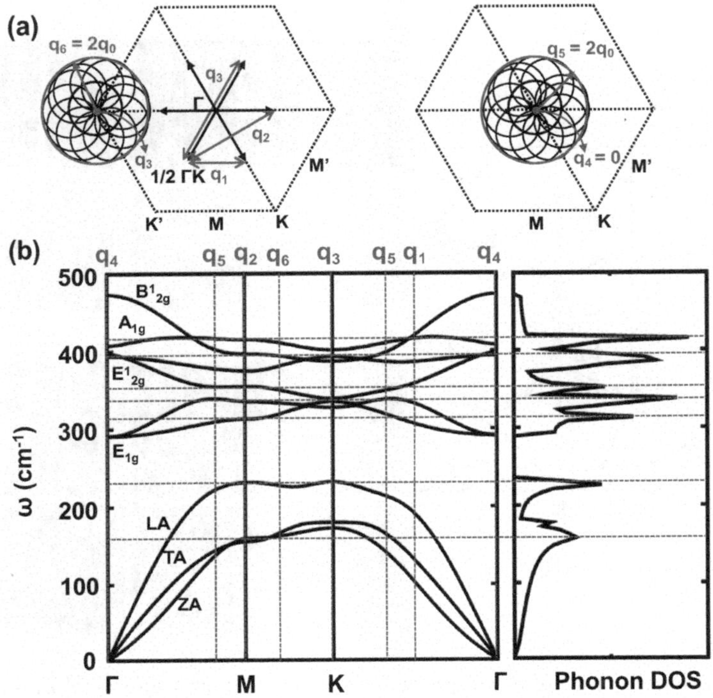

Fig. 7.13 Phonons contributing to the second order Raman scattering process. (**a**) Typical phonon vectors arising from both intervalley and intravalley electron-phonon interactions. Arrow/circle represents wave vector having large amplitude of optical absorption, gray arrow/circle represents phonon wave vector out of electron-phonon scattering process. (**b**) Phonon dispersion relation and density of states of monolayer MoS_2 [10]

phonon wave vectors $\boldsymbol{q}$. The intervalley electron scattering between six $\frac{1}{2}\overrightarrow{\Gamma K}$ vector points leads to three different $\boldsymbol{q}$ vectors, $\boldsymbol{q}_1$ $(= \frac{1}{2}\overrightarrow{\Gamma K})$, $\boldsymbol{q}_2$ $(= \overrightarrow{\Gamma M})$, and $\boldsymbol{q}_3$ $(= \overrightarrow{\Gamma K})$, as shown by gray double-ended arrow. For the intravalley electron-phonon interaction process, two different phonon vectors $\boldsymbol{q}_4$ $(= 0)$ and $\boldsymbol{q}_5$ $(= 2\boldsymbol{q}_0)$, in which $\boldsymbol{q}_0$ is the radius of amplitude circle for large optical absorption, can be obtained.

To assign the phonon modes corresponding to the experimental Raman shift excited by the 354 nm laser line in Fig. 7.8b, we calculated the phonon dispersion relation and the phonon density of states (PhDOS) in Fig. 7.13b. The assigned phonon wave vectors, all of which contribute to the first and second order Raman scattering process, are highlighted by the dashed gray lines. Since we can see some saddle points of the phonon energy bands appear at the M point which gives VHS's in the PhDOS, as indicated by the horizontal dash lines. This suggests that the M point phonons with phonon vector $\boldsymbol{q}_2$ dominate over the Γ $(\boldsymbol{q}_4)$, K point $(\boldsymbol{q}_3)$ and also $\frac{1}{2}\overrightarrow{\Gamma K}$ phonons $(\boldsymbol{q}_1)$ in the second order Raman scattering.

Table 7.3 Phonon assignment of the first and second order Raman scattering spectra of monolayer MoS_2 excited by the 354 nm laser (Fig. 7.8b). A comparison is given between the experimentally unassigned modes and the calculated second order phonon spectrum at both the M and Γ points [10]

	ω_1	ω_2	ω_3	ω_4	ω_5
Exp.	384.4	403.0	467.4	519.2	564.5
Cal.	393.2	407.6	463.4	510.7	566.7
Mode	$E^1_{2g}(\Gamma)$	$A_{1g}(\Gamma)$	2LA(M)	E^1_{2g}+TA(M)	E_{1g}+LA(M)
	ω_6	ω_7	ω_8	ω_9	ω_{10}
Exp.	596.0	625.7	675.8	707.3	725.9
Cal.	586.4	626.6	670.0	709.4	727.7
Mode	E^1_{2g}+LA(M)	$2E_{1g}$(M)	$2E_{1g}$(M)	$2E^1_{2g}$(M)	A_{1g}+E_{1g}(M)
	ω_{11}	ω_{12}	ω_{13}		
Exp.	756.2	780.4	819.3		
Cal.	749.0	786.4	814.0		
Mode	$2E^1_{2g}$(M)	$2E^1_{2g}(\Gamma)$	$2A_{1g}(\Gamma)$		

In Table 7.3, we list the experimental and calculated phonon frequencies at the singular points in the first and second order Raman scattering spectra of monolayer MoS_2 and phonon mode assignments as well. Most phonon frequency differences between the experiment and the theory are within a few cm^{-1}. Besides the combination and overtone modes assigned at the M point, some overtone modes are also assigned at the Γ point. To note that an opposite trend of the intensity can be found between the first order $A_{1g}(\Gamma)$ (decreasing) and overtone $2A_{1g}(\Gamma)$ (increasing) modes for both 354 and 266 nm laser lines in the experiment [10]. We expect that there exists a branch of the Raman scattering processes for a given photo excited electron into the first order and second order Raman processes at the Γ point. The salient aspect of the A_{1g} peak intensity at 532, 488, and 257 nm laser lines shows that the first order $A_{1g}(\Gamma)$ mode is dominant to the second order $2A_{1g}(\Gamma)$ mode.

In Fig. 7.8c, a weak phonon peak at around 450 cm^{-1} is found dispersive with the laser energy increasing from 2.33 eV (532 nm) to 3.50 eV (354 nm), indicating that this is the second order phonon mode. However, a very small JDOS and absorption probability α for both 2.33 eV (532 nm) and 2.54 eV (488 nm) laser lines in Fig. 7.12b, c show that both the electron-photon and electron-phonon resonance processes shall occur at the K point, suggesting that 450 cm^{-1} peaks at 2.33 and 2.54 eV are the 2LA(K) overtone mode. The phonon vector $\boldsymbol{q}_6$ (=$2\boldsymbol{q}_0$(K) with $\boldsymbol{q}_0$(K) extracted from Fig. 7.13a) arises from the intervalley electronic scattering process between the K and K' points. The blue-shift of 2LA(K) mode from 532 to 488 nm can be analyzed by the phonon vector $\boldsymbol{q}_6$ along the M-K direction (with large PhDOS) other than along the Γ-K direction (with small PhDOS) of the phonon dispersion relation in Fig. 7.13. The phonon vector $\boldsymbol{q}_6$ gets closer to the M point with increasing E_L and the frequency of the 2LA(K) mode increases. For the 354 nm laser, the overtone mode appears right at the M point (2LA(M)) with a higher frequency than that of the 2LA(K) modes. This is indeed the case for the experiment shown in Fig. 7.8b.

The low amplitude of Raman scattering spectra above 500 cm^{-1} for both 2.33 eV (532 nm) and 2.54 eV (488 nm) laser lines in Fig. 7.8c can be understood by the low absorption probability around the K point. In this respect, due to a large absorption probability as calculated for both 266 nm (∼4.66 eV) and 257 nm (∼4.82 eV), we would expect an enhancement of the intensity for the Raman peaks above 500 cm^{-1} at these two deep-ultraviolet laser lines, similar to those at 354 nm (∼3.50 eV) laser line. However, this is not the case as shown in Fig. 7.8b. We speculate this phenomenon could be due to the low density of states of phonon for the related phonon modes. In Fig. 7.12c, the two laser lines (4.66 and 4.82 eV) have six spots in the BZ with moderately high absorption amplitude beyond the brightest circle around the Γ point. These spots lie close to the K point with a length of approximately $\frac{2}{3}\Gamma K$ and contribute to several possible intervalley electronic scattering by phonons with $\boldsymbol{q} = \frac{2}{3}\overrightarrow{\Gamma K}, \frac{2}{3}\overrightarrow{\Gamma M}$, all of which do not have a large PhDOS by the M point phonons for the 354 nm laser line. As a result, the intensity of the second order phonon modes excited by the deep-ultraviolet laser line is expected to be weak, which is in agreement with the measurements.

7.5 Summary and Discussion

In this chapter, we discussed double resonance Raman spectra of two dimensional materials such as graphene and transition metal dichalcogenides. In the second order Raman process, if two of the three intermediate states are real electronic states, the corresponding Raman intensity can be enhanced significantly which is known as double resonance effect. In particular, the double resonance Raman spectra is enhanced in the case of 2D materials mainly by the following three reasons; (1) an exciton that makes optical absorption and emission large compared with single particle excitation of an electron can exist in 2D materials even at the room temperature, (2) 2D Van Hove singularity of joint density of states at the saddle point of energy dispersion in the $\boldsymbol{k}$ space gives many possible states that contribute to the Raman processes in a narrow energy region, and (3) electron-phonon interaction for a particular $\boldsymbol{q}$ is relatively large compared with the case of general $\boldsymbol{q}$. Thus when laser excitation energy E_{L} matches either the exciton states or VHS of the 2D materials, we can observe a stronger double resonance Raman intensity than that of first order Raman spectra. For example, the G′ band of monolayer graphene (Case (3)) or double resonance peaks of monolayer MoS_2 at $E_{\mathrm{L}} = 2.04$ eV (Case (1)) $E_{\mathrm{L}} = 3.50$ eV (Case (2)) are, respectively, stronger than the G band or the first order Raman peaks of MoS_2.

Thus the analysis of double resonance Raman spectra probes solid state properties of 2D materials such as (A) the phonon energy dispersion as a function of non-zero phonon wavevector $\boldsymbol{q}$ for each phonon mode, (B) electronic energy information if we have many E_{L}'s, and (C) electron phonon interaction as a function of $\boldsymbol{k}$ and $\boldsymbol{q}$. By changing E_{L}, both the electronic wavevector $\boldsymbol{k}$ and the phonon wavevector $\boldsymbol{q}$ that satisfy the double resonance condition changes accordingly, which is useful for not only probing the phonon dispersion but also assigning the Raman spectra as double resonance Raman spectra.

The first order Raman spectra of solid is frequently used for characterizing the structure of the materials combined with the phonon calculation at the Γ point in the Brillouin zone. Similarly, the second order Raman spectra can be used as additional information for the characterization of much complicated properties. For example, we can distinguish, by double resonance Raman spectra, the ABA and ABC stacking of trilayer graphene [54] or number of graphene layers [30].

However, the phonon assignment for double resonance Raman peaks are not possible only by experimental observation. Theoretical assignment that is discussed in this chapter is needed for the phonon assignment which is not always straightforward. It is because that the phonon wavevector of double resonance Raman spectra is not $\boldsymbol{q} = 0$ and thus the group theory can not be used in double resonance Raman peaks while Raman tensor analysis by group theory can be used for first order Raman peaks combined with polarized Raman spectroscopy [20]. Even though we need many theoretical works for phonon assignment of double resonance peaks, the analysis will be powerful especially for 2D materials in which double resonance Raman peaks are significant. We hope that many experimentalists will be interested in not only first order Raman peaks but also double resonance Raman peaks that are generally weak for a general E_{L}'s.

So far we have only considered the electron-involved double resonance process, we have not discussed the hole-involved double resonance process, nor the multiple resonance effect suggested by Venezuela et al. [55]. As Venezuela et al. [55] pointed out, the enhancement by the double resonance is strong in graphene for one-electron and one hole scattering, since the energy denominators of the Raman intensity formula become zero for the symmetric π and π^* bands of graphene (triple resonance). In the case of TMDs, however, a similar effect is not expected since the conduction band and the valence band are not symmetric around the Fermi energy [48, 56].

Before closing this chapter, let us point out the missing points in this chapter that can be improved in the future. For an experimental point of view, the polarized Raman spectra of double resonance Raman spectra is not yet performed. For given electronic wavevector $\boldsymbol{k}$, electron-photon (or exciton-photon) matrix element depends on the polarization direction of the incident/scattered light which reflects the symmetry of the initial and final electronic wavefunctions. In particular for few layers 2D materials, both the valence and conduction bands are split into few energy bands with different symmetries. Thus for a given E_{L}, when the number of layers is changed, the symmetry of the final states can be changed by the splitting of energy bands, which gives a different polarized Raman spectra even for the same phonon mode. Such phenomena are already observed in the first order Raman spectroscopy of black phosphorene or GaTe 2D materials [57, 58]. Such number of layer dependence on polarized Raman spectra for a given symmetry phonon can not be explained by classical theory of Raman tensor and we need quantum theory of Raman spectroscopy [20]. If we can see a similar phenomena on polarized Raman spectra for double resonance Raman peaks, it would be more informative since we can confirm that the theoretical assignment of the symmetry of the final states. It is because that we can use the common electron-photon matrix elements for first order and second order Raman intensity formula. Further, by comparing polarized Raman spectra of first and second order Raman spectra, we can discuss the polarization dependence of the electron-phonon interaction, though the analysis will be much complicated.

In the theoretical calculation, we can calculate electronic energy bands, phonon dispersion and electron-photon matrix elements that are discussed in this chapter. However, we do not finish to make a computational programs of calculating first order and second order, resonance Raman spectra by first principles calculation combining with programs of electron-phonon interaction, exciton, and exciton-photon and exciton-phonon calculation. It should be mentioned that non-resonant Raman spectra for Raman active modes can be calculated by existing program of first principles calculation (See for example of $MoTe_2$ [59] and GaTe [58]). Since we have developed an electron-phonon matrix element calculation that is used for indirect ARPES measurement [60] and electron-photon matrix element calculation that is used for optical valley polarization [53], the goal of calculating resonance Raman spectra by first principles calculation will be possible in the near future. As for the exciton calculation, although there is a package of exciton calculation such as Yambo, it requires a huge computational time to obtain Raman spectra. Thus it is required for using localized wavefunction for calculating the excitonic states such as tight-binding wavefunctions [2, 17, 18].

Acknowledgements All authors sincerely acknowledge Professor Mildred S. Dresselhaus who passed away on February 20th, 2017, before finishing this article. We all thank her for supervising us Raman spectroscopy of nano carbons and 2D materials. R.S. acknowledges JSPS KAKENHI Grant Numbers JP25286005, JP225107005, JP15K21722 and JP18H01810. T.Y. acknowledges the Major Program of Aerospace Advanced Manufacturing Technology Research Foundation NSFC and CASC, China (No. U1537204) and National Basic Research Program (No.2017YFA0206301) of China. H.H.G. acknowledges the support by the Liaoning Province Doctor Startup Fund (Grant 201601325) and Liaoning Shihua University Grant 2016XJJ-044. S.H. and L.Z. acknowledge financial support by STC Center for Integrated Quantum Materials, NSF Grant No. DMR-1231319, EFRI 2-DARE(EFMA-1542815), NSF grant DMR-1507806, and the U.S. Army Research Office through the MIT Institute for Soldier Nanotechnologies (Grant No. 023674).

References

1. R. Saito, M. Hofmann, G. Dresselhaus, A. Jorio, M.S. Dresselhaus, Adv. Phys. **60**, 413 (2011)
2. M.S. Dresselhaus, G. Dresselhaus, R. Saito, A. Jorio, Phys. Rep. **409**, 47 (2005)
3. R. Saito, A. Grüneis, G.G. Samsonidze, V.W. Brar, G. Dresselhaus, M.S. Dresselhaus, A. Jorio, L.G. Cançado, C. Fantini, M.A. Pimenta, A.G. Souza Filho, New J. Phys. **5**, 157.1 (2003)
4. M.A. Pimenta, G. Dresselhaus, M.S. Dresselhaus, L.G. Cançado, A. Jorio, R. Saito, Phys. Chem. Chem. Phys. **9**, 1276 (2007)
5. A. Jorio, M.A. Pimenta, A.G. Souza Filho, R. Saito, G. Dresselhaus, M.S. Dresselhaus, New J. Phys. **5**, 139.1 (2003)

6. A. Grüneis, R. Saito, G.G. Samsonidze, T. Kimura, M.A. Pimenta, A. Jorio, A.G.S. Filho, G. Dresselhaus, M.S. Dresselhaus, Phys. Rev. B **67**, 165402 (2003)
7. J. Jiang, R. Saito, A. Grüneis, G. Dresselhaus, M.S. Dresselhaus, Chem. Phys. Lett. **392**, 383 (2004)
8. J. Jiang, R. Saito, G.G. Samsonidze, S.G. Chou, A. Jorio, G. Dresselhaus, M.S. Dresselhaus, Phys. Rev. B **72**, 235408 (2005)
9. L.V. Hove, Phys. Rev. **89**, 1189 (1953)
10. H. Liu, H. Guo, T. Yang, Z. Zhang, Y. Kumamoto, C. Shen, Y. Hsu, R. Saito, S. Kawata, Phys. Chem. Chem. Phys. **17**, 14561 (2015)
11. R. Saito, A.R.T. Nugraha, E.H. Hasdeo, S. Siregar, H. Guo, T. Yang, Phys. Status Solidi B **252**, 2363 (2015)
12. R. Saito, A. Jorio, A.G. Souza Filho, G. Dresselhaus, M.S. Dresselhaus, M.A. Pimenta, Phys. Rev. Lett. **88**, 027401 (2002)
13. P. Giannozzi, S. Baroni, N. Bonini, M. Calandra, R. Car, C. Cavazzoni, D. Ceresoli, G.L. Chiarotti, M. Cococcioni, I. Dabo, et al., J. Phys. Condensed Matter **21**(39), 395502 (2009)
14. F. Aryasetiawan, O. Gunnarsson, Reports on Progress in Physics **61**(3), 237 (1998)
15. S. Albrecht, L. Reining, R. Del Sole, G. Onida, Phys. Rev. Lett. **80**, 4510 (1998)
16. M. Rohlfing, S.G. Louie, Phys. Rev. Lett. **81**, 2312 (1998)
17. J. Jiang, R. Saito, G.G. Samsonidze, A. Jorio, S.G. Chou, G. Dresselhaus, M.S. Dresselhaus, Phys. Rev. B **75**, 035407 (2007)
18. J. Jiang, R. Saito, K. Sato, J.S. Park, G.G. Samsonidze, A. Jorio, G. Dresselhaus, M.S. Dresselhaus, Phys. Rev. B **75**, 035405 (2007)
19. S. Baroni, S. de Gironcoli, A. Dal Corso, P. Giannozzi, Rev. Mod. Phys. **73**, 515 (2001)
20. R. Saito, Y. Tatsumi, S. Huang, X. Ling, M.S. Dresselhaus, J. Phys. Cond. Matt. **28**(35), 353002 (2016)
21. P. Ayria, A.R.T. Nugraha, E.H. Hasdeo, T.R. Czank, S. Tanaka, R. Saito, Phys. Rev. B **92**, 195148 (2015)
22. J. Noffsinger, F. Giustino, B.D. Malone, C.H. Park, S.G. Louie, M.L. Cohen, Comput. Phys. Commun. **181**, 2140 (2010)
23. A. Jorio, M.S. Dresselhaus, R. Saito, G. Dresselhaus, *Raman Spectroscopy in Graphene Related Systems* (Wiley-VCH Verlag GmbH & Co KGaA, WeinHeim, Germany, 2010)
24. A.C. Ferrari, J.C. Meyer, V. Scardaci, C. Casiraghi, M. Lazzeri, F. Mauri, S. Piscanec, D. Jiang, K.S. Novoselov, S. Roth, A.K. Geim, Phys. Rev. Lett. **97**, 187401 (2006)
25. C. Thomsen, S. Reich, Phys. Rev. Lett. **85**, 5214 (2000)
26. J.S. Park, A. Reina Cecco, R. Saito, J. Jiang, G. Dresselhaus, M.S. Dresselhaus, Carbon **47**, 1303 (2009)
27. L.G. Cançado, M.A. Pimenta, R. Saito, A. Jorio, L.O. Ladeira, A. Grüneis, A.G. Souza Filho, G. Dresselhaus, M.S. Dresselhaus, Phys. Rev. B **66**, 035415 (2002)
28. P.H. Tan, C.Y. Hu, J. Dong, W.C. Shen, B.F. Zhang, Phys. Rev. B **64**, 214301 (2001)
29. J. Maultzsch, S. Reich, C. Thomsen, Phys. Rev. B **70**, 155403 (2004)
30. C. Cong, T. Yu, R. Saito, G.F. Dresselhaus, M.S. Dresselhaus, ACS Nano **5**, 1600 (2011)
31. V.W. Brar, G.G. Samsonidze, G. Dresselhaus, M.S. Dresselhaus, R. Saito, A.K. Swan, M.S. Ünlü, B.B. Goldberg, A.G. Souza Filho, A. Jorio, Phys. Rev. B **66**, 155418 (2002)
32. L.M. Malard, J. Nilsson, D.C. Elias, J.C. Brant, F. Plentz, E.S. Alves, A.H. Castro Neto, M.A. Pimenta, Phys. Rev. B **76**, 201401 (2007)
33. P.H. Tan, W.P. Han, W.J. Zhao, Z.H. Wu, K. Chang, H. Wang, Y.F. Wang, N. Bonini, N. Marzari, N. Pugno, G. Savini, A. Lombardo, A.C. Ferrari, Nat. Mater. **11**(4), 294 (2012)
34. P.H. Tan, J.B. Wu, W.P. Han, W.J. Zhao, X. Zhang, H. Wang, Y.F. Wang, Phys. Rev. B **89**, 235404 (2014)
35. R. Saito, G. Dresselhaus, M.S. Dresselhaus, *Physical Properties of Carbon Nanotubes* (Imperial College Press, London, 1998)
36. K.A.N. Duerloo, Y. Li, E.J. Reed, Nat. Commun. **5**, 5214 (2014)
37. C. Ruppert, O.B. Aslan, T.F. Heinz, Nano Lett. **14**(11), 6231 (2014)

38. N.R. Pradhan, D. Rhodes, S. Feng, Y. Xin, S. Memaran, B.H. Moon, H. Terrones, M. Terrones, L. Balicas, ACS Nano **8**(6), 5911 (2014)
39. M. Yamamoto, S.T. Wang, M.Y. Ni, Y.F. Lin, S.L. Li, S. Aikawa, W.B. Jian, K. Ueno, K. Wakabayashi, K. Tsukagoshi, ACS Nano **8**(4), 3895 (2014)
40. H. Guo, T. Yang, M. Yamamoto, L. Zhou, R. Ishikawa, K. Ueno, K. Tsukagoshi, Z. Zhang, M. Dresselhaus, R. Saito, Phys. Rev. B **91**, 205415 (2015)
41. S.Y. Chen, T. Goldstein, D. Venkataraman, A. Ramasubramaniam, J. Yan, Nano Lett. **16**(9), 5852 (2016)
42. K.F. Mak, J. Shan, T.F. Heinz, Phys. Rev. Lett. **104**, 176404 (2010)
43. A. Splendiani, L. Sun, Y. Zhang, T. Li, J. Kim, C.Y. Chim, G. Galli, F. Wang, Nano Lett. **10**(4), 1271 (2010)
44. M.A. Pimenta, E. del Corro, B.R. Carvalho, C. Fantini, L.M. Malard, Acc. Chem. Res. **48**(1), 41 (2014)
45. B. Chakraborty, H.S.S.R. Matte, A.K. Sood, C.N.R. Rao, J. Raman Spectrosc. **44**, 92 (2013)
46. A. Kumar, P. Ahluwalia, Eur. Phys. J. B **85**(6), 186 (2012)
47. Y. Ding, Y.L. Wang, J. Ni, L. Shi, S.Q. Shi, W.H. Tang, Physica B Condensed Matter **406**(11), 2254 (2011)
48. D. Xiao, G.B. Liu, W. Feng, X. Xu, W. Yao, Phys. Rev. Lett. **108**, 196802 (2012)
49. Y. Cheng, U. Schwingenschlögl, *MoS_2: A First-Principles Perspective* (Ed. Zhiming M. Wang, Springer, Berlin, 2014)
50. D.Y. Qiu, F.H. da Jornada, S.G. Louie, Phys. Rev. Lett. **111**, 216805 (2013)
51. H.L. Liu, C.C. Shen, S.H. Su, C.L. Hsu, M.Y. Li, L.J. Li, Appl. Phys. Lett. **105**(20), 201905 (2014)
52. J.W. Park, H.S. So, S. Kim, S.H. Choi, H. Lee, J. Lee, C. Lee, Y. Kim, J. Appl. Phys. **116**(18), 183509 (2014)
53. Y. Tatsumi, K. Ghalamkari, R. Saito, Phys. Rev. B **94**, 235408 (2016)
54. C. Cong, T. Yu, K. Sato, J. Shang, R. Saito, G. Dresselhaus, M.S. Dresselhaus, ACS Nano **5**, 8760 (2011)
55. P. Venezuela, M. Lazzeri, F. Mauri, Phys. Rev. B **84**, 035433 (2011)
56. Z.Y. Zhu, Y.C. Cheng, U. Schwingenschlögl, Phys. Rev. B **84**, 153402 (2011)
57. X. Ling, S. Huang, E.H. Hasdeo, L. Liang, W.M. Parkin, Y. Tatsumi, A.R.T. Nugraha, A.A. Puretzky, P.M. Das, B.G. Sumpter, D.B. Geohegan, J. Kong, R. Saito, M. Drndic, V. Meunier, M.S. Dresselhaus, Nano Lett. **16**(0), 2260 (2016)
58. S. Huang, Y. Tatsumi, X. Ling, H. Guo, Z. Wang, G. Watson, A.A. Puretzky, D.B. Geohegan, J. Kong, J. Li, T. Yang, R. Saito, M.S. Dresselhaus, ACS Nano **10**(9), 8964 (2016)
59. L. Zhou, S. Huang, Y. Tatsumi, L. Wu, H. Guo, Y. Bie, K. Ueno, T. Yang, Y. Zhu, J. Kong, R. Saito, M.S. Dresselhaus, J. Am. Chem. Soc. (2017)
60. P. Ayria, S. Tanaka, A.R.T. Nugraha, M.S. Dresselhaus, R. Saito, Phys. Rev. B. **94**, 075429 (2016)

Chapter 8
Raman Signatures of Surface and Interface Effects in Two-Dimensional Layered Materials: Theoretical Insights

Sandhya Chintalapati, Xin Luo, and Su Ying Quek

Abstract Raman spectroscopy is a non-destructive and versatile method of identifying materials through their Raman "fingerprints". To this end, first principles calculations are essential to predict the Raman spectra of different materials. First principles calculations, together with parametrized models, can also give atomic scale insights into the origins of Raman shifts and Raman intensities, thus providing a guide to experiments. In this chapter, we will discuss some insights we have gained through our theoretical modeling of Raman spectra in 2D materials and their heterostructures. In particular, we show that surface and interface effects in 2D materials can give rise to observable changes in the Raman spectra. For example, we show that the formation of a surface in the 2D material leads to larger interatomic force constants at the surface, which results in experimentally observed anomalous frequency trends of the E_{2g}^1 mode in MoS_2, and the E_{2g}^1 and B_{2g}^1 modes in WSe_2. We further show that the Raman intensities of the interlayer shear modes in 2D layered materials can be simply predicted based on the stacking sequence.

8.1 Introduction

The 1930 Nobel prize was awarded to Sir C.V. Raman for his work on light scattering in materials, now known as Raman scattering. Raman discovered that inelastic scattering of light in materials resulted in a difference in energy between

S. Chintalapati
Centre for Advanced 2D Materials, National University of Singapore, Singapore, Singapore

X. Luo (✉)
Department of Applied Physics, The Hong Kong Polytechnic University, Hong Kong, People's Republic of China
e-mail: xin.xi.luo@polyu.edu.hk

S. Y. Quek (✉)
Centre for Advanced 2D Materials, National University of Singapore, Singapore, Singapore

Department of Physics, National University of Singapore, Singapore, Singapore
e-mail: phyqsy@nus.edu.sg

P.-H. Tan (ed.), *Raman Spectroscopy of Two-Dimensional Materials*, Springer Series in Materials Science 276, https://doi.org/10.1007/978-981-13-1828-3_8

the incident and scattered photons, that was characteristic of the material. The incident light can lose energy to fundamental excitations in the system, typically phonons, or can gain energy from the same excitations. These two processes lead to Stokes and anti-Stokes Raman spectra, respectively. The Raman scattering process obeys conservation of energy and momentum. Due to the very small wave vector of light compared to the size of the Brillouin Zone in periodic crystals, only gamma point (zone center) phonons can be detected in first-order Raman processes. If the incident light energy is not resonant with any electronic excitation, the intensity of the peaks in the Raman spectra depends on the degree to which the polarizability of the material changes during the scattering process. This is known as non-resonant Raman scattering and has been analyzed in detail using group theory, where phonon modes with certain symmetries are Raman inactive if the phonon displacements do not lead to any change in polarizability of the material. Resonant Raman scattering can result in a breakdown of this selection rule. The line-width of the Raman peaks gives information on phonon lifetimes.

Today, almost a century after the Nobel prize was awarded to Sir Raman, Raman scattering is still evolving as a powerful, non-destructive spectroscopic tool. Surface-enhanced Raman spectroscopy can be used to make sensitive chemical sensors for molecules. Raman spectroscopy is also commonly used to obtain detailed structural information about 2D materials, where direct imaging techniques can be inaccessible. Advances in experimental techniques have also allowed for the detection of ultralow frequency phonon modes. In this chapter, we show that by using theoretical modeling in combination with experiment, Raman spectroscopy in 2D materials not only can provide structural information, but can also provide physical insights into the nature and strength of individual bonds, at the atomic level. The contents of this chapter are organized as follows. In Sect. 8.2, we discuss the importance and nature of interlayer interactions in 2D layered materials. In Sect. 8.3, we show how these interlayer interactions can be probed using the ultralow frequency phonon modes in Raman spectra, in combination with ab initio calculations and force constants models. In Sect. 8.4, we show how the stacking sequence influences the frequency trends of observed interlayer shear modes, using a physically intuitive bond polarizability model for the Raman intensities. In Sect. 8.5, we show that the interlayer interactions discussed so far can lead to different force constants when a neighboring layer is removed, at the surface of a thin film. We demonstrate that this so-called surface effect can explain the well-known anomalous frequency trend of the E^1_{2g} mode in 2H transition metal dichalcogenides (TMDs), as well as other interesting features in Raman spectra. Finally, the chapter concludes with a discussion of the outlook of the potential of Raman spectroscopy in 2D materials from a theoretical perspective.

8.2 Interlayer Interactions in 2D Layered Materials

The stability of 2D materials generally requires that the energy needed to form a surface is small compared to the bulk cohesive energies. Typically, 2D materials are derived from materials that are layered in the bulk, with sufficiently weak interlayer interactions. Although the interlayer interactions are weak, they can have significant implications. For example, as we will discuss below, the absence of interlayer interactions going from bilayer to monolayer results in the well-known indirect-to-direct band gap transition for MoS_2 [1–3]. We will also show that the creation of a surface in thin films (i.e., the absence of interlayer interactions for the surface layer) results in observable changes in the frequencies of phonon modes [4].

Interlayer interactions can be detected experimentally via the interlayer force constants, i.e., the second derivative of the total energy with respect to interlayer displacements. Interlayer phonon modes and kinetic friction between layers are both measures of the interlayer force constants. Interlayer force constants are important quantities for tribology applications. Layered materials, such as graphite and transition metal dichalcogenides (TMDs) have long been used as solid lubricants, made possible by their small interlayer force constants [3, 5]. The fact that these interlayer force constants remain small even at high temperature makes these layered materials especially useful to replace liquid lubricants, which are ejected at high temperature. On the other hand, nanoscale friction has also generated significant attention in recent years [6], and it is interesting to consider how these interlayer force constants would change when going from 3D to 2D.

It is generally accepted in the literature that the interlayer interactions in many layered materials, such as graphite and MoS_2, are weak vdW interactions. While this is certainly true for graphite, the interlayer interactions in TMDs, and in other layered materials such as black phosphorus, are not simply of the weak vdW type. For example, in TMDs, Coulombic interactions play an important role in the interlayer interactions, because of the partially ionic intralayer bonds between the transition metal and the chalcogen. An important consequence is the indirect-to-direct band gap transition for MoS_2 [1–3]. The direct band gap at K remains constant as the thickness of MoS_2 changes because the valance and conduction band states at K are derived mainly from transition metal states. Except for monolayer MoS_2, the band gap of MoS_2 is indirect, with valence band maximum states at Γ and conduction band minimum states at K. However, the valence states at Γ are derived from both chalcogen p_z orbitals and metal d_z^2 orbitals. In the case of monolayer MoS_2, the absence of coulomb repulsion between p_z orbitals in neighboring layers stabilizes the valence band state at Γ, resulting in a larger indirect band gap from Γ to K, thus leading to an indirect to direct band gap transition [2, 3].

The presence of Coulombic interactions between layers is also evident in the anomalous Davydov splitting in 2H-MoS_2 [7, 8]. The phonons at Γ point in bulk 2H-MoS_2 can be decomposed into 12 irreducible representations, which can be further divided into six conjugate pairs. The phonon modes in each pair differs from its partner by an interlayer phase shift of 180°. For instance the vibrations of neighbouring layers of conjugate pair (E_{1u}^2,E_{2g}^1) are shown in Fig. 8.1. The difference in frequency between the two modes of a conjugate pair is called Davydov splitting. As the E_{2g}^1 (E_{1u}^2)mode has sulphur atoms on adjacent layers moving out of phase (in phase), one would expect that the additional bond stretch involving sulphur atoms in adjacent layers would result in the E_{2g}^1 mode having a higher frequency than the E_{1u}^2 mode. However, the opposite is found in experiment.

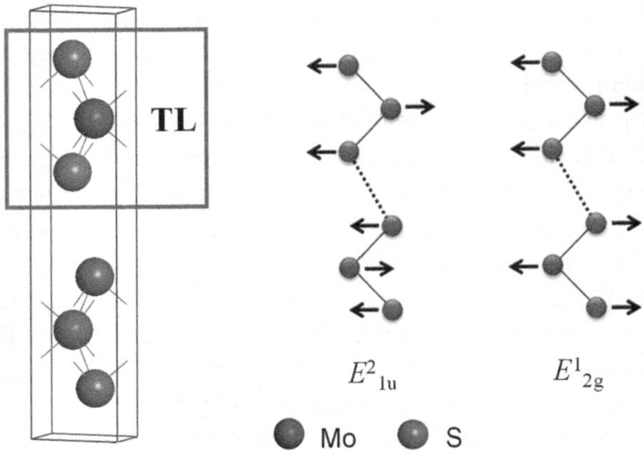

Fig. 8.1 Schematic diagram of MoS_2 with trilayer (TL) and the schematic illustration of the phonon displacements in the (E^2_{1u},E^1_{2g}) Davydov doublet (conjugate pair) in a unit cell of bulk 2H-MoS_2. (This figure is adapted with permissions from Ref. 4)

It turns out that, when second-nearest neighbour atomic interactions are included, taking into account the Coulomb interactions between a sulphur atom in one layer and the Mo atom in the adjacent layer, the experimentally observed Davydov splitting can be obtained. First principles calculations show that the force constant involving such a Coulomb interaction is ~40% larger in magnitude than that involving sulphur atoms in adjacent layers [4].

We remark that interlayer interactions of the vdW type are nonlocal. This means that the exchange correlation functional in density functional theory (DFT) calculations must be carefully tested. Even though the local density approximation (LDA) to the exchange correlation functional does not include any non-local van der Waals interactions, LDA predicts both the low [9, 10] and high-frequency [4, 11, 12] phonon modes in MoS_2, in good agreement with experiment. Surprisingly, the predicted frequencies obtained with van der Waals-DFT functionals, and with generalized gradient approximation with dispersion correction (GGA + D), have significantly larger discrepancies with experiment [4, 9]. While most treatments of vdW effects in DFT can successfully predict structural properties and binding energies, our results suggest that these treatments cannot properly describe the force constants in layered materials.

8.3 Probing Interlayer Interactions

In this section, we show how Raman spectroscopy, together with theoretical modeling, has uncovered key insights into the nature of interlayer interactions in layered materials. We first begin with a discussion of phonon modes in layered materials, since Raman spectroscopy is primarily a probe of phonon frequencies. Compared to a single layer, an N-layer system has N times as many phonon modes. Each phonon mode in a monolayer will evolve into N modes in N layers, with slightly different frequencies, depending on the relative phase shifts between layers. This phenomenon is a generalization of the Davydov doublets (conjugacy pairs)

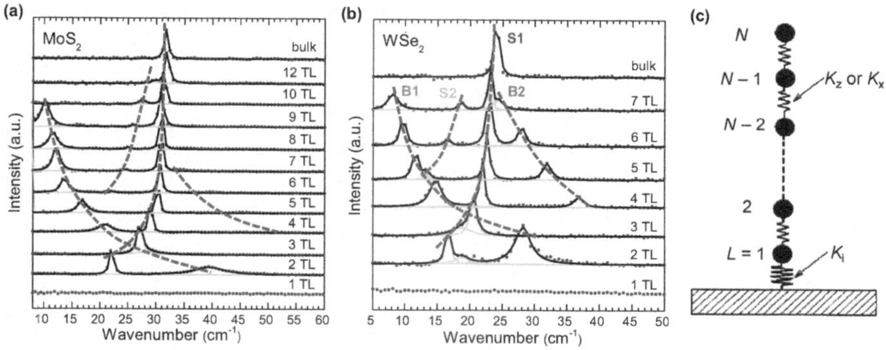

Fig. 8.2 Low-frequency Raman spectrum evolutions as a function of trilayer number in MoS_2 (**a**) and WSe_2 (**b**) measured along the $\bar{z}(xx)z$ polarization configuration. The dashed lines are fits to a linear chain model. (**c**) Illustration of the linear chain model with substrate. (Figures (**a** and **b**) are adapted with permissions from Ref. [9]. Figure (**c**) is adapted with permissions from Ref. [21])

described in Sect. 8.2. When all the atoms in each layer are moving in phase, the intralayer atomic force constants do not affect the phonon frequencies. Instead, only the interlayer force constants are important, and the corresponding phonon modes are called interlayer phonon modes. These interlayer modes are derived from the three acoustic modes of the monolayer. Thus, inclusive of the acoustic modes, there are 3N interlayer modes in an N layer system. Among these modes, there are $(N-1)$ interlayer breathing modes, and $2\,N-2$ interlayer shear modes. In the case of multi-layer graphene and 2H-TMDs, which have a hexagonal lattice, each shear mode is two-fold degenerate, leading to $(N-1)$ two-fold degenerate interlayer shear modes. Due to the weak interlayer interactions, the frequencies of these interlayer modes are generally in the ultralow frequency regime (<55 cm^{-1}) and can be difficult to distinguish from Rayleigh scattering. However, advancements in experimental techniques have made it possible to observe these low frequency modes [9, 13–15]. As discussed in Sect. 8.4 below, typically, only one or two of the $(N-1)$ breathing or shear modes can be observed, leading to characteristic low frequency Raman peaks that shift in frequency as the thickness of the film is changed. Interestingly, our first principles calculations show that the interlayer modes with the largest Raman intensity are the lowest frequency breathing modes, and the lowest or highest frequency shear modes. An example is shown in Fig. 8.2a, b, where two shear modes and two breathing modes are observed in few-layer MoS_2 and WSe_2. In each case, one predominant shear mode and one predominant breathing mode are clearly present in the Raman spectrum.

Before going into the details of the shear and breathing modes, we highlight the link between the Raman frequencies and the interlayer force constants discussed in the previous section. It has been found that the experimental Raman frequencies can be fitted remarkably well to a simple linear chain model (Fig. 8.2c) [9, 15], that considers each layer as a single entity (a ball in Fig. 8.2c), connected to adjacent, nearest neighbor layers by springs with force constants corresponding to effective

Table 8.1 Force constants per unit area derived from fitting of the linear chain model

		DFT (LDA)	Experiment
MoS_2	$K_z(10^{19}\ Nm^{-3})$	9.26	8.62
	$K_x(10^{19}\ Nm^{-3})$	3.51	2.72
WSe_2	$K_z(10^{19}\ Nm^{-3})$	8.38	8.63
	$K_x(10^{19}\ Nm^{-3})$	3.41	3.07
Few layer graphene	$K_z(10^{19}\ Nm^{-3})$	9.63	9.86
	$K_x(10^{19}\ Nm^{-3})$	1.40	1.28
Black phosphorus	$K_z(10^{19}\ Nm^{-3})$	14.1	12.7

The numbers in this table are obtained from Ref. [9] (TMD materials), Ref. [17] (DFT, FLG), Ref. [15] (K_x, exp., FLG), Ref. [18] (K_z, exp., FLG) and Ref. [19] (black phosphorus)

interlayer force constants in the material. Results from such a fit are shown by the dashed lines in Fig. 8.2a, b. Out-of-plane force constants K_z would give rise to the interlayer breathing modes, while in-plane force constants K_x would give rise to the shear modes. Within such a linear chain model, and ignoring the substrate effect (*i.e.*, $K_i = 0$ N/m^3), the phonon frequencies (unit: cm^{-1}) can be represented as [15]:

$$\omega_\alpha = \sqrt{\frac{K}{2\mu\pi^2c^2}\left(1 - \cos\left(\frac{(\alpha-1)\pi}{N}\right)\right)}, \tag{8.1}$$

with N being the number of layers, K being the force constant per unit area, μ the mass per unit area (*e.g.*, $\mu = 8.2 \times 10^{-6}$ kg/m^2 for Bi_2Te_3, $\mu = 3.1 \times 10^{-6}$ kg/m^2 for MoS_2), and c is the speed of light. α is a parameter that determines if the corresponding mode is a highest frequency mode, second highest frequency mode, etc., and $\alpha = 1$ corresponds to the acoustic mode. The fact that this simple linear chain model describes the experimental observations so well tells us several important things about the interlayer interactions. Firstly, nearest neighbour interactions are sufficient, despite the nonlocal nature of the interlayer interactions. This apparent contradiction stems from the fact that the force constants in the model are really *effective* force constants, and not interatomic force constants. Secondly, the effective interlayer force constants do not depend on N (allowing for a single K that does not change with N). In fact, our calculations explicitly show that these effective interlayer force constants are also the same in thin film as in the bulk [9]. Thirdly, it also appears that interactions with the substrate do not change even the low interlayer mode frequencies. Indeed, experiments have shown that the Raman spectra taken from substrate-supported and suspended few-layered MoS_2 and WSe_2 show almost no difference, indicating that there is negligible interaction between the substrate and sample in those instances [9, 16]. Using the linear chain model, it is possible to derive the effective interlayer force constants directly from Raman spectra. A comparison between interlayer force constants derived thus from experimental Raman frequencies, and those from DFT (LDA)-calculated frequencies, show that the agreement is reasonably good (Table 8.1).

The effect of interactions with the substrate can also be considered in the linear chain model by adding a force constant K_i, describing the interaction between the substrate and the bottommost layer of the film (Fig. 8.2c) [20, 21]. Although this is not important for MoS_2 and WSe_2 [9, 16], a non-zero force constant K_i was required to explain the low frequency interface (I) mode in CVD-grown Bi_2Te_3 nanoplates on SiO_2/Si. Through the linear chain model with $K_i/K_z = 0.5$, the calculated frequency trend is in agreement with experimental observations. Furthermore, the I peak is relatively more sensitive to substrate, which supports its origin of a substrate introduced I mode [21].

We now discuss the details of the shear and breathing modes, which are made possible using DFT calculations and group theory analysis [9]. The MoS_2 and WSe_2 considered in this work are of the most common 2H type. The bulk form belongs to the D_{6h}^4 (P63/mmc) space group [22]. The primitive cell in the bulk consists of two trilayers (TLs) as shown in Fig. 8.1, resulting in 18 Brillouin zone center (Γ) phonons. Based on group theory, the irreducible representations of the phonon modes can be classified as $\Gamma_{bulk} = A_{1g} + 2A_{2u} + B_{1u} + 2B_{2g} + E_{1g} + 2E_{1u} + E_{2u} + 2E_{2g}$, among which $2E_{2g}$, E_{1g}, and A_{1g} are Raman active modes. The two fold degenerate E_{2g} modes represents the in-plane vibrations (shear modes) and A mode represents the out-of-plane vibrations (breathing modes). In N layer films, the symmetry depends on whether N is even or odd. Even-layer films have inversion symmetry and belong to the D_{3d}^3(P$\bar{3}$m1) space group, and the corresponding irreducible representations for the zone center phonons are: $\Gamma_{even} = (3N/2)(A_{1g} + A_{2u} + E_g + E_u)$, $N = 2, 4, 6$, etc. Odd-layer films have no inversion symmetry and belong to the D_{3h}^1 (P$\bar{6}$m2) space group, and $\Gamma_{odd} = ((3N - 1)/2)(A'_1 + E'') + ((3N + 1)/2)(A''_2 + E')$, $N = 1, 3, 5, \ldots$. In the bulk, there are two interlayer optical phonon modes; that is, the Raman active E_{2g}^2 (shear mode) and the optically inactive B_{2g}^2 (breathing mode), where two adjacent TLs vibrate out-of-phase in-plane and out-of plane, respectively (Fig. 8.3a). When N is odd, the interlayer breathing modes are either Raman-active (A'_1) or IR-active (A''_2), while the interlayer shear modes are either Raman-active (E'') or both Raman-active and IR-active (E'). When N is even, the interlayer shear modes are either Raman-active (E_g) or IR-active (E_u) and the interlayer breathing modes are either Raman-active (A_{1g}) or IR-active (A_{2u}).

DFT calculations revealed that experimentally observed S1 peak in the NTL systems corresponds to the highest frequency shear mode, while the peak B1 corresponds to the lowest frequency interlayer out-of-plane breathing mode. The normal vibrational displacements of the S1 and B1 modes in the NTL crystals ($2 <= N <= 9$) are schematically displayed in Fig. 8.3b, c. In the S1 mode, adjacent TLs are distinctly out-of phase, while in the B1 mode, the TLs can be divided into two groups, with TLs in each group being approximately in-phase with one another. This picture is consistent with the fact that S1 is the highest frequency shear mode while B1 is the lowest frequency-breathing mode, because the frequency of an interlayer phonon mode is larger if the adjacent TLs have more out-of-phase displacement. Furthermore, in the case of S1, a larger N implies more out-of-phase displacement between adjacent layers, leading to higher frequencies for larger N, as

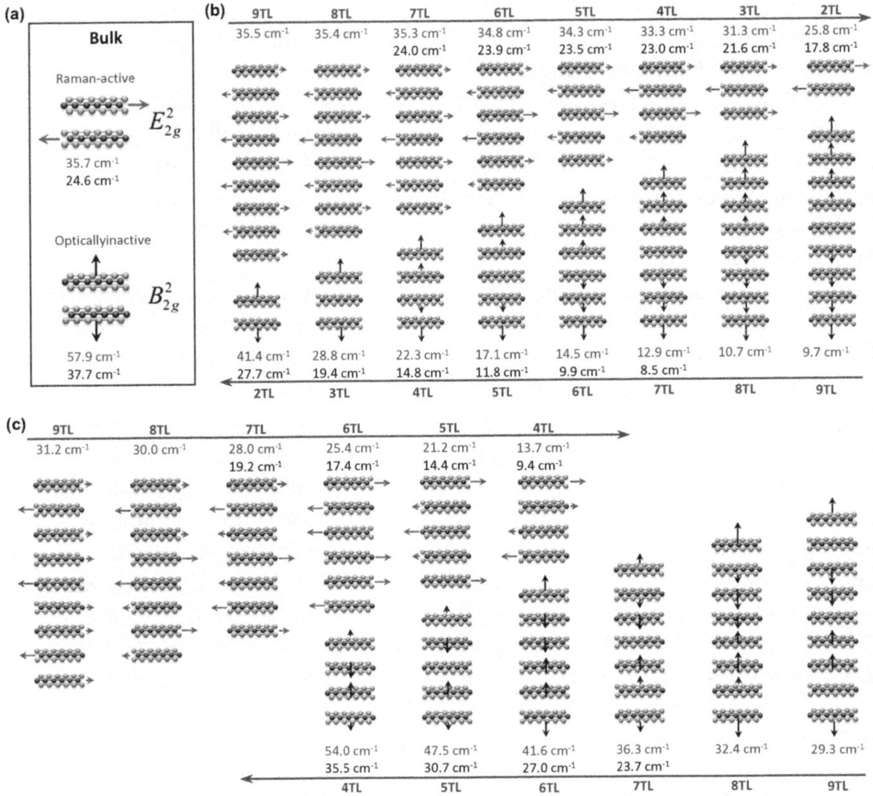

Fig. 8.3 Vibrational normal modes of the interlayer shear and breathing modes in MoS_2/WSe_2. (**a**) The vibrational normal modes of the interlayer shear (E_{2g}^2) and breathing modes (B_{2g}^2) in bulk 2H-MoS_2/WSe2. The shear mode is Raman-active, and the breathing mode is optically inactive. (**b**) Vibrational normal modes of the highest frequency shear mode S1 (top) and the lowest frequency breathing mode B1 (bottom) from 2TL to 9TL. (**c**) Vibrational normal modes of the minor shear mode S2 (top) and the minor breathing mode B2 (bottom) from 4TL to 9TL. The arrows indicate the direction of motion of the whole TL, and the length of the arrows represents the magnitude. The denoted frequencies are results of the first principles calculations for both MoS_2 (in gray) and WSe_2 (in black). (This figure is adapted with permissions from Ref. [9])

shown also in Figs. 8.2b and 8.3b. In the case of B1, a larger N implies a greater proportion of approximately in phase displacement, leading to lower frequencies. Similar arguments can be made for S2 and B2, which are, respectively, the shear modes with the third highest frequencies, and breathing modes with the third lowest frequencies (Fig. 8.3c). Comparing the S1 mode with the bulk E_{2g}^2 mode, it is clear that the S1 mode evolves to the Raman-active E_{2g}^2mode in the bulk, although the vibration amplitudes of the surface layers are slightly smaller than for layers in the middle of the few-layer materials. The reason for this difference is that the TLs at the surface have only one nearest neighbour TL, in contrast to TLs in the bulk that have

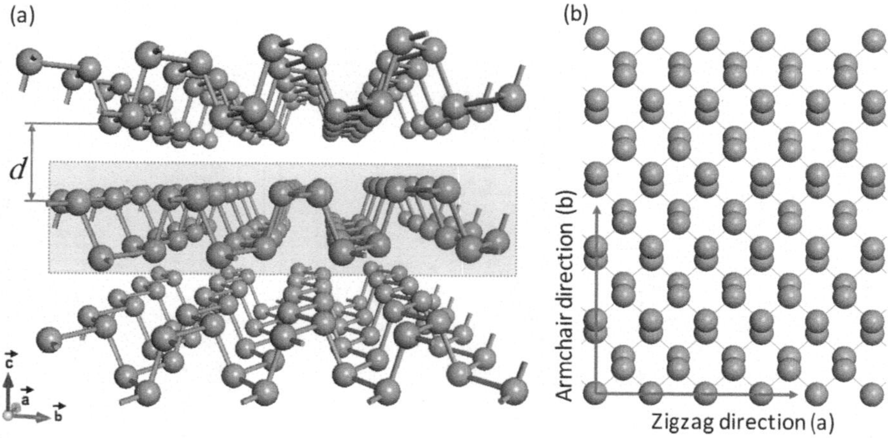

Fig. 8.4 (**a**) Structural illustration of black phosphorus (BP). Phosphorene is a single layer of the puckered sheet as depicted by the dash rectangle. (**b**) Top view of phosphorene, with the zigzag and armchair directions indicated by arrows. (This figure is adapted with permissions from Ref. [19])

two nearest neighbour layers. In this way, the first principles calculations based on DFT successfully revealed that the experimentally observed S1 and S2 modes are found to be interlayer shear modes with the highest and third highest frequencies for each *N*TL system, while B1 and B2 modes are interlayer breathing modes with the lowest and third lowest frequencies. This result is consistent with the fact that $\alpha = N, N-2, 2, 4$ were used in the linear chain model to obtain the frequencies for S1, S2, B1, and B2 modes, respectively.

The interlayer modes in another 2D material, few layer black phosphorous (BP) (Fig. 8.4), have also been studied with DFT calculations and Raman spectroscopy [19]. Similar to the case of 2H-MoS_2 and WSe_2, DFT calculations showed that the lowest frequency interlayer breathing mode had the largest Raman intensity ($A^3{}_g$ in Fig. 8.5). However, interestingly, this interlayer breathing mode exhibits a significantly larger frequency change with film thickness, compared to MoS_2 and WSe_2. These predictions were confirmed in experiment [19]. Using eq. 8.1 (linear chain model), this implies that the effective interlayer force constants are significantly larger in black phosphorus, which was confirmed through explicit calculations (Table 8.1).

These Raman studies drew attention to the unusual nature of the interlayer interactions in BP. Unlike TMDs, BP only consists of one type of atom – phosphorus. Thus, interatomic Coulomb interactions is unlikely to affect the phonon frequencies in the same way as in the TMDs. However, for BP, only three out of five valence electrons are used to form intralayer bonds, leaving an electron lone pair directed into the interlayer region. Based on DFT calculations using LDA and vdW functionals, it was found that these lone pairs interacted with one another to form covalent interactions between layers [19]. However, quantum Monte Carlo calculations indicated that the electrons are pushed by Pauli repulsion into the

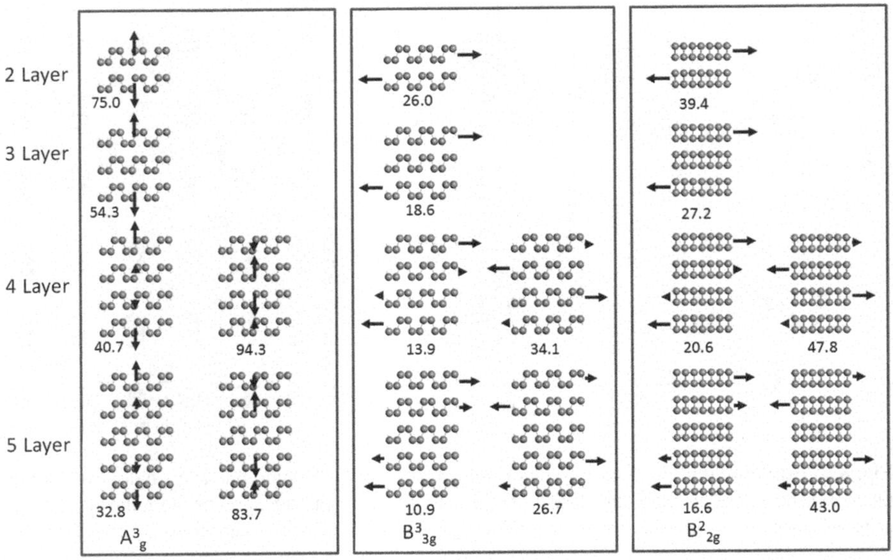

Fig. 8.5 Atomic displacements for the Raman active interlayer vibration modes in 2–5 layer BP; the frequency labeled below the displacement pattern is in the unit of cm^{-1}. (This figure is adapted with permissions from Ref. [19])

intralayer regions, resulting in a different type of interlayer interaction [23]. On the other hand, we note that based on the quantum Monte Carlo results, the derived force constants underestimate the experimental interlayer force constant by 28%, in contrast to the LDA frequencies which overestimate the experimental force constant by 14% [24]. The reason for these differences is at present unknown, but regardless of the exact nature of the interlayer interactions, it is interesting that these interlayer interactions are significant, and also anisotropic in nature. These large anisotropic interlayer interactions play an essential role in the formation of a Dirac cone in few-layer black phosphorous when the band gap is closed, by an electric field [25] or by Li adsorption [26]. Likewise, it was suggested that formation of topological phases in bulk BP under pressure may be due to large interlayer interactions [27].

8.4 Stacking Sequence Probed by Raman Spectroscopy

Besides providing insights into interlayer interactions, the ultralow frequency Raman spectrum can also give information on the stacking sequence in 2D layered materials, i.e. how the different layers in a 2D material are stacked relative to one another. The properties of layered materials can be very sensitive to the stacking sequence. For instance, AB (Bernal)-stacked three-layer graphene (3LG) is a semimetal, whereas ABC-stacked 3LG has a band gap with symmetry-breaking ground states [28]. Broken symmetry quantum hall states have been observed in dual-gated AB-stacked 3LG [29]. On the other hand, circularly polarized photoluminescence measurements show an enhancement of valley polarization in non-centrosymmetric ABC-stacked 3R few layered MoS_2 compared to the more common AB-stacked 2H MoS_2 [30]. Due to limited resolution, it can be difficult to unambiguously identify the stacking sequence in 2D layered materials using imaging techniques. In the case of few-layer graphene, AB and ABC stacking were identified by different line shapes in their Raman spectra [31, 32]. However, this method is very specific to few-layer graphene.

During our calculations of the Raman spectra for interlayer modes in different 2D layered materials, we noticed a very interesting phenomenon. Earlier, we had mentioned that there would typically be one interlayer shear mode and one interlayer breathing mode with larger Raman intensity than the others, and this would correspond to a particular α in the linear chain model (Eq. 8.1), such as the lowest frequency or highest frequency shear/breathing mode. We noticed that curiously, the highest frequency shear mode had the largest Raman intensity in 2H-MoS_2, 2H-WSe_2, and Bernal-stacked few-layer graphene, but it was the lowest frequency shear mode that had the largest Raman intensity in Bi_2Se_3 and Bi_2Te_3 [15, 21]. This is despite the fact that Bi_2Se_3 and Bi_2Te_3 share the same space group ($D^3{}_{3d}$) as even-layered Bernal-stacked graphene. The space group determines the Raman activity, but of all the Raman active modes, the relative Raman intensities (in non-resonant scattering) is determined by how much the polarizability of the material changes during the scattering process. At this point, we had a strong hunch that the reason for the stark contrast between Bi_2Se_3/Bi_2Te_3 and 2H-MoS_2/2H-WSe_2/Bernal-stacked few-layer graphene was that the former are ABC-stacked, while the latter are AB-stacked, and this stacking difference was very likely to strongly affect the change in polarizability during the Raman process. In Fig. 8.6, we show a schematic of the phonon displacements for the interlayer shear mode with the largest Raman intensity in AB- and ABC-stacked systems. It is plausible to imagine that the polarizability of AB-stacked systems changes the most when adjacent layers are moving out-of-phase with one another, as shown in Fig. 8.6a. At the same time, the polarizability of ABC-stacked systems will change the most when the whole film is stretched like a stack of cards, as shown in Fig. 8.6b. Such displacements in turn correspond respectively to the highest frequency (maximally out-of-phase) and lowest frequency (maximally in-phase) interlayer shear modes. We tested our hypothesis of the connection between stacking sequence and Raman intensity through explicit calculations of the Raman spectra of ABC-stacked few-layer graphene, where we found that, like ABC-stacked Bi_2Se_3 and Bi_2Te_3, and in contrast to AB-stacked few-layer graphene, the lowest frequency shear mode had the largest Raman intensity [17].

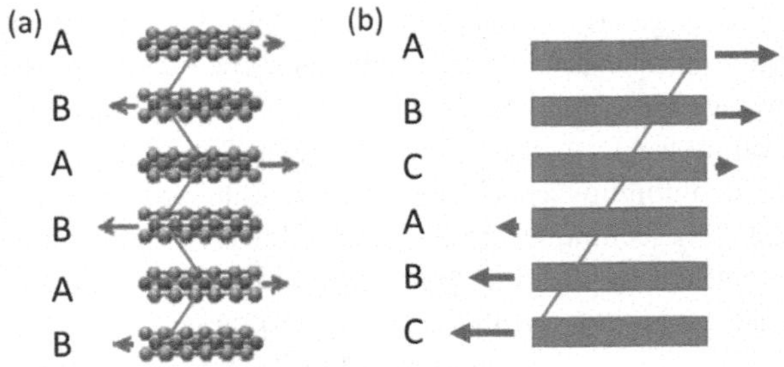

Fig. 8.6 Schematic illustration of stacking order and displacement of layers for AB- and ABC-stacked systems

It was necessary to quantify the above ideas. To do so, we used a bond polarizability model for the interlayer shear modes [17]. The bond polarizability model has previously been used to study the Raman intensity of fullerenes, graphene ribbons and boron nitride nanotubes [33–35]. In this model, the polarizability of the material is written as a sum of individual bond polarizabilities, which are assumed to be independent of the chemical environment, i.e. the polarizability tensor is approximated as:

$$P_{\alpha\beta} = \frac{1}{2}\sum_{l,B}\left\{\frac{1}{3}\left(\alpha_{\parallel} + 2\alpha_{\perp}\right)\delta_{\alpha\beta} + \left(\alpha_{\parallel} - \alpha_{\perp}\right)\left(R_{\alpha}R_{\beta} - \frac{1}{3}\delta_{\alpha\beta}\right)\right\} \quad (8.2)$$

where $\mathbf{R}(l, B)$ is the bond vector from atom l to one of its nearest neighbor atoms l' connected by bond B, the vector being normalized to unity. $\alpha_{\parallel}$ and $\alpha_{\perp}$ are the static longitudinal and perpendicular bond polarizability, respectively. Within the Placzek approximation, the non-resonant Raman intensity for phonon mode k, I^k, can be written as a function of the polarizability as follows:

$$I^k \propto \frac{(n_k + 1)}{\omega_k}\left|\sum_{\alpha\beta}\eta_{\alpha}\eta'_{\beta}P_{\alpha\beta,k}\right|^2 \quad (8.3)$$

where η and η' are the unit vectors for the polarization of the incident and scattered light, ω_k and $n_k = [\exp(\hbar\omega_k/k_BT) - 1]^{-1}$ are the frequency and the Boltzmann distribution function of phonon mode k, respectively. $P_{\alpha\beta,k} = \sum_{l\gamma}\left[\frac{\partial P_{\alpha\beta}}{\partial u_{l\gamma}}\right]_0 \chi^k_{l\gamma}$ is the derivative of the electronic polarizability tensor $P_{\alpha\beta}$ with respect to the atomic displacements in mode k. $u_{l\gamma}$ is the displacement of lth atom in the γ direction for normal mode k and $\chi^k_{l\gamma}$ is the γth (x, y, or z) component of the eigenvector of phonon mode k. To compute the Raman intensities within the bond polarizability model, we use Eq. 8.3 with Eq. 8.2 as the expression for the polarizability. Since most of the experiments are done in the back scattering geometry, we derived the derivative of the bond polarizability for the widely used $\overline{z}(xx)z$ configuration:

$$P_{xx,k} = -\sum_{lB}$$

$$\left\{\frac{\alpha'_{\parallel} + 2\alpha'_{\perp}}{3}\mathbf{R}_0 \cdot \vec{\chi}_l^k + \left(\alpha'_{\parallel} - \alpha'_{\perp}\right) R_{0x}{}^2\mathbf{R}_0 \cdot \vec{\chi}_l^k - \frac{1}{3}\left(\alpha'_{\parallel} - \alpha'_{\perp}\right)\mathbf{R}_0 \cdot\right.$$

$$\left.\vec{\chi}_l^k + 2R_{0x}\chi_{lx}^k \frac{\alpha_{\parallel} - \alpha_{\perp}}{R_0} - 2\frac{\alpha_{\parallel} - \alpha_{\perp}}{R_0} R_{0x}{}^2\mathbf{R}_0 \cdot \vec{\chi}_l^k\right\} \tag{8.4}$$

where R_0 is the bond length at equilibrium, and R_{0x} is the x component of $\mathbf{R_0}$, which is the bond vector normalized to unity. The "primes" on the bond polarizabilities refer to radial derivatives with respect to the bond lengths, and χ_l^k is the eigenvector of the phonon mode k. While this expression looks complicated, for the purposes of our study, the bond polarizabilities and their derivatives can be treated as unknown constants. Furthermore, the presence of symmetry renders many of the terms zero. For example, in systems with hexagonal symmetry, the expression reduces to [17]

$$P_{xx,k} \propto \sum_{lB} R_{0x}^3 \vec{\chi}_{lx}^k. \tag{8.5}$$

Let us consider AB- and ABC-stacked 3-layer graphene (3LG) as an example. There are two shear modes in 3LG as shown in Fig. 8.7. The Raman intensities derived by applying Eq. 8.5 are in good agreement with our DFT calculations [17] as well as the experimental observation that the highest and lowest frequency shear mode is observed, resepectively, in AB and ABC stacked 3LG [36]. Thus, the bond polarizability model quantifies and confirms our hypothesis in Fig. 8.6, that relates the Raman intensities of interlayer shear modes with the stacking sequence. We note that in order for our hypothesis and the bond polarizability model to be applicable to interlayer modes, there should be some directional bonding between layers that is sensitive to the stacking sequence. Our DFT calculations show that indeed, there are weak directional covalent bonds between the layers (Fig. 8.8). The charge accumulated in these weak covalent bonds can produce dipole moments in the presence of an applied electric field imposed by incident radiation, resulting in Raman intensities that depend on the stacking order. We note that the experimentally observed sensitivity of Raman intensities to the stacking order in fact confirms the presence of weak covalency in the interlayer interactions.

We have shown above that the non-resonant Raman intensity of the interlayer shear modes depends directly on the stacking sequence of the layered material. For AB-stacked materials, the highest frequency shear mode is observed, while for ABC-stacked materials, the lowest frequency shear mode is observed. This is very useful because these two different shear modes have different frequency

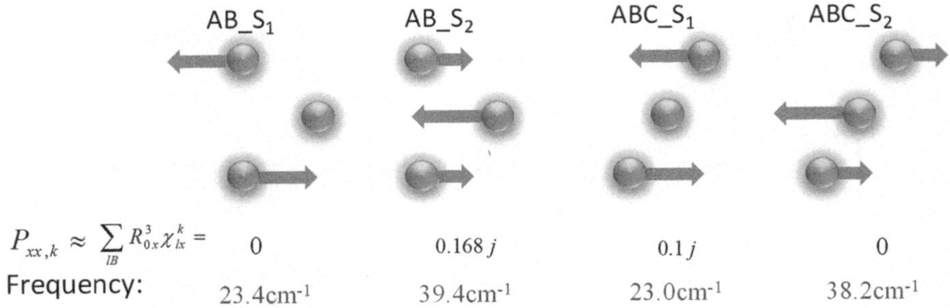

Fig. 8.7 The atomic displacements of the two shear modes in AB- and ABC-stacked 3LG. The Raman intensities derived from the bond polarizability are also given, where j is a constant related with the bond length

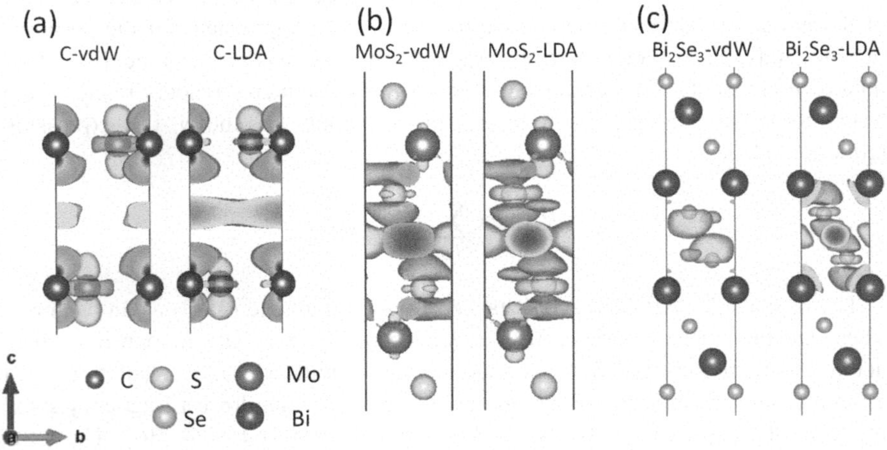

Fig. 8.8 Charge accumulation in the interlayer regions of few-layer (**a**) graphene, (**b**) MoS_2 and (**c**) Bi_2Se_3, as computed using DFT calculations with vdW-functional ((vdW-DF) with Cooper's exchange) and with the LDA exchange correlation functional. (This figure is adapted with permissions from Ref. [17])

trends as a function of film thickness. As the number of layer decreases, the highest frequency shear mode (prominent in AB-stacked materials) red shifts, while the lowest frequency shear mode (prominent in ABC-stacked materials) blue shifts. In this way, the ultralow frequency Raman spectra can provide a clear signature of the stacking order in layered materials. We note that the stacking order does not significantly change the interlayer mode frequencies or the Raman intensities of breathing modes. Another advantage of the bond polarizability model is that it can be easily applied to more complicated stacked sequences, such as ABACA stacking in five-layer $MoSe_2$ [37].

8.5 Surface Effects Probed by Raman Spectroscopy

So far, we have discussed how the interlayer interactions in 2D layered materials are not just of the weak vdW type. We have shown how the effective interlayer force constants can be probed using Raman spectroscopy, and also, that slight covalency in the interlayer interactions results in a remarkable effect on the frequency trends of the interlayer shear modes. In this section, we discuss what happens when an adjacent layer is removed at a surface of a thin film. Are the interlayer interactions important enough, that the absence of an adjacent layer creates observable effects? Before we started this project, we had thought naively that since 2D layered materials can be easily exfoliated, even with scotch tape, from bulk layered materials, the creation of a surface in a thin film should not have significant observable effects. This would be in contrast to the creation of a silicon surface, where dramatic atomic rearrangements take place to make up for the loss of neighboring atoms. However, it turns out that in fact, the surface of a thin film, even in 2D layered materials, is not the same as in the bulk. Specifically, we find that larger interatomic force constants are present at the surface of layered TMDs. We refer to this as the surface effect. In this section, we will discuss how the surface effect leads to several interesting features in Raman spectra.

Our discussion begins with trying to understand the so-called anomalous frequency trend in the E_{2g}^1 mode of few layer 2H-MoS_2 [4]. With decreasing thickness, the A_{1g} mode of MoS_2 red shifts to lower frequencies while the E_{2g}^1 mode blue shifts toward higher frequencies (Fig. 8.9). In both A_{1g} and E_{2g}^1 modes, the sulphur atoms in adjacent layers are moving out of phase. Thus, one would expect that for fewer layers, there will be less friction, and accordingly, lower frequencies. We call this the thickness effect, and the frequency shift of the A_{1g} mode is consistent with the thickness effect. However, the frequency trend of the E_{2g}^1 mode is not consistent with this thickness effect, and was thus called "anomalous". In order to understand this anomalous trend, we examine the interatomic force constants up to the second nearest neighbour, as shown in Fig. 8.10. We highlight two trends from this figure. Firstly, the force constants between atomic pairs at or close to the surface are different from those in the interior of thin films. Except for the second nearest neighbour compressive force constants, the force constants at the surface are larger in magnitude than that in the interior of the thin films. We call this the surface effect. For instance, in the six TL system, the force constant $k^S{}_{Mo\text{-}S\text{-}1}$ of the Mo-S bond at the surface is ~2.6% larger than that in the middle. This observation is in fact consistent with a previous experimental finding that the Mo-S bonds at the surface of MoS_2 are slightly smaller than those in the bulk [38]. Secondly, in some cases, the force constants in the bulk are different from those in the interior of the thin films. This difference could be due to a difference in dielectric screening between the bulk and the thin films, and we call this the screening effect.

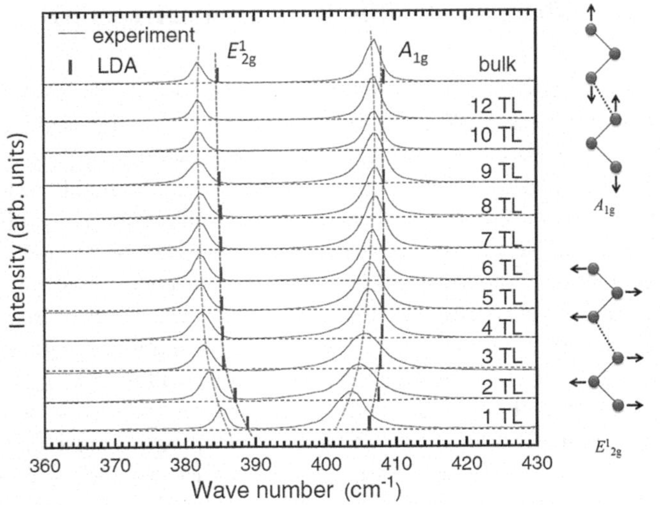

Fig. 8.9 Experimental and LDA-calculated frequencies for the E^1_{2g} and A_{1g} modes in few-TL and bulk MoS_2. The E^1_{2g} modes blue shift and the A_{1g} modes red shift as the thickness of the MoS_2 film decreases. (This figure is adapted with permissions from Ref. [4])

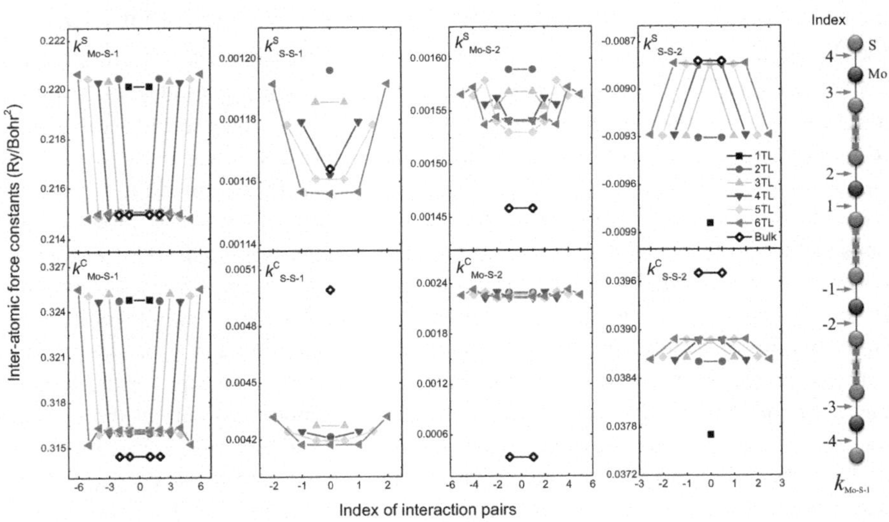

Fig. 8.10 Spatial variations of the magnitude of LDA interatomic force constants in bulk and N-TL MoS_2. The legend is in the top right panel. The superscript S or C in the force constant refers to the in-plane "shear" or out-of-plane "compressive" force constant, respectively. The subscript refers to the elements of the atoms in the interatomic pair for which the force constant is computed, followed by a number indicating whether these atoms are first-nearest [1] or second-nearest [2] neighbours. The index of interaction pairs (horizontal axis) is counted from the bottom TL to the surface TL, with the middle set to 0. An illustration of the index of $k_{Mo\text{-}S\text{-}1}$ in 4-TL MoS_2 in the linear chain representation is shown on the right-hand side. For example, an index of -4 refers to the force constant between the bottommost S and Mo atoms in 4-TL MoS_2. (This figure is adapted with permissions from Ref. [4])

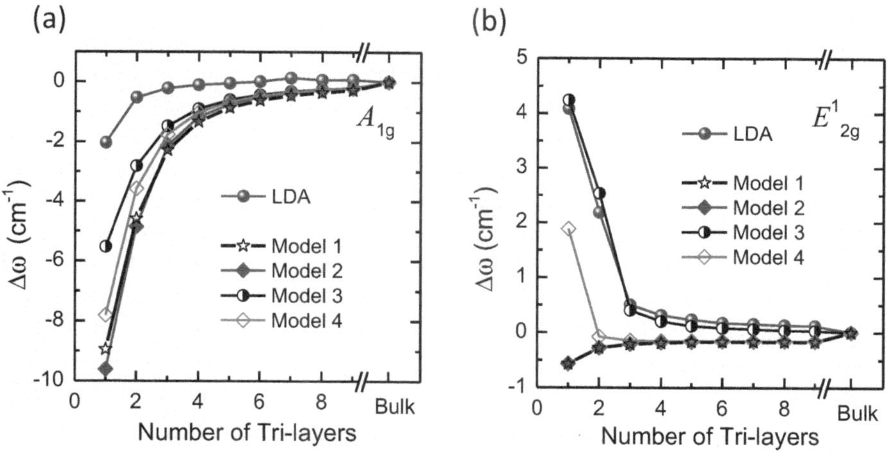

Fig. 8.11 The force constants model results for A_{1g} and E^1_{2g} modes are shown in (**a**) and (**b**). Only the frequency differences with respect to their corresponding bulk modes are displayed. Model 1 uses parameters from the bulk, Model 2 uses parameters from the interior of thin films, Model 3 is the same as Model 2 but taking into account modified surface force constants on both surfaces, while Model 4 is like Model 3 but with surface force constants on one surface only. (This figure is adapted with permissions from Ref. [4])

To distinguish the different physical effects (thickness, surface and screening effects), we computed the frequencies of the E^1_{2g} and A_{1g} modes using three different force constants models. The models are: model 1 – force constants from the bulk (thickness effect only); model 2 – force constants from the interior of the thin films (thickness effect and screening effect only); model 3 – force constants from Fig. 8.10, with different force constants for the interior and the surface of the thin films (thickness, screening and surface effects); model 4 – same as model 3 but with surface force constants on one surface only [4]. The results are compared to the DFT LDA frequencies and shown in Fig. 8.11. All the models give the correct red shift of the A_{1g} mode with decreasing thickness, just like the DFT and experimental result. However, for the E^1_{2g} mode, only models 3 and 4 give the correct blue shift with decreasing thickness. These results clearly demonstrate that the anomalous frequency trend for the E^1_{2g} mode arises from the surface effect. We note that the surface effect is more apparent for the E^1_{2g} mode than for the A_{1g} mode, partly because the thickness effect (model 1) has a smaller effect on the frequency trend in the E^1_{2g} mode than in the A_{1g} mode, due to the smaller shear force constants than compressive force constants.

Thus, we have shown that the absence of interlayer interactions at the surface lead to larger interatomic force constants at the surface, which in turn result in an anomalous frequency trend in the E^1_{2g} mode. Due to the opposite frequency trends of the E^1_{2g} and A_{1g} modes, the frequency difference of the two modes can be used to determine the thickness of thin film layer (<6 L). However, this frequency difference is also slightly sensitive to the way the few-layer MoS_2 sample is prepared. A frequency difference of 19 cm^{-1} is obtained for the mechanically exfoliated monolayer MoS_2, while a slightly smaller value of 16 cm^{-1} was reported for the micromechanically exfoliated monolayer MoS_2 and a larger value of 20 cm^{-1} was reported in the laser- and thermally thinned monolayer. The discrepancy mainly arose from the different frequencies of the E^1_{2g} mode across the different samples. The discrepancy reported here is not surprising as these different samples can have different surface environments, giving slightly different surface force constants.

The red shift of the A_{1g} mode and blue shift of the E^1_{2g} mode are also observed in other 2H stacked group VI TMDs, including WS_2, WSe_2, $MoSe_2$ and $MoTe_2$ [2, 39–43]. The surface effect can also affect the frequency trend of the ${B_{2g}}^1$ mode, that is observed in N layer ($N > 2$) $MoTe_2$ and WSe_2 due to the lower symmetry.[16, 43] The adjacent planes are moving out of phase in B^1_{2g}, very similar to that of the E^1_{2g} but in the out of plane direction. Both the E^1_{2g} and B^1_{2g} modes exhibit a blue shift frequency trend with decreasing thickness. However, in contrast to the E^1_{2g} mode, the magnitude of the atomic displacement at the surface is larger than that in the interior for the B^1_{2g} mode. Thus, the frequency trend of the B^1_{2g} mode does not depend significantly on the film thickness. On the other hand, the frequency of the E^1_{2g} mode changes significantly going from 3 L to 1 L due to the larger proportion of interatomic vibrations at the surface for 1 L and 2 L films.

Froehlicher et al. [44] have also used the same force constants model to perform a detailed investigation of the optical phonon modes in few layer 2H-$MoTe_2$. This model was able to reproduce the experimental results reasonably well. In particular, it was found that the surface effect was important for explaining very small anomalous Davydov splittings for the high frequency modes.

8.6 Conclusion and Outlook

In the above discussion, we have shown that DFT calculations, together with remarkably simple theoretical models, and experiment, can provide detailed atomic-scale information about the interlayer and surface effects in 2D layered materials. We show that the effective interlayer force constants can be derived using a force constants model, while the stacking sequence can be identified by the frequency trends of interlayer shear modes. On the other hand, larger force constants at the surface can result in anomalous frequency trends. Several of these findings relied on the ability to detect ultralow frequency phonon modes, which can still be challenging due to the need to filter away the large peak from Rayleigh (elastic) scattering. The development of cheaper and better light filters will thus help make this experimental measurement accessible to a larger research community. At the same time, even phonon modes with higher frequencies can be used to tease out information on interatomic force constants, such as in the analysis of the surface effect in Sect. 8.5.

Looking forward, one major area that calls for further theoretical input is the study of resonant Raman excitations [45]. These are typically very computational expensive calculations, and the development of simpler, yet predictive, models, would be very useful. Furthermore, it would be interesting to apply the force constants model and bond polarizability model to more complicated interfaces involving twist angles, such as in the reported twisted bilayer $MoS_2/MoSe_2$ structures [46, 47]. Such a study would allow one to obtain the effective interlayer force constants for different twist angles. The interaction strength between the substrate and the 2D material can also be assessed from the ultralow frequency modes [21]. Interestingly, most reports do not find a large influence of the substrate on the ultralow frequency modes – a point that is worthy of further investigation.

It would also be interesting to study the validity of the phonon confinement effect [48] in 2D layered materials. The phonon confinement effect uses information about the phonon band structure from the corresponding 3D material, tracing a particular phonon in the 2D film to a corresponding phonon band in the 3D material, but allowing for wave vectors away from the Gamma point. This would correspond to the different layers in the 2D film having out-of-phase displacements. The phonon confinement effect has been used widely in the field of silicon nanoparticles [48] and has also been used to explain the frequency trends in Bi_2Se_3 thin films [49]. In our recent review article [24], however, we showed that the phonon confinement effect is not applicable for understanding the anomalous frequency trend of the E^1_{2g} mode in MoS_2. Likewise, in nanoclusters or ribbons of 2D materials, edge-activated modes have been reported in monolayer TMD materials [13, 50–52], and have been attributed to corresponding the ZA phonon at the M point due to the breaking of translational symmetry. However, whether the phonon confinement effect applies or not remains to be tested with first principles calculations, which can give atomic-level insight into these edge-activated modes.

Finally, we wish to add that interlayer force constants as detected with experimental Raman spectroscopy provide a valuable data set that can be used to assess the quality of the exchange-correlation functional in DFT. Much effort has been undertaken in recent years to develop exchange-correlation functionals that include van der Waals dispersive interactions. However, in our benchmark studies on interlayer phonon frequencies, we have found that these van der Waals functionals perform poorly compared to the local density approximation (LDA) to the exchange-correlation functional. This poor performance tells us that there is much room for further improvement of the van der Waals functionals to describe not only bond lengths and bond energies accurately, but also force constants. Thus, the synergy between theory and experiment in Raman spectroscopy not only gives new physical insights into 2D materials, but can also pave the way for new and exciting developments applicable to other fields.

References

1. K.F. Mak, C. Lee, J. Hone, J. Shan, T.F. Heinz, Atomically thin MoS2: a new sirect-gap semiconductor. Phys. Rev. Lett. **105**, 136805 (2010)
2. S. Tongay, J. Zhou, C. Ataca, K. Lo, T.S. Matthews, J. Li, J.C. Grossman, W.J. Thermally Driven, Crossover from indirect toward direct bandgap in 2D semiconductors: MoSe2 versus MoS2. Nano Lett. **12**, 5576–5580 (2012)
3. V. Sorkin, H. Pan, H. Shi, S.Y. Quek, Y.W. Zhang, Nanoscale transition metal dichalcogenides: structures, properties, and applications. Crit. Rev. Solid State Mater Sci **39**, 319–367 (2014)
4. X. Luo, Y.Y. Zhao, J. Zhang, Q.H. Xiong, S.Y. Quek, Anomalous frequency trends in MoS2 thin films attributed to surface effects. Phys. Rev. B **88**, 75320 (2013)
5. L. Rapoport, Y. Bilik, Y. Feldman, M. Homyonfer, S.R. Cohen, R. Tenne, Hollow nanoparticles of WS2 as potential solid-state lubricants. Nature **387**, 791–793 (1997)
6. C. Lee, Q. Li, W. Kalb, X.Z. Liu, H. Berger, R.W. Carpick, J. Hone, Frictional characteristics of atomically thin sheets. Science **328**, 76–80 (2010)
7. C. Lee, H. Yan, L.E. Brus, T.F. Heinz, J. Hone, S. Ryu, Anomalous lattice vibrations of single- and few-layer MoS2. ACS Nano **4**, 2695–2700 (2010)
8. T. Wieting, J. Verble, Infrared and Raman studies of long-wavelength optical phonons in hexagonal MoS2. Phys. Rev. B **3**, 4286 (1971)
9. Y. Zhao, X. Luo, H. Li, J. Zhang, P.T. Araujo, C.K. Gan, J. Wu, H. Zhang, S.Y. Quek, M.S. Dresselhaus, Q. Xiong, Interlayer breathing and shear modes in few-trilayer MoS2 and WSe2. Nano Lett. **13**, 1007 (2013)
10. H. Zeng, B. Zhu, K. Liu, J. Fan, X. Cui, Q.M. Zhang, Low-frequency Raman modes and electronic excitations in atomically thin MoS2 films. Phys. Rev. B **86**, 241301 (2012)
11. C. Ataca, M. Topsakal, E. Aktürk, S. Ciraci, A comparative study of lattice dynamics of three- and two-dimensional MoS2. J. Phys. Chem. C **115**, 16354–16361 (2011)
12. A. Molina-SÃ¡nchez, L. Wirtz, Phonons in single-layer and few-layer MoS2 and WS2. Phys. Rev. B **84**, 155413 (2011)
13. X. Zhang, X.-F. Qiao, W. Shi, J.-B. Wu, D.-S. Jiang, P.-H. Tan, Phonon and Raman scattering of two-dimensional transition metal dichalcogenides from monolayer, multilayer to bulk material. Chem. Soc. Rev. **44**, 2757–2785 (2015)
14. X. Zhang, W.P. Han, J.B. Wu, S. Milana, Y. Lu, Q.Q. Li, A.C. Ferrari, P.H. Tan, Raman spectroscopy of shear and layer breathing modes in multilayer MoS2. Phys. Rev. B **87**, 115413 (2013)
15. P. Tan, W. Han, W. Zhao, Z. Wu, K. Chang, H. Wang, Y. Wang, N. Bonini, N. Marzari, N. Pugno, The shear mode of multilayer graphene. Nat. Mater. **11**, 294–300 (2012)
16. X. Luo, Y. Zhao, J. Zhang, M. Toh, C. Kloc, Q. Xiong, S.Y. Quek, Effects of lower symmetry and dimensionality on Raman spectra in two-dimensional WSe 2. Phys. Rev. B **88**, 195313 (2013)
17. X. Luo, X. Lu, C. Cong, T. Yu, Q. Xiong, S. Ying Quek, Stacking sequence determines Raman intensities of observed interlayer shear modes in 2D layered materials – a general bond polarizability model. Sci. Rep. **5**, 14565 (2015)
18. C.H. Lui, T.F. Heinz, Measurement of layer breathing mode vibrations in few-layer graphene. Phys. Rev. B **87**, 121404 (2013)
19. X. Luo, X. Lu, G.K.W. Koon, A.H.C. Neto, B. Özyilmaz, Q. Xiong, S.Y. Quek, Large frequency change with thickness in interlayer breathing mode – significant interlayer interactions in few layer black phosphorus. Nano Lett. **15**, 3931 (2015)
20. N.S. Luo, P. Ruggerone, J.P. Toennies, Theory of surface vibrations in epitaxial thin films. Phys. Rev. B **54**, 5051–5063 (1996)
21. Y. Zhao, X. Luo, J. Zhang, J. Wu, X. Bai, M. Wang, J. Jia, H. Peng, Z. Liu, S.Y. Quek, Q. Xiong, Interlayer vibrational modes in few-quintuple-layer Bi2Te3 and Bi2Se3 two-dimensional crystals: Raman spectroscopy and first-principles studies. Phys. Rev. B **90**, 245428 (2014)

22. J.A. Wilson, A.D. Yoffe, The transition metal dichalcogenides discussion and interpretation of the observed optical, electrical and structural properties. Adv. Phys. **18**, 193–335 (1969)
23. L. Shulenburger, A.D. Baczewski, Z. Zhu, J. Guan, D. Tomanek, The nature of the interlayer interaction in bulk and few-layer phosphorus. Nano Lett. **15**, 8170–8175 (2015)
24. X. Lu, X. Luo, J. Zhang, S.Y. Quek, Q. Xiong, Lattice vibrations and Raman scattering in two-dimensional layered materials beyond graphene. Nano Res. **9**, 3559–3597 (2016)
25. K. Dolui, S.Y. Quek, Quantum-confinement and structural anisotropy result in electrically-Tunable Dirac cone in few-layer black phosphorous (vol 5, 11699, 2015). Sci. Rep. **6**, 11699 (2016)
26. J. Kim, S.S. Baik, S.H. Ryu, Y. Sohn, S. Park, B.-G. Park, J. Denlinger, Y. Yi, H.J. Choi, K.S. Kim, Observation of tunable band gap and anisotropic Dirac semimetal state in black phosphorus. Science **349**, 723–726 (2015)
27. R.X. Fei, V. Tran, L. Yang, Topologically protected Dirac cones in compressed bulk black phosphorus. Phys. Rev. B **91**, 195319 (2015)
28. W. Bao, L. Jing, J. Velasco Jr., Y. Lee, G. Liu, D. Tran, B. Standley, M. Aykol, S.B. Cronin, D. Smirnov, Stacking-dependent band gap and quantum transport in trilayer graphene. Nat. Phys. **7**, 948–952 (2011)
29. Y. Lee, J. Velasco, D. Tran, F. Zhang, W. Bao, L. Jing, K. Myhro, D. Smirnov, C.N. Lau, Broken symmetry quantum hall states in dual-gated ABA trilayer graphene. Nano Lett. **13**, 1627–1631 (2013)
30. R. Suzuki, M. Sakano, Y.J. Zhang, R. Akashi, D. Morikawa, A. Harasawa, K. Yaji, K. Kuroda, K. Miyamoto, T. Okuda, K. Ishizaka, R. Arita, Y. Iwasa, Valley-dependent spin polarization in bulk MoS2 with broken inversion symmetry. Nat. Nanotechnol. **9**, 611–617 (2014)
31. C.H. Lui, Z. Li, Z. Chen, P.V. Klimov, L.E. Brus, T.F. Heinz, Imaging stacking order in few-layer graphene. Nano Lett. **11**, 164–169 (2011)
32. C. Cong, T. Yu, K. Sato, J. Shang, R. Saito, G.F. Dresselhaus, M.S. Dresselhaus, Raman characterization of ABA- and ABC-stacked trilayer graphene. ACS Nano **5**, 8760–8768 (2011)
33. R. Saito, M. Furukawa, G. Dresselhaus, M. Dresselhaus, Raman spectra of graphene ribbons. J. Phys. Condens. Matter **22**, 334203 (2010)
34. S. Guha, J. Menendez, J. Page, G. Adams, Empirical bond polarizability model for fullerenes. Phys. Rev. B **53**, 13106 (1996)
35. L. Wirtz, M. Lazzeri, F. Mauri, A. Rubio, Raman spectra of BN nanotubes: Ab initio and bond-polarizability model calculations. Phys. Rev. B **71**, 241402 (2005)
36. C.H. Lui, Z. Ye, C. Keiser, E.B. Barros, R. He, Stacking-dependent shear modes in trilayer graphene. Appl. Phys. Lett. **106**, 041904 (2015)
37. X. Lu, M.I.B. Utama, J. Lin, X. Luo, Y. Zhao, J. Zhang, S.T. Pantelides, W. Zhou, S.Y. Quek, Q.H. Xiong, Rapid and nondestructive identification of polytypism and stacking sequences in few-layer molybdenum diselenide by Raman spectroscopy. Adv. Mater. **27**, 4502–4508 (2015)
38. B.J. Mrstik, R. Kaplan, T.L. Reinecke, M. Van Hove, S.Y. Tong, Surface-structure determination of the layered compounds MoS2 and NbSe2 by low-energy electron diffraction. Phys. Rev. B **15**, 897–900 (1977)
39. P. Tonndorf, R. Schmidt, P. Böttger, X. Zhang, J. Börner, A. Liebig, M. Albrecht, C. Kloc, O. Gordan, D.R. Zahn, Photoluminescence emission and Raman response of monolayer MoS_2, $MoSe_2$, and WSe_2. Opt. Express **21**, 4908–4916 (2013)
40. W. Zhao, Z. Ghorannevis, A.K. Kumar, J.R. Pang, M. Toh, X. Zhang, C. Kloc, P.H. Tan, G. Eda, Lattice dynamics in mono-and few-layer sheets of WS2 and WSe2. arXiv:13040911 (2013)
41. S.-Y. Chen, C. Zheng, M.S. Fuhrer, J. Yan, Helicity-resolved Raman scattering of MoS2, MoSe2, WS2, and WSe2 atomic layers. Nano Lett. **15**, 2526–2532 (2015)
42. C. Ruppert, O.B. Aslan, T.F. Heinz, Optical properties and band gap of single-and few-layer MoTe2 crystals. Nano Lett. **14**, 6231–6236 (2014)
43. M. Yamamoto, S.T. Wang, M. Ni, Y.-F. Lin, S.-L. Li, S. Aikawa, W.-B. Jian, K. Ueno, K. Wakabayashi, K. Tsukagoshi, Strong enhancement of Raman scattering from a bulk-inactive vibrational mode in few-layer MoTe2. ACS Nano **8**, 3895–3903 (2014)

44. G. Froehlicher, E. Lorchat, F. Fernique, C. Joshi, A. Molina-Sanchez, L. Wirtz, S. Berciaud, Unified description of the optical phonon modes in N-layer $MoTe_2$. Nano Lett. **15**, 6481–6489 (2015)
45. H.P.C. Miranda, S. Reichardt, G. Froehlicher, A. Molina-Sánchez, S. Berciaud, L. Wirtz, Quantum interference effects in resonant Raman spectroscopy of single- and triple-layer MoTe2 from first-principles. Nano Lett. **17**, 2381–2388 (2017)
46. S. Huang, L. Liang, X. Ling, A.A. Puretzky, D.B. Geohegan, B.G. Sumpter, J. Kong, V. Meunier, M.S. Dresselhaus, Low-frequency interlayer Raman modes to probe interface of twisted bilayer MoS2. Nano Lett. **16**, 1435–1444 (2016)
47. A.A. Puretzky, L. Liang, X. Li, K. Xiao, B.G. Sumpter, V. Meunier, D.B. Geohegan, Twisted MoSe2 bilayers with variable local stacking and interlayer coupling revealed by low-frequency raman spectroscopy. ACS Nano **10**, 2736–2744 (2016)
48. H. Richter, Z.P. Wang, L. Ley, The one phonon Raman spectrum in microcrystalline silicon. Solid State Commun. **39**, 625–629 (1981)
49. J. Zhang, Z. Peng, A. Soni, Y. Zhao, Y. Xiong, B. Peng, J. Wang, M.S. Dresselhaus, Q. Xiong, Raman spectroscopy of few-quintuple layer topological insulator Bi2Se3 nanoplatelets. Nano Lett. **11**, 2407–2414 (2011)
50. S. Mignuzzi, A.J. Pollard, N. Bonini, B. Brennan, I.S. Gilmore, M.A. Pimenta, D. Richards, D. Roy, Effect of disorder on Raman scattering of single-layer Mo S 2. Phys. Rev. B **91**, 195411 (2015)
51. W. Shi, M.-L. Lin, Q.-H. Tan, X.-F. Qiao, J. Zhang, P.-H. Tan, Raman and photoluminescence spectra of two-dimensional nanocrystallites of monolayer WS2 and WSe2. 2D Materials **3**, 025016 (2016)
52. J.B. Wu, H. Zhao, Y. Li, D. Ohlberg, W. Shi, W. Wu, H. Wang, P.H. Tan, Monolayer molybdenum disulfide nanoribbons with high optical anisotropy. Adv. Opt. Mater. **4**, 756–762 (2016)

Chapter 9
Resonant Raman Spectroscopy of Two Dimensional Materials Beyond Graphene

Hyeonsik Cheong and Jae-Ung Lee

Abstract The resonance Raman effects in two dimensional materials including transition metal dichalcogenides and black phosphorus are reviewed. The Raman intensities of high-frequency intra-layer vibration modes are enhanced near resonance with exciton states. Some Raman peaks that are either forbidden or weak in non-resonant cases show strong enhancement near resonances. In the low-frequency Raman spectra, some unusual features, in addition to shear and breathing modes, appear near resonance with exciton states. Some intra-layer vibration modes exhibit Davydov splitting due to inter-layer interactions when the excitation energy is close to resonances. The polarization behaviors of some Raman modes in anisotropic two-dimensional materials have peculiar dependences on the excitation energy, which is related to the resonance effect.

9.1 Introduction

9.1.1 Electronic Band Structure of Two Dimensional Materials Beyond Graphene

Raman spectroscopy is widely used as an important characterization tool for two dimensional materials beyond graphene such as transition metal dichalcogenides (TMD's) and black phosphorus. Unlike graphene, which has a rather simple band structure, these two dimensional materials have complex band structures as summarized in Chap. 7. Most TMD's have a direct band gap for monolayers and an indirect gap for few-layers or bulk [1–5], whereas black phosphorus has a direct gap of ~0.3 to ~2.0 eV depending on the thickness [5, 6]. Strong electron-hole interaction due to reduced screening of Coulomb interaction [7, 8] leads to formation of robust exciton states even at room temperature, which dominate the optical absorption

H. Cheong (✉) · J.-U. Lee
Department of Physics, Sogang University, Seoul, Korea
e-mail: hcheong@sogang.ac.kr

P.-H. Tan (ed.), *Raman Spectroscopy of Two-Dimensional Materials*, Springer Series in Materials Science 276, https://doi.org/10.1007/978-981-13-1828-3_9

spectra. Figure 9.1 shows room-temperature differential reflection spectra, which approximately mimic the corresponding absorption spectra, for several monolayer TMD's [9, 10]. Three or four peaks are observed. The A and B excitons are bound states of the electrons in the lowest energy conduction band and holes in the highest-energy valence band or the spin-orbit split band at the K or K' points of the Brillouin zone, respectively. The C exciton state involves higher energy conduction bands. The spectrum for WSe_2 show four peaks, which are labelled A, B, A′, and B′. The two higher energy peaks are often interpreted as excited states of A and B excitons [9, 11–14]. Other researchers assign these peaks to independent higher energy exciton states and label them as C and D instead of A′ and B′ [10, 15, 16]. Owing to the large exciton binding energies the photoluminescence is strong, especially for the direct-gap monolayer case. The photoluminescence peak usually coincides with the A exciton peak in reflectance or absorption. The Raman scattering process is profoundly affected by the exciton states when the excitation laser energy is resonant with the exciton states. The resonance leads not only to the enhancement of the Raman intensities but also to several peculiar phenomena that are not observed in resonance Raman scattering of conventional semiconductors.

9.1.2 Resonance Effects in Raman Scattering

In graphene, the main features of the Raman spectrum do not exhibit strong dependence on the excitation energy except that the 2D (G′) band shifts with the excitation energy due to the nature of the double resonance Raman scattering [17]. This is because graphene has almost constant optical absorption over a wide energy range that covers most lasers used in Raman spectroscopy [18]. On the other hand, the absorption spectra of TMD's show rich features that originate from the complex band structure and the excitonic states [9, 10]. Due to the reduced screening in 2-dimensional materials, the electrons and holes are strongly bound to form excitons with binding energies on the order of 100 s of meV [8, 19–22]. When the excitation laser energy matches an allowed optical transition energy, the resonance effect dominates the Raman scattering process [23]. The Raman scattering intensity given by Eq. (9.1) of Chap. 7 can be rewritten in the following form

$$I_{ph}(\omega_s) \approx \left(\frac{2\pi}{\hbar}\right)\left|\frac{\langle 0|\, H_{eR}(\omega_s)\,|a\rangle \langle a|\, H_{e-ph}\,|a\rangle \langle a|\, H_{eR}(\omega_i)\,|0\rangle}{[E_a - \hbar\omega_i - i\Gamma_a][E_a - \hbar\omega_i + \hbar\Omega - i\Gamma_a]} + C\right|^2. \quad (9.1)$$

Here, $\hbar\Omega$ is the energy of the phonon involved, $\hbar\omega_i$ the excitation laser energy, $|a\rangle$ the intermediate state involved in the resonance process with an energy E_a above the ground state $|0\rangle$, H_{eR} and $H_{e\text{-ph}}$ electron-radiation and electron-phonon interaction Hamiltonians, respectively, and Γ_a the damping constant for the resonance [24, 25]. The constant C represents the contribution from nonresonant scattering processes. In general, there are two resonances, an incoming and an outgoing, at $\hbar\omega_i = E_a$ and $\hbar\omega_i = E_a + \hbar\Omega$, respectively. The most prominent resonance effect is an

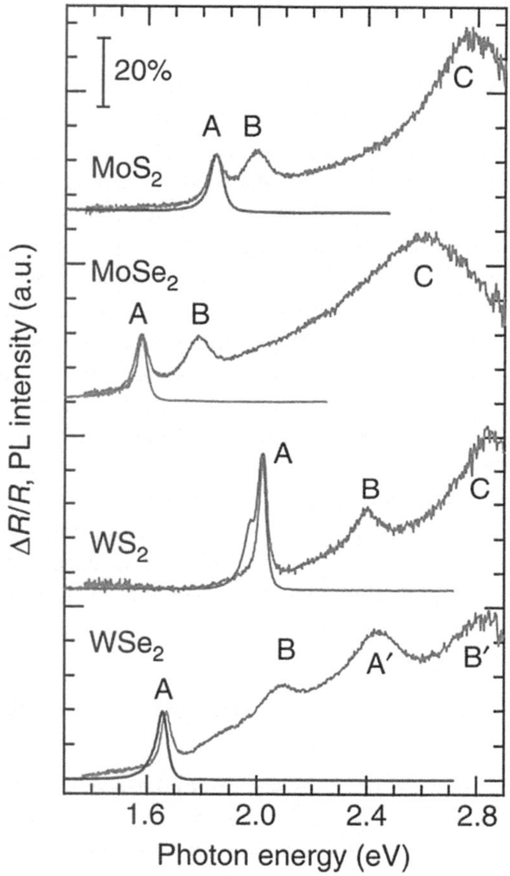

Fig. 9.1 Differential reflectance spectra which mimic absorption spectra of monolayer TMD's on quartz substrates. Spectra features ascribed to excitonic transitions are indicated. PL spectra measured with excitation at the C (A′ for WSe_2) peak are also shown [9]

increase of the Raman signal near these resonances. But there are further effects due to the symmetry of the states involved. If the intermediate state $|a\rangle$ is a Bloch state, the translational symmetry is preserved, and so is the momentum conservation. However, if the intermediate state is not a Bloch state, the momentum conservation is relaxed, and some forbidden Raman peaks may appear near resonances with such states [25, 26]. Since an exciton is a bound state of an electron and a hole, it is not a Bloch state, especially in TMD's where the strong coulomb interaction results in tightly localized exciton wave functions [7, 8, 22]. In such cases, the relaxation of the momentum conservation becomes pronounced, and several peculiar resonance effects are observed [27–35].

9.2 Resonance Effects on High-Frequency Raman Spectra

Figure 9.2 shows Raman spectra of several TMD materials and their dependence on the number of layers. The excitation energy for each material is chosen to avoid resonance effect as much as possible. Because the point group symmetry depends on the number of layers, the mode notations are different for similar vibration modes depending on the number of layers. However, the bulk mode notations are often used for convenience regardless of the number of layers. The *intra-layer* vibration modes E_{2g}^1 and A_{1g} correspond to vibrations of the metal atom against the chalcogen atoms in the in-plane and out-of-plane directions, respectively [36–38], whereas rigid vibrations of the layers against each other correspond to *inter-layer* vibration modes of shear and breathing modes. The separation between the E_{2g}^1 and A_{1g} peaks increases with the number of layers and can be used to estimate the number of layers of MoS_2 [33, 36, 39], $MoSe_2$ [40–42], WS_2 [42, 43], and WSe_2 [42–44]. The inter-layer modes show more pronounced dependence on the number of layers and are better fingerprints of the layer numbers. In addition to the main Raman modes, some peaks due to multiple phonon scattering are observed. For example, a peak originating from two-phonon scattering involving zone-boundary longitudinal acoustic phonons (2LA) is frequently observed as shown in Fig. 9.2.

9.2.1 *Resonance Effects on Main Modes*

Figure 9.3 compares several Raman spectra of 4-layer MoS_2 [33], $MoSe_2$ [41], WS_2 [45], and WSe_2 [46], taken with several excitation energies. Dramatic changes due to resonance effects are observed. In addition to the variation of the intensities of the main E_{2g}^1 and A_{1g} modes, strong enhancement of several peaks that are not observed for nonresonant excitations are observed. For example, in the case of MoS_2, several peaks that are observed only for the 1.96 eV excitation are even stronger than the main $E_{2g}{}^1$ and A_{1g} peaks.

Figure 9.4 presents the normalized intensities of some of the Raman peaks from monolayer WS_2 as a function of the excitation laser energy. Both $E_{2g}{}^1$ and A_{1g} peaks as well as the 2LA signal exhibit resonance enhancement near 2.0, 2.4, and 2.7 eV, corresponding to the A, B, and C exciton energies, respectively, determined from optical absorption spectra. It is clear that the resonance enhancement is dominated by resonance with the exciton states. On the other hand, closer inspection reveals that the relative enhancement of the Raman peaks differs for each exciton energy. For example, in Fig. 9.4, the A_{1g} mode is enhanced more than the $E_{2g}{}^1$ mode at the A exciton energy, but the $E_{2g}{}^1$ mode is more enhanced for other exciton energies. Similar effects are often observed in other TMD materials [11, 31, 33, 41, 47, 48]. Different resonance behaviors of the $E_{2g}{}^1$ and A_{1g} phonon modes are often interpreted in terms of the correlation between the atomic displacements of the phonons and the symmetry of the electronic orbitals that contribute to the

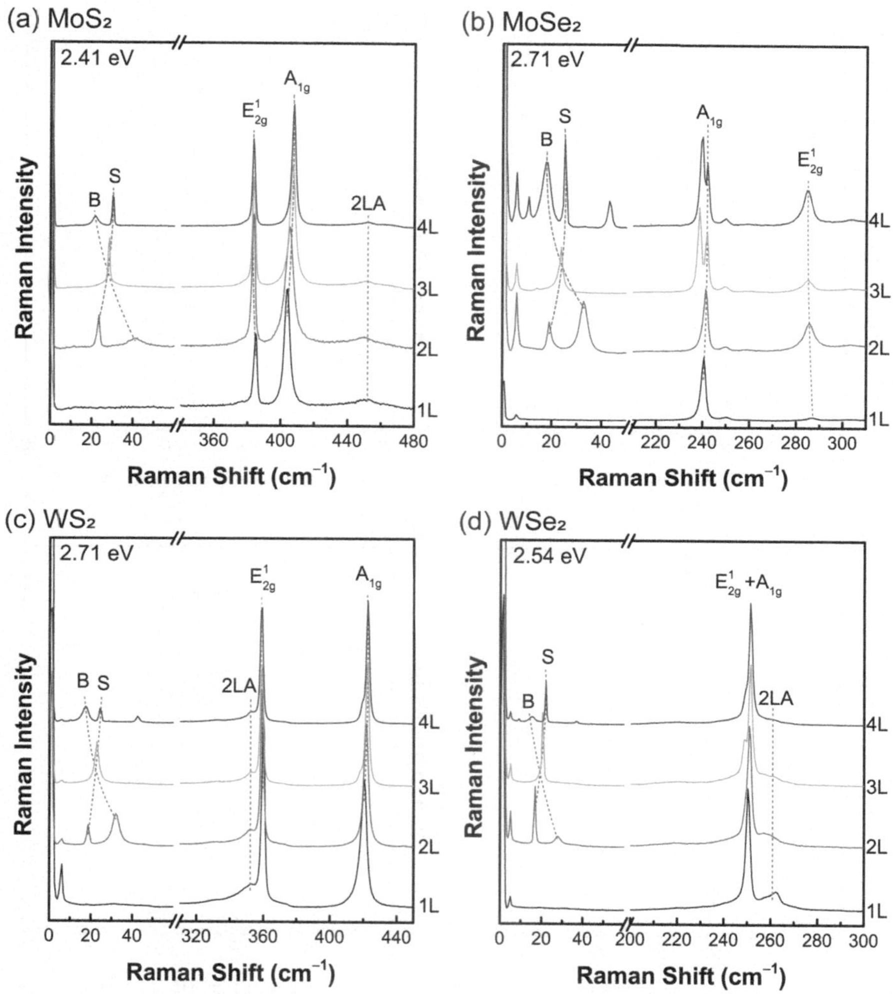

Fig. 9.2 Raman spectra of few-layer TMD materials. The excitation energy for each material is indicated. The inter-layer shear (S) and breathing (B) modes as well as intra-layer modes of $E_{2g}{}^{1}$ and A_{1g} modes and the two-phonon scattering signal involving two longitudinal acoustic phonons (2LA) are indicated

exciton states [32, 49]. On the other hand, Miranda *et al.*, suggested that quantum interference effects between contributions of electronic transitions from different parts of the Brillouin zone are responsible for the observed differences between the phonon modes [50]. More studies are needed to clarify this issue.

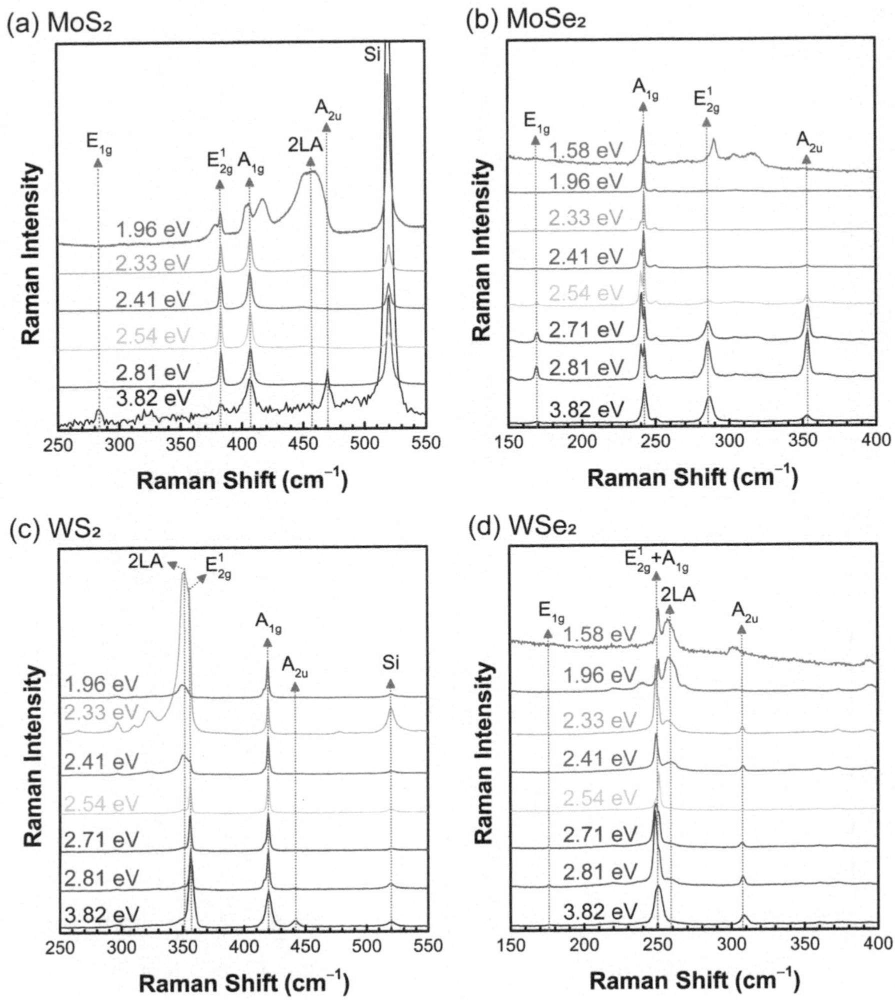

Fig. 9.3 Excitation energy dependence of Raman spectra for several 4-layer TMD materials

9.2.2 Resonance Effects on Forbidden or Multi-phonon Peaks

Figure 9.5 shows the Raman spectra of 7-layer MoS_2, measured with several excitation energies. As noted in the previous section, the spectrum for the 1.96 eV (633 nm) excitation exhibits many peculiar features. This excitation energy is close to the A (~1.8 eV) and B (~2.0 eV) exciton states. Some of the peaks that are either forbidden or weak in non-resonant cases are strongly enhanced depending on the excitation energy [33]. The small, sharp peak at 285 cm^{-1} is observed for all thicknesses and is strongly enhanced for excitation energies larger than 2.5 eV [33].

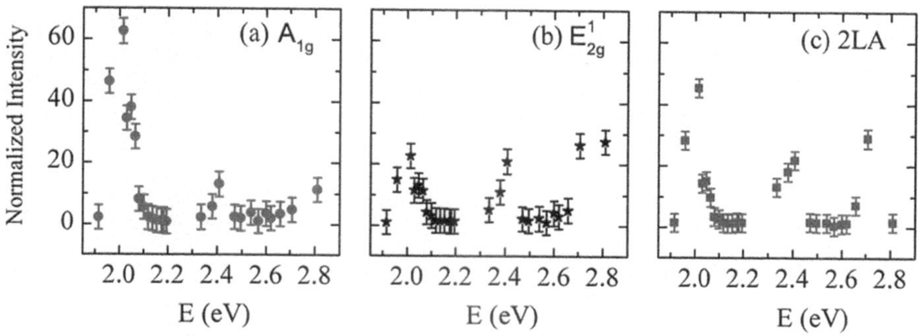

Fig. 9.4 Resonance profiles of the Raman peaks A_{1g}, $E_{2g}{}^1$, and 2LA of monolayer WS_2. Each peak exhibits enhancements near 2.0, 2.4, and 2.7 eV, corresponding to the A, B, and C exciton energies, respectively [47]

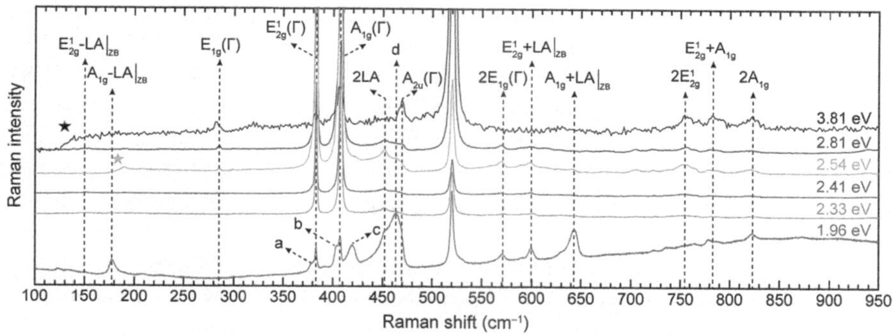

Fig. 9.5 Raman spectra of 7-layer MoS_2 measured with several excitation energies indicated [33]

In bulk MoS_2, this peaks was assigned as the E_{1g} mode, in which the Mo atoms are stationary and the S atoms above and below the Mo atom move in opposite directions in the lattice plane. However, this mode in *bulk* MoS_2 is forbidden in backscattering geometry. The corresponding mode is E'' in odd number of layers and E_g in even number of layers [37, 38, 51]. Group theoretical considerations predict that the E'' mode is forbidden and the E_g mode is allowed in backscattering geometry. The fact that this mode appears for all thicknesses and only for high excitation energies seems to imply that the appearance of this peak is closely related to the resonance with the C exciton state [33]. The peak at 469 cm^{-1} shows a trend that is very similar to the E''or E_g peak at 285 cm^{-1}: it is well resolved as an isolated peak for the 3.81-eV excitation but appears as a shoulder for low energy excitations. This peak is tentatively assigned as a mode corresponding to the bulk $A_{2u}(\Gamma)$ mode, which is forbidden but appears near resonance with the C exciton [33].

More dramatic enhancements are observed for the peak *c* at ~418 cm^{-1}, a broad asymmetric peak at ~460 cm^{-1}, and a series of peaks in the 570–650 cm^{-1} range [33]. These features are attributed to multi-phonon scattering through various combinations of phonon modes [29, 33, 34, 52]. As explained in Chap. 7, these peaks are enhanced due to combination of the increased optical absorption, the resonance Raman effect, and the large density of states for these phonon bands near particular Brillouin zone boundaries. Similar enhancement of forbidden or multi-phonon peaks near excitonic resonances has been observed for $MoSe_2$ as well [41, 48, 53].

9.2.3 Davydov Splitting

Davydov splitting is the splitting of bands in the electronic or vibrational spectra of crystals due to the presence of more than one equivalent entities in the unit cell that interact with each other [54, 55]. In layered materials, the weak *inter*-layer interaction causes splitting of the *intra*-layer vibration modes. Since this splitting is directly related to the *inter*-layer interaction, Davydov splitting can be used to estimate the *inter*-layer interaction as well as the number of layers [41, 56, 57]. Davydov splitting has been observed in $MoTe_2$ [56, 58–60], $MoSe_2$ [40, 41], WS_2 [61], WSe_2 [46], and black phosphorus [62]. In most cases, Davydov splitting is observed only for excitation energies specific to the material. For example, it is most prominently observed for 1.96-eV excitation in the case of $MoTe_2$ [56, 58–60] whereas it is observed in $MoSe_2$ only for excitation energies greater than 2.33 eV [41].

Figure 9.6a shows Davydov splitting of the main A_{1g} mode of 1 to 6-layer $MoSe_2$ for several excitation energies [41]. The number of peaks increase with the number of layers, and multiple Lorentzian fits can be used to deconvolute the peaks. For 1 and 2 layers, the spectra can be fitted with a single Lorentzian function except for the one for 2-layer $MoSe_2$ measured with the 1.58 eV excitation. Two peaks are resolved for 3 and 4 layers, and 3 peaks for 5 and 6 layers, and so on. Figure 9.6b illustrates the out-of-plane vibration modes in 2 to 4-layer $MoSe_2$ [41]. The vibration modes are within each layer, but the relative phases between the layers are different. If a vibration mode has an inversion center (for even number of layers) or a mirror symmetry plane (for odd number of layers), it is Raman active, and IR active otherwise [37, 41, 58]. In 2-layer $MoSe_2$, there are a Raman active mode (A_{1g}) in which the two layers vibrate in phase and an infrared active mode (A_{2u}) in which the layers vibrate out of phase. Therefore, normally only one A_{1g} peak should be observed for 2-layer $MoSe_2$. The small signal on the lower- wavenumber side of the A_{1g} peak for the 1.58 eV excitation is ascribed to the out-of-phase A_{2u} mode which becomes partially active due to resonance with the A exciton in $MoSe_2$ [41].

For 3 layers, the two $A_1{}'$ modes are Raman active, and the $A_2{}''$ mode is infrared active. All the layers vibrate in phase for the $A_1{}'(1)$ mode, whereas one layer is out of phase for the $A_1{}'(2)$ mode. Because the (nonresonant) Raman scattering cross section is proportional to the derivative of the susceptibility tensor with respect to displacement, the contributions of out-of-phase vibrations would tend to cancel each other. Hence the $A_1{}'(1)$ mode is observed for all excitation energies whereas the $A_1{}'(2)$ mode appears only for some excitation energies near resonances [41].

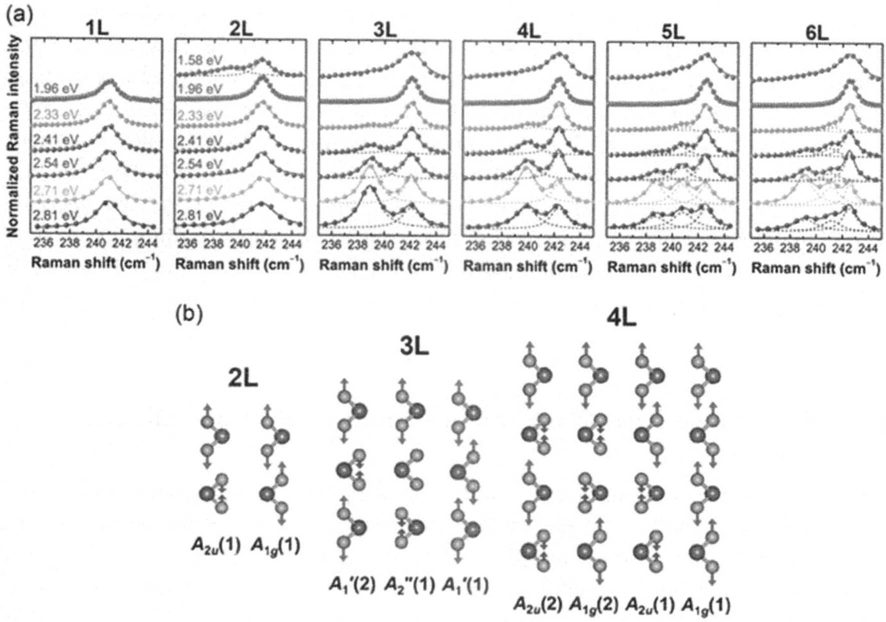

Fig. 9.6 (**a**) Davydov splitting of A_{1g} mode in 1 to 6-layer $MoSe_2$ measured with several excitation energies. (**b**) Schematics of out-of-plane vibrational modes for 2 to 4-layer $MoSe_2$ [41]

Similarly for 4 layers, the all-in-phase $A_{1g}(1)$ mode is always observed. Such excitation-energy-dependent relative enhancement of Davydov-split Raman modes has been ascribed to quantum interference effects [50]. Recent measurements on the temperature dependence of the relative intensities of Davydov-split peaks seem to indicate that the interplay between resonant and nonresonant contributions to the Raman scattering is playing an important role [60]. For 3 and 4 layers, the infrared active modes are not resolved even for resonant excitations, probably because they are overwhelmed by relatively stronger Raman active modes with similar frequencies [41, 42, 56, 61]. Davydov splitting also affects the in-plane E modes [56], but because the splitting is relatively smaller, the split peaks are often difficult to resolve. In the case of $MoSe_2$, a small splitting of the E_{1g} mode can be extracted from its asymmetric line shape, but no such splitting is resolved for the $E_{2g}{}^1$ mode [41]. Davydov splitting can explain why some forbidden modes appear near resonances. When an infrared-active but Raman-forbidden peak in few-layer TMD is Davydov split, some of the split peaks become Raman-active due to symmetry considerations as explained in the above. In that case, the Raman-active components would appear in the Raman spectrum. If the splitting is small, it would appear that the 'forbidden' peak becomes active near resonances in 2 layers or thicker materials.

By fitting the peak positions of the *intra-* and *inter*-layer Raman modes including the Davydov splitting to the linear chain model, one can estimate the strength of the inter-layer interaction [41, 46, 56]. Although the thickness dependence of the shear and breathing modes can be modelled by considering only the nearest-neighbor inter-layer interaction [37, 38], the Davydov splitting cannot be fitted without considering the second-nearest-neighbor interaction, the interaction between the transition metal atoms and the chalcogen atoms in adjacent layers. The second-nearest-neighbor inter-layer interaction turns out to be nonnegligible, about 30% of the nearest-neighbor interaction for both in-plane and out-of-plane vibrations in $MoSe_2$ [41] and also for out-of-plane vibrations in WSe_2 [46].

9.3 Resonance Effects on Low-Frequency Raman Spectra

The resonance has a strong impact on the low-frequency Raman spectrum as well. Figure 9.7 shows the low-frequency Raman spectra of 2-layer WS_2 measured with six different excitation energies [45]. The overall shape of the spectrum varies greatly depending on the excitation energy, although the peak positions are fixed. Some of the peaks are observed only for some excitation energies. The shear mode (S_1) at 19 cm^{-1} and the breathing mode (B_1) at 32 cm^{-1} are strongly enhanced for higher energy excitations. Two additional features, labelled X_1 at 28 cm^{-1} and X_2 at 46 cm^{-1}, are observed for some excitation energies. These peaks seem to be enhanced near resonance with exciton states, although they are not always resolved. These two peaks appear in all thicknesses, including monolayer, and their positions do not have dependence on the thickness (see Fig. 9.8). A similar peak was reported for MoS_2 when the excitation energy was close to the A and B exciton states [33]. These peaks cannot be due to phonons at high-symmetry points in the Brillouin zone because no high-symmetry point phonon in monolayer WS_2 has energy this small except for the acoustic phonons at the zone center which has zero energy. Also, they are not related to the excitons in a simple way because the exciton energy depends on the thickness, but these peaks do not. There is a suggestion that these peaks originate from acoustic phonons with a specific finite momentum, but it is not clear why a specific momentum is preferentially chosen [63]. Another possibility is that these peaks are due to some complex excitation which is a combination of low-energy acoustic phonon(s) and some low-energy quasi-particle with a momentum related to the excitons. More studies are needed to resolve this puzzle.

In addition to the sharp peaks, there is a rather broad peak centered at the origin when the excitation energy is near the A (1.96 eV) or B (2.33 or 2.41 eV) exciton resonances (see Fig. 9.8). This 'central peak' was also observed for MoS_2 and was ascribed to acoustic phonon scattering mediated by the exciton states broadened by the inhomogeneity in the sample [33]. In Fig. 9.8, the spectra measured with the 1.96-eV excitation show a series of small peaks marked 'unknown'. These peaks do not correspond to allowed shear or breathing modes. Comparison of their positions with a linear-chain-model calculation revealed that they correspond to shear modes

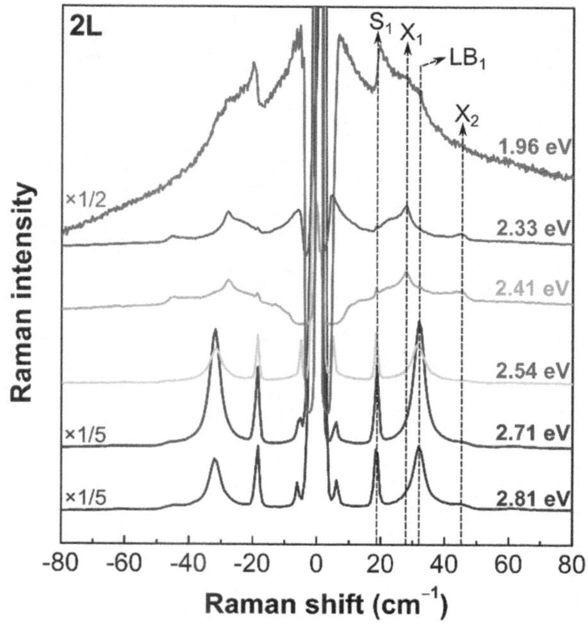

Fig. 9.7 Low-frequency Raman spectra of 2-layer WS_2 measured with six different excitation energies [45]

that are forbidden in back scattering (Fig. 9.9). Their appearance is ascribed to excitonic resonance, but it is not clear why they are almost as strong as 'allowed' shear modes. A similar series of forbidden shear modes have been observed in MoS_2 as well [33].

9.4 Excitation Energy Dependence of Polarized Raman Scattering in Anisotropic Layered Materials

Polarized Raman spectroscopy is often a convenient tool to determine the crystallographic orientation of anisotropic layered materials such as ReS_2 [64–66], $ReSe_2$ [67, 68], WTe_2 [69], and black phosphorus [70–72] crystals. However, because the polarization dependences of the main Raman peaks often exhibit peculiar excitation energy dependences [68–70, 72], determination of the crystallographic orientation based on polarized Raman measurements could yield erroneous results if such excitation energy dependences are not carefully accounted for. This peculiar phenomena may originate from anisotropy of the electron-photon or electron-phonon interactions which should be related to the electronic band structure of the material.

Figure 9.10a shows the crystal structure of black phosphorus, with the principal axes indicated. The x axis is set along the zigzag direction. Figure 9.10b shows typical polarized Raman spectra of bulk black phosphorus measured with the 441.6-nm excitation. The analyzer is set parallel to the incident laser polarization (parallel

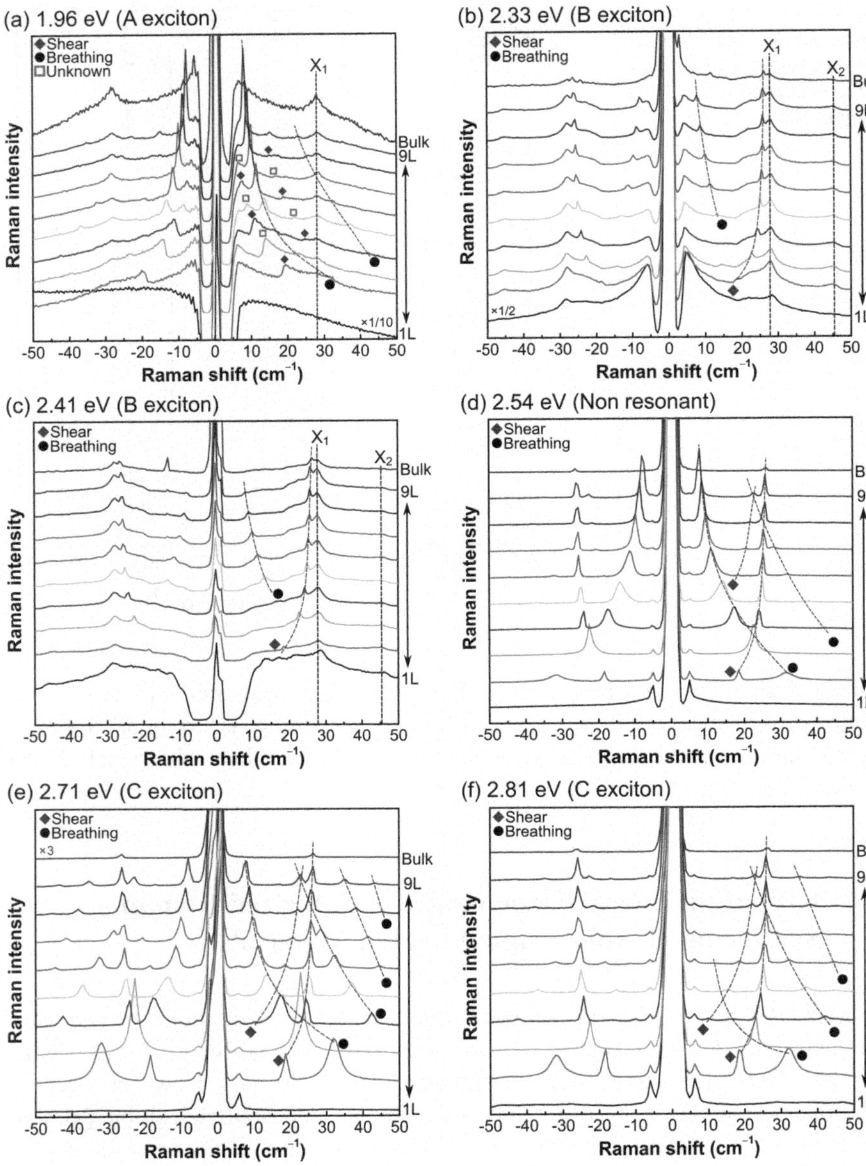

Fig. 9.8 Dependence of low-frequency Raman spectra of WS_2 on number of layers for six different excitation energies. In the spectra taken with 1.96 eV excitation, a series of 'unknown' peaks that do not correspond to allowed shear or breathing modes also appear [45]

polarization). Three major peaks are observed. The peak at 360 cm^{-1} is strongest at 0° and disappears at 90°. The peak at 437 cm^{-1}, on the other hand, disappears at 0° and 90°. The strongest peak at 464 cm^{-1} is maximum at 90° and minimum at

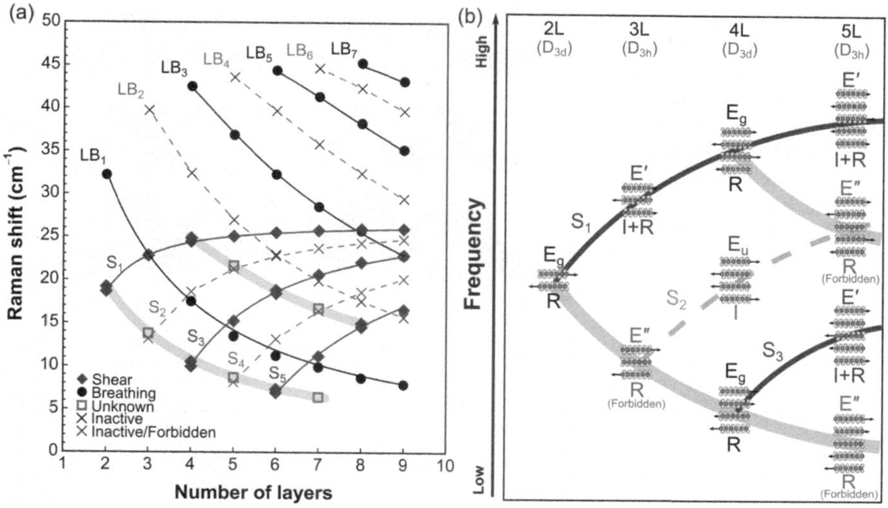

Fig. 9.9 (**a**) Position of inter-layer vibration modes of WS_2 as a function of the number of layers. The experimentally observed Raman active modes are shown in solid symbols, and Raman inactive or forbidden modes obtained from calculations are indicated by '×'. Open squares are experimentally observed peaks that do not correspond to 'allowed' modes. (**b**) Schematic diagram of shear modes for 2–5-layer WS_2. The mode designations for each thickness are shown. E_g modes are Raman active, E_u infrared active (Raman inactive), E' Raman and infrared active, and E'' Raman active but forbidden in backscattering [45]

0°, which is orthogonal to the one at 360 cm^{-1}. The polarization dependence of the peaks can be compared with group theoretical analysis to assign the modes as A_g^1, B_{2g}, and A_g^2, respectively, and the vibrational modes are shown schematically.

Figure 9.11 summarizes the polarization dependence of the A_g^1, A_g^2, and B_{2g} modes for excitation wavelengths of 441.6, 488, 514.5, 532, and 632.8 nm in parallel polarization. The polarization dependence of the A_g modes dramatically changes with the excitation wavelength. The B_{2g} mode, on the other hand, has the same polarization dependence regardless of the excitation wavelength. Similar measurements on black phosphorus samples with different thicknesses revealed that the polarization dependence of the A_g modes vary also with the thickness of the sample, which can be explained in terms of the birefringence and the interference effect [70]. However, the dependence of the polarization behavior on the excitation energy for the same thickness cannot be explained in a similar manner. Because of the anisotropy in the band structure, it is reasonable to expect that the Raman tensor elements vary with the excitation wavelength. If the excitation energy dependence of the tensor elements are not identical, the polarization behaviors, which reflect the relative ratio of tensor elements, would depend on the excitation energy. On the other hand, because the Raman tensor of the B_{2g} mode has only one element e, the functional form of the polarization dependence of the B_{2g} mode will not change even if the magnitude of the tensor element changes. This would explain the observed

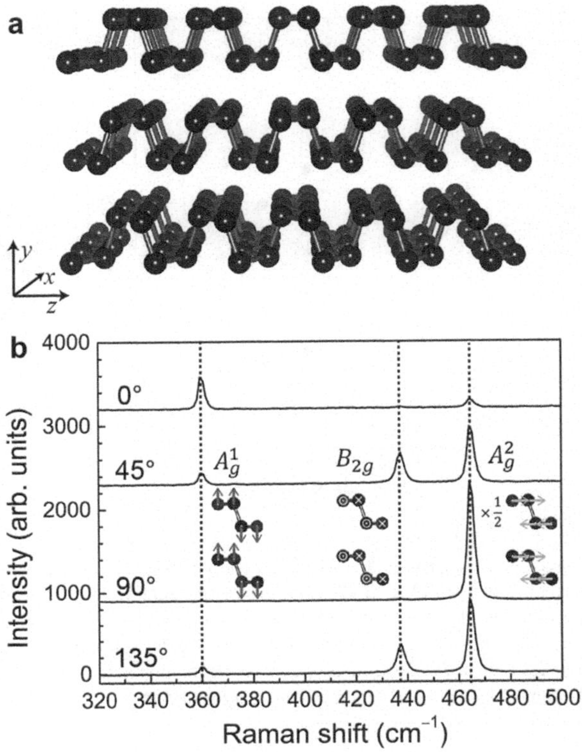

Fig. 9.10 (**a**) Crystal structure of black phosphorus. The principal axes are shown. (**b**) Polarized Raman spectra (parallel polarization) of black phosphorus measured with the 441.6-nm excitation. The angle between the incident polarization and the zigzag direction for each spectrum is indicated [70]

difference between the A_g modes and the B_{2g}mode [70]. More studies are needed to find direct correlation between the excitation energy dependence of the tensor elements with the observed polarization behaviors. Similar anomalous polarization behavior depending on the excitation wavelength or the sample thickness was observed in few-layer WTe_2 [69].

9.5 Summary and Outlook

Understanding the resonance effects is critically important if one wants to use Raman spectroscopy to study physics in two dimensional materials beyond graphene. The excitonic resonances play a very important role in Raman scattering owing to the large exciton binding energy. In addition to usual enhancement of Raman signals, several peculiar phenomena such as appearance of forbidden modes, enhancement of multiple-phonon scattering peaks, emergence of unidentified new peaks, Davydov splitting, and unusual polarization behaviors are observed. The knowledge of the resonance Raman effects and the data for excitation dependence of the Raman spectra would be immensely helpful in characterizing these materials

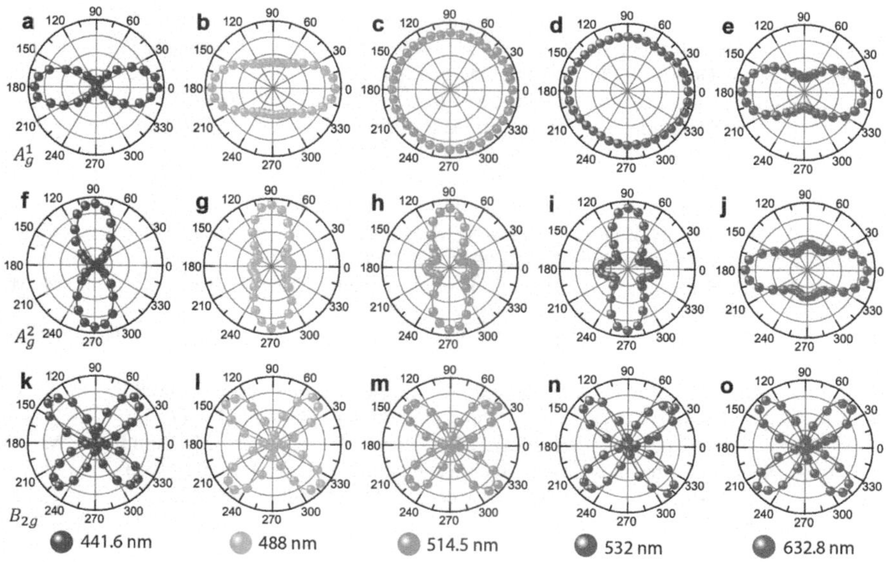

Fig. 9.11 Polarization dependence of Raman modes of black phosphorus for different excitation wavelengths. Each row shows polarization dependence of A_g^1, A_g^2 and B_{2g}modes, respectively, taken with excitation wavelengths of 441.6, 488, 514.5, 532, and 632.8 nm as indicated. The polarization behaviors of the A_g modes exhibit strong dependence on the excitation wavelength, whereas the B_{2g}mode does not show such dependence [70]

and understanding the unique physics in two dimensions. The origin of the so-called 'X' peaks and the excitation energy dependence of the polarization behaviors in some of the Raman peaks in black phosphorus and WTe_2 are still poorly understood and call for further studies. Theoretical investigations of the electron-photon and electron-phonon interactions, in comparison with experimental results, might shed light on these issues.

Acknowledgements This work was supported by the National Research Foundation (NRF) grant funded by the Korean government (MSIP) (NRF-2016R1A2B3008363) and by a grant (No. 2011-0031630) from the Center for Advanced Soft Electronics under the Global Frontier Research Program of MSIP. The authors also acknowledge the contributions of K. Kim, M. Kim, J. Kim, J. Yang, J. Park, S. Lim, D. Nam, S. Han, and S. Kim.

References

1. Q.H. Wang, K. Kalantar-Zadeh, A. Kis, J.N. Coleman, M.S. Strano, Nat. Nanotechnol. **7**, 699 (2012)
2. K.F. Mak, J. Shan, Nat. Photonics **10**, 216 (2016)
3. A. Splendiani, L. Sun, Y. Zhang, T. Li, J. Kim, C.Y. Chim, G. Galli, F. Wang, Nano Lett. **10**, 1271 (2010)

4. K.F. Mak, C. Lee, J. Hone, J. Shan, T.F. Heinz, Phys. Rev. Lett. **105**, 136805 (2010)
5. F. Xia, H. Wang, D. Xiao, M. Dubey, A. Ramasubramaniam, Nat. Photonics **8**, 899 (2014)
6. A. Castellanos-Gomez, J. Phys. Chem. Lett. **6**, 4280 (2015)
7. D.Y. Qiu, F.H. da Jornada, S.G. Louie, Phys. Rev. Lett. **111**, 216805 (2013)
8. A. Chernikov, T.C. Berkelbach, H.M. Hill, A. Rigosi, Y. Li, O.B. Aslan, D.R. Reichman, M.S. Hybertsen, T.F. Heinz, Phys. Rev. Lett. **113**, 76802 (2014)
9. D. Kozawa, R. Kumar, A. Carvalho, K. Kumar Amara, W. Zhao, S. Wang, M. Toh, R.M. Ribeiro, A.H. Castro Neto, K. Matsuda, G. Eda, Nat. Commun. **5**, 4543 (2014)
10. Y. Li, A. Chernikov, X. Zhang, A. Rigosi, H.M. Hill, A.M. van der Zande, D.A. Chenet, E.-M. Shih, J. Hone, T.F. Heinz, Phys. Rev. B **90**, 205422 (2014)
11. E. del Corro, H. Terrones, A. Elias, C. Fantini, S. Feng, M.A. Nguyen, T.E. Mallouk, M. Terrones, M.A. Pimenta, ACS Nano **8**, 9629 (2014)
12. W. Zhao, Z. Ghorannevis, L. Chu, M. Toh, C. Kloc, P.-H. Tan, G. Eda, ACS Nano **7**, 791 (2013)
13. R.A. Bromley, R.B. Murray, A.D. Yoffe, J. Phys. C Solid State Phys. **5**, 759 (2001)
14. A.R. Beal, J.C. Knights, W.Y. Liang, J. Phys. C Solid State Phys. **5**, 3540 (1972)
15. R. Schmidt, I. Niehues, R. Schneider, M. Drüppel, T. Deilmann, M. Rohlfing, S.M. de Vasconcellos, A. Castellanos-Gomez, R. Bratschitsch, 2D Mater. **3**, 21011 (2016)
16. R. Frisenda, Y. Niu, P. Gant, A.J. Molina-Mendoza, R. Schmidt, R. Bratschitsch, J. Liu, L. Fu, D. Dumcenco, A. Kis, D.P. De Lara, A. Castellanos-Gomez, J. Phys. D. Appl. Phys. **50**, 74002 (2017)
17. T.A. Nguyen, J.-U. Lee, D. Yoon, H. Cheong, Sci. Rep. **4**, 4630 (2014)
18. R.R. Nair, P. Blake, A.N. Grigorenko, K.S. Novoselov, T.J. Booth, T. Stauber, N.M.R. Peres, A.K. Geim, Science **320**, 1308 (2008)
19. Z. Ye, T. Cao, K. O'Brien, H. Zhu, X. Yin, Y. Wang, S.G. Louie, X. Zhang, Nature **513**, 214 (2014)
20. M.M. Ugeda, A.J. Bradley, S.-F. Shi, F.H. da Jornada, Y. Zhang, D.Y. Qiu, W. Ruan, S.-K. Mo, Z. Hussain, Z.-X. Shen, F. Wang, S.G. Louie, M.F. Crommie, Nat. Mater. **13**, 1091 (2014)
21. K. He, N. Kumar, L. Zhao, Z. Wang, K.F. Mak, H. Zhao, J. Shan, Phys. Rev. Lett. **113**, 1 (2014)
22. A.R. Klots, A.K.M. Newaz, B. Wang, D. Prasai, H. Krzyzanowska, J. Lin, D. Caudel, N.J. Ghimire, J. Yan, B.L. Ivanov, K.A. Velizhanin, A. Burger, D.G. Mandrus, N.H. Tolk, S.T. Pantelides, K.I. Bolotin, Sci. Rep. **4**, 6608 (2014)
23. M. Cardona, *Light Scattering in Solids I* (Springer Berlin Heidelberg, Berlin/Heidelberg, 1983)
24. P.Y. Yu, M. Cardona, *Fundamentals of Semiconductors* (Springer Berlin Heidelberg, Berlin/Heidelberg, 2010)
25. H.M. Cheong, Y. Zhang, A. Mascarenhas, J.F. Geisz, Phys. Rev. B **61**, 13687 (2000)
26. C. Trallero-Giner, A. Cantarero, M. Cardona, M. Mora, Phys. Rev. B **45**, 6601 (1992)
27. B. Chakraborty, H.S.S.R. Matte, A.K. Sood, C.N.R. Rao, J. Raman Spectrosc. **44**, 92 (2013)
28. J.-H. Fan, P. Gao, A.-M. Zhang, B.-R. Zhu, H.-L. Zeng, X.-D. Cui, R. He, Q.-M. Zhang, J. Appl. Phys. **115**, 53527 (2014)
29. K. Gołasa, M. Grzeszczyk, P. Leszczyński, C. Faugeras, A.A.L. Nicolet, A. Wysmołek, M. Potemski, A. Babiński, Appl. Phys. Lett. **104**, 92106 (2014)
30. K. Gołasa, M. Grzeszczyk, R. Bozek, P. Leszczyński, A. Wysmołek, M. Potemski, A. Babiński, Solid State Commun. **197**, 53 (2014)
31. B.R. Carvalho, L.M. Malard, J.M. Alves, C. Fantini, M.A. Pimenta, Phys. Rev. Lett. **114**, 136403 (2015)
32. B.R. Carvalho, L.M. Malard, J.M. Alves, C. Fantini, M.A. Pimenta, Phys. Rev. Lett. **116**, 89904 (2016)
33. J.-U. Lee, J. Park, Y.-W. Son, H. Cheong, Nanoscale **7**, 3229 (2015)
34. T. Livneh, J.E. Spanier, 2D Mater. **2**, 35003 (2015)
35. M.A. Pimenta, E. del Corro, B.R. Carvalho, C. Fantini, L.M. Malard, Acc. Chem. Res. **48**, 41 (2015)

36. C. Lee, H. Yan, L.E. Brus, T.F. Heinz, J. Hone, S. Ryu, ACS Nano **4**, 2695 (2010)
37. Y. Zhao, X. Luo, H. Li, J. Zhang, P.T. Araujo, C.K. Gan, J. Wu, H. Zhang, S.Y. Quek, M.S. Dresselhaus, Q. Xiong, Nano Lett. **13**, 1007 (2013)
38. X. Zhang, W.P. Han, J.B. Wu, S. Milana, Y. Lu, Q.Q. Li, A.C. Ferrari, P.H. Tan, Phys. Rev. B **87**, 115413 (2013)
39. X. Luo, Y. Zhao, J. Zhang, Q. Xiong, S.Y. Quek, Phys. Rev. B **88**, 75320 (2013)
40. P. Tonndorf, R. Schmidt, P. Böttger, X. Zhang, J. Börner, A. Liebig, M. Albrecht, C. Kloc, O. Gordan, D.R.T. Zahn, S. Michaelis de Vasconcellos, R. Bratschitsch, Opt. Express **21**, 4908 (2013)
41. K. Kim, J.-U. Lee, D. Nam, H. Cheong, ACS Nano **10**, 8113 (2016)
42. S.-Y. Chen, C. Zheng, M.S. Fuhrer, J. Yan, Nano Lett. **15**, 2526 (2015)
43. W. Zhao, Z. Ghorannevis, K.K. Amara, J.R. Pang, M. Toh, X. Zhang, C. Kloc, P.H. Tan, G. Eda, Nanoscale **5**, 9677 (2013)
44. X. Luo, Y. Zhao, J. Zhang, M. Toh, C. Kloc, Q. Xiong, S.Y. Quek, Phys. Rev. B **88**, 195313 (2013)
45. J. Yang, J.-U. Lee, H. Cheong, FlatChem **3**, 64 (2017)
46. S. Kim, K. Kim, J.-U. Lee, H. Cheong, 2D Materials **4**, 045002 (2017)
47. E. del Corro, A. Botello-Méndez, Y. Gillet, A.L. Elias, H. Terrones, S. Feng, C. Fantini, D. Rhodes, N. Pradhan, L. Balicas, X. Gonze, J.-C. Charlier, M. Terrones, M.A. Pimenta, Nano Lett. **16**, 2363 (2016)
48. P. Soubelet, A.E. Bruchhausen, A. Fainstein, K. Nogajewski, C. Faugeras, Phys. Rev. B **93**, 155407 (2016)
49. B. Chakraborty, A. Bera, D.V.S. Muthu, S. Bhowmick, U.V. Waghmare, A.K. Sood, Phys. Rev. B **85**, 161403 (2012)
50. H.P.C. Miranda, S. Reichardt, G. Froehlicher, A. Molina-Sánchez, S. Berciaud, L. Wirtz, Nano Lett. **17**, 2381 (2017)
51. J. Ribeiro-Soares, R.M. Almeida, E.B. Barros, P.T. Araujo, M.S. Dresselhaus, L.G. Cançado, A. Jorio, Phys. Rev. B **90**, 115438 (2014)
52. B.R. Carvalho, Y. Wang, S. Mignuzzi, D. Roy, M. Terrones, C. Fantini, V.H. Crespi, L.M. Malard, M.A. Pimenta, Nat. Commun. **8**, 14670 (2017)
53. D. Nam, J.-U. Lee, H. Cheong, Sci. Rep. **5**, 17113 (2015)
54. A.S. Davydov, Sov. Phys. Uspekhi **7**, 145 (1964)
55. C.G. Wermuth, C.R. Ganellin, P. Lindberg, L.A. Mitscher, Pure Appl. Chem. **51**, 1129 (1979)
56. G. Froehlicher, E. Lorchat, F. Fernique, C. Joshi, A. Molina-Sánchez, L. Wirtz, S. Berciaud, Nano Lett. **15**, 6481 (2015)
57. J.-U. Lee, K. Kim, S. Han, G.H. Ryu, Z. Lee, H. Cheong, ACS Nano **10**, 1948 (2016)
58. Q.J. Song, Q.H. Tan, X. Zhang, J.B. Wu, B.W. Sheng, Y. Wan, X.Q. Wang, L. Dai, P.H. Tan, Phys. Rev. B **93**, 115409 (2016)
59. M. Grzeszczyk, K. Gołasa, M. Zinkiewicz, K. Nogajewski, M.R. Molas, M. Potemski, A. Wysmołek, A. Babiński, 2D Mater. **3**, 25010 (2016)
60. K. Gołasa, M. Grzeszczyk, M.R. Molas, M. Zinkiewicz, Ł. Bala, K. Nogajewski, M. Potemski, A. Wysmołek, A. Babiński, Nanophotonics **6**, 1281 (2017)
61. M. Staiger, R. Gillen, N. Scheuschner, O. Ochedowski, F. Kampmann, M. Schleberger, C. Thomsen, J. Maultzsch, Phys. Rev. B **91**, 195419 (2015)
62. A.-L. Phaneuf-L'Heureux, A. Favron, J.-F. Germain, P. Lavoie, P. Desjardins, R. Leonelli, R. Martel, S. Francoeur, Nano Lett. **16**, 7761 (2016)
63. Q.-H. Tan, Y.-J. Sun, X.-L. Liu, Y. Zhao, Q. Xiong, P.-H. Tan, J. Zhang, 2D Mater. **4**, 031007 (2017)
64. D.A. Chenet, O.B. Aslan, P.Y. Huang, C. Fan, A.M. van der Zande, T.F. Heinz, J.C. Hone, Nano Lett. **15**, 5667 (2015)
65. R. He, J. Yan, Z. Yin, Z. Ye, G. Ye, J. Cheng, J. Li, C.H. Lui, Nano Lett. **16**, 1404 (2016)
66. X.-F. Qiao, J.-B. Wu, L.-W. Zhou, J.-S. Qiao, W. Shi, T. Chen, X. Zhang, J. Zhang, W. Ji, P.-H. Tan, Nanoscale **8**, 8324 (2016)

67. H. Zhao, J. Wu, H. Zhong, Q. Guo, X. Wang, F. Xia, L. Yang, P. Tan, H. Wang, Nano Res. **8**, 3651 (2015)
68. E. Lorchat, G. Froehlicher, S. Berciaud, ACS Nano **10**, 2752 (2016)
69. M. Kim, S. Han, J.H. Kim, J.-U. Lee, Z. Lee, H. Cheong, 2D Mater. **3**, 34004 (2016)
70. J. Kim, J.-U. Lee, J. Lee, H.J. Park, Z. Lee, C. Lee, H. Cheong, Nanoscale **7**, 18708 (2015)
71. J. Wu, N. Mao, L. Xie, H. Xu, J. Zhang, Angew. Chem. **127**, 2396 (2015)
72. H.B. Ribeiro, M.A. Pimenta, C.J.S. de Matos, R.L. Moreira, A.S. Rodin, J.D. Zapata, E.A.T. de Souza, A.H. Castro Neto, ACS Nano **9**, 4270 (2015)

Chapter 10
Ultralow-Frequency Raman Spectroscopy of Two-dimensional Materials

Miao-Ling Lin and Ping-Heng Tan

Abstract In two-dimensional materials (2DMs), atoms within one layer (in-plane) are joined by covalent bonds, whereas van der Waals (vdW) interactions keep the layers together. Raman spectroscopy is a powerful tool for measuring the lattice vibrational modes in 2DMs, including the intralayer and interlayer vibrations, and has shown great potential for the characterizations of the layer number, interlayer coupling and layer-stacking configurations in 2DMs via the ultralow-frequency (ULF) interlayer vibrational modes. This chapter begins with an introduction of how the monolayer 2DMs stack to assemble a large family of two-dimensional systems (Section 10.1), which are likely to exhibit modified interlayer coupling and thus various ULF mode behaviours. In sequence, Section 10.2 provides a detailed description of the physical origins of the interlayer vibrations and the linear chain model (LCM) to depict their layer-number dependent frequencies. Subsequently, two popular Raman setups are introduced to perform the ULF modes measurements (Section 10.3). Then, we provide a review of the ULF Raman spectroscopy of various types of 2DMs, including: (1) layer-number dependent (Section 10.4.1) and (2) stacking-order dependent (Section 10.4.2) ULF Raman spectroscopy in isotropic 2DMs; (3) ULF Raman spectroscopy in anisotropic 2DMs(Section 10.4.3); and (4) ULF Raman modes in twisted 2DMs (Section 10.4.4) and heterostructures (Section 10.4.5).

The original version of this chapter was revised. A correction to this chapter is available at https://doi.org/10.1007/978-981-13-1828-3_12

M.-L. Lin
State Key Laboratory of Superlattices and Microstructures, Institute of Semiconductors, Chinese Academy of Sciences, Beijing, China

College of Materials Science and Opto-Electronic Technology, University of Chinese Academy of Science, Beijing, China

P.-H. Tan (✉)
State Key Laboratory of Superlattices and Microstructures, Institute of Semiconductors, Chinese Academy of Sciences, Beijing, China

Center of Materials Science and Opto-Electronics Engineering & CAS Center of Excellence in Topological Quantum Computation, University of Chinese Academy of Sciences, Beijing, China
e-mail: phtan@semi.ac.cn

P.-H. Tan (ed.), *Raman Spectroscopy of Two-Dimensional Materials*, Springer Series in Materials Science 276, https://doi.org/10.1007/978-981-13-1828-3_10

10.1 Introduction

Two dimensional (2D) materials can usually be exfoliated from bulk layered materials (LMs), which form strong in-plane chemical bonds, but have weak out-of-plane van der Waals (vdW) interactions [1, 2]. The enormous interest in 2D materials (2DMs) is fuelled by the desire to investigate the unique optoelectronic properties of these ultra-thin flakes. Actually, 2DMs constitute a large family [1, 3–7]. The diverse and novel physical properties of these 2DMs are usually ascribed to their different symmetries and various interlayer coupling [3, 5, 7]. Thus, 2DMs can be classified according to their symmetry [2]. Large advances have been made on the ultrathin 2DMs whose bulk forms belong to D_{6h} symmetry, such as graphite and TMDs. Recently, much attention has been paid to the 2DMs with low symmetry due to their in-plane anisotropy. 2DMs consist of one or more rigid layers. Each rigid layer can be regarded as one individual *layer* in 2DMs or LMs. The stacking of these layers along the c axis can form the corresponding multilayer (ML) 2DMs, and N layer (NL) 2DMs contains N rigid layers in the flakes. The symmetries in ML 2DMs are likely to be reduced because the out-of-plane translational symmetry will break. For example, the D_{6h} in graphite is reduced to D_{3h} and D_{3d} for ML graphenes (MLGs) with odd and even layer numbers (ONLs and ENLs), respectively, whereas it remains D_{6h} for monolayer graphene (1LG).

The stacking order of the layers in 2DMs can be different, which leads their symmetries as well as the properties to become more diverse. Taking the in-plane isotropic MoS_2 as a simple example, as shown in Fig. 10.1a, monolayer MoS_2 belongs to D_{3h} symmetry, referred to as 1H-MoS_2. The stacking of 1H-MoS_2 along the c axis can generate two common polytypes, 2H (H represents hexagonal) and 3R (R represents rhombohedral), as shown in Fig. 10.1b. The obvious difference between 2H and 3R begins from the second layer. The stacking along the c axis in 3R only undergoes a layer shift, whereas the Nth layer of 2H stacking is rotated by 180° with respect to the $(N-1)$th layer. Thus, the 3R-stacking MoS_2 from the monolayer to bulk preserves noncentrosymmetry. However, in 2H stacking, ENL MoS_2 possesses an inversion symmetry point, but ONL MoS_2 is noncentrosymmetric. Furthermore, the metal atoms can also behave in an octahedral coordination in which the spatial inversion symmetry is preserved in the monolayer MoS_2, forming the new polytype 1T stacking, as depicted in right panel of Fig. 10.1b. The different symmetries of these three stackings lead to different interlayer couplings and electronic properties [8]. The symmetries and interlayer couplings in polytypic MoS_2 and other 2DMs can be easily distinguished by ultralow-frequency (ULF) Raman spectroscopy [2, 6, 9–11]. In addition, during the mechanical exfoliation of 2DM flakes, one thin flake with layer number n (denoted as nL-2DM, $n = 1$, 2,...) may be randomly folded onto another flake with a layer number m (denoted as mL-2DM, $m = 1$, 2,...) to form a twisted-2DM (t($m+n$)L-2DM). For example, 1LG can be stacked onto another 1LG and 2LG with a twist angle θ_t to form a twisted 2LG and 3LG, denoted as t(1 + 1)LG and t(1 + 2)LG, as shown in Fig. 10.1c, d, respectively. These twisted systems have been achieved in graphene [12–14],

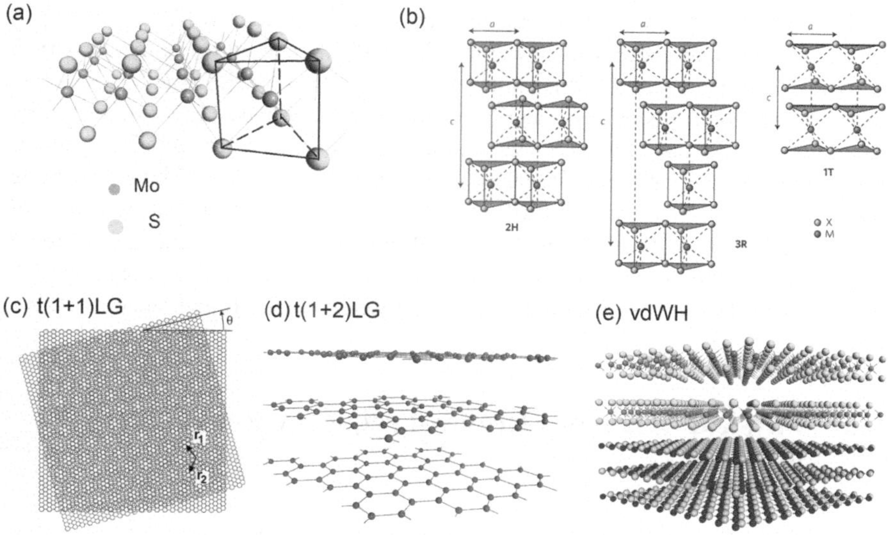

Fig. 10.1 (**a**) Schematic diagram of the three-dimensional structure of 1L-MoS_2. The triangle indicated by black lines display the trigonal prismatic coordination of Mo atoms. (**b**) Schematic diagrams of the three typical structural polytypes of MX_2: 2H, 3R and 1T. a and c represent the in-plane and out-of-plane lattice constants, respectively. (Reproduced with permission from Ref. [4]). (**c**) Schematics of a rotationally stacked bilayer graphene with the slant one sitting on the top. The top and bottom layers are rotated with respect to each other by a generic angle θ_t, generating a periodic Moiré pattern. $\mathbf{r}_1$ and $\mathbf{r}_2$ are the direct vectors defining the supercell. The schematic of (**d**) t(1 + 2)LG with $\theta_t = 21.8°$ and (**e**) 2D heterostructures from the side view

MoS_2, $MoSe_2$ [15–17] and so on. For a given layer number N, the choices of m, n and the relative twist angles between two individual constituents leads to another large family of 2DMs with a variety of optical and electronic properties. The recent research frontier has also advanced to investigate hybrid systems of 2DMs. The flat and inert surfaces of these systems enable us to assemble stacks of different 2D heterostructures in a chosen sequence, coupled vertically only by vdW interactions (Fig. 10.1e), which can offer large opportunities for designing the functionalities of these vdW heterostructures (vdWHs). The interfacial interactions between two atomic layers from adjacent constituents of vdWHs can dramatically influence the properties of vdWHs and induce remarkable phenomena that are absent in individual constituents [18–20].

Thus, developing a direct way to detect the interlayer and interfacial coupling in all types of the above 2DMs and related vdWHs is critically needed to investigate their physical and chemical properties. Raman scattering is an important and versatile tool for probing lattice vibrations, including the intralayer and interlayer modes [2, 6, 9–11]. In contrast to the intralayer vibrations, which are driven by the chemical bonds in the plane, interlayer vibrations, i.e., the shear and layer-breathing modes, are determined by the interlayer coupling. It should be noted that the shear mode is referred to as the C mode in multilayer graphenes (MLGs) because it provides a direct measurement of the interlayer $Coupling$ and was first observed in AB-stacked MLGs [21]. Other notations for the shear modes, such as SM, and for layer-breathing modes, such as the B modes or LBM have been introduced by various research groups [10]. As a general notation for interlayer vibrational modes in layered materials, in this chapter, we denote the shear mode and layer breathing mode as the S and LB modes, [10], respectively. Because the restoring forces of the interlayer modes are weak due to the vdW interactions, the frequencies of these modes are expected to arise in the ULF region. In the following sections, we will provide a broad overview of ULF Raman spectroscopy in 2DMs and related vdWHs.

10.2 Physical Origin of Ultralow-Frequency Raman Modes in 2DMs

In ML 2DMs, the relative motions of the atoms in each monolayer result in the so-called intralayer Raman modes, which have high frequencies, whreas the interlayer vibrations in the ULF region are induced by the relative motions of adjacent layers (Fig. 10.2). As referred to in Sect. 10.1, there are two typical interlayer Raman modes, the S and LB modes, which are perpendicular or parallel to their normals, respectively. All of these lattice vibrations at the Brillouin zone (BZ) centre (Γ) can be expressed by the irreducible representations based on their symmetries, and the number of the vibrational branches ($3\times n$) depends on the number of atoms (n) in the unit cell. For example, there are two atoms in the unit cell of 1LG (D_{6h}), so there are six vibrations (three acoustic phonons and three optical phonons), which can be represented by: $\Gamma = E_{2g} + E_{1u} + B_{2g} + A_{2u}$. The E_{1u} and A_{2u} are assigned to the acoustic phonons, whereas the others correspond to the optical phonons at Γ. Because there is only one layer in 1LG, no interlayer Raman modes are expected. For graphite, all the modes become doublets: $\Gamma = 2(E_{2g} + E_{1u} + B_{2g} + A_{2u})$. The interlayer vibrations can be denoted E_{2g} (for the S mode) and B_{2g} (for the LB mode) [21]. Due to the different symmetries between ENL graphenes (ENLGs) and ONL graphenes (ONLGs), the vibrations can be represented by $\Gamma = NE_g + NE_u + NA_{1g} + NA_{2u}$ for ENLGs and $\Gamma = (N-1)A_1' + (N+1)A_2'' + (N+1)E' + (N-1)E''$ for ONLGs, respectively. For a NLG, there are $N-1$ LB modes and $N-1$ pairs of doubly degenerate S modes. Hence, the interlayer vibrations in NLG can be represented by $\frac{N}{2}(A_{1g}+E_g) + (\frac{N}{2}-1)(A_{2u}+E_u)$ for ENLG and $\frac{(N-1)}{2}(E'+E''+A_1'+A_2'')$ for ONLG, as shown in Fig. 10.2a, b. For the in-plane isotropic N-layer 2DMs (denoted as NL-2DMs), each S mode is doubly degenerate and belong to irreducible representation E, whereas the LB modes belong to irreducible representation A or B. Therefore, the S modes are independent on the in-plane polarization, whereas the LB modes exhibit high dependence on the excitation polarization direction, with the result that the LB

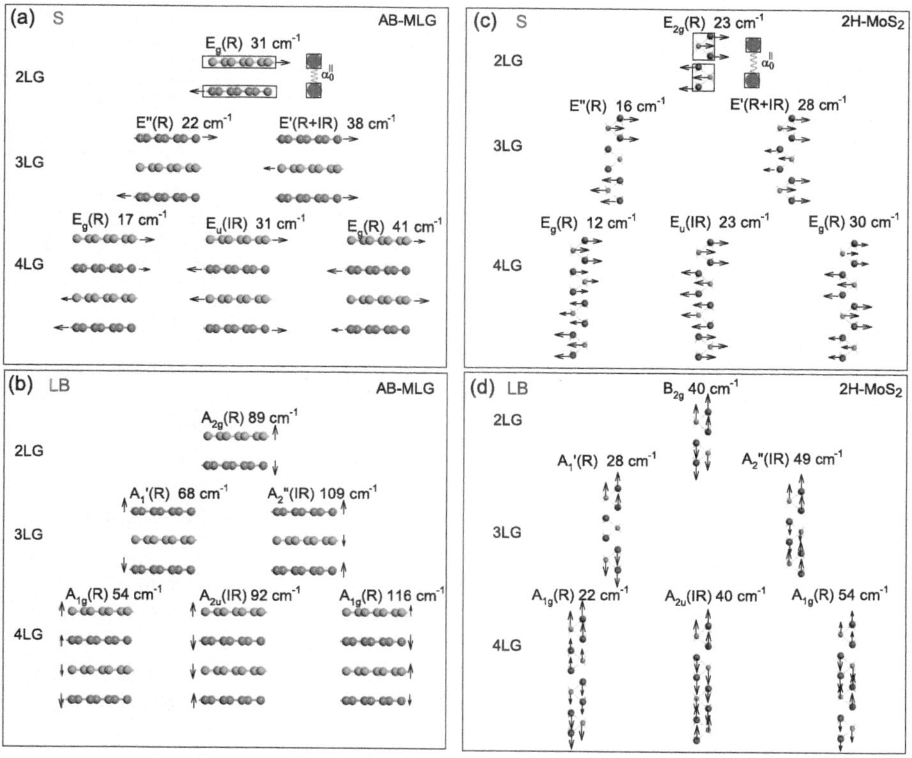

Fig. 10.2 Symmetry, frequency, Raman activity and normal mode displacements for the S (**a**) and LB (**b**) modes in AB-(2-4)LG and the S (**c**) and LB (**d**) modes in 2-4L 2H-MoS_2. R represents Raman active, and IR represents infrared active. The frequencies are calculated from the LCM. The arrows indicate the vibrational directions of the corresponding layers and the length represents the amplitude of the displacement

modes cannot be detected in the cross polarization (HV). This implies that the LB and S modes can be distinguished via the polarized Raman spectroscopy. The assignments of interlayer vibrations in MoS_2 and other TMDs are similar to those in MLGs (Fig. 10.2c, d) due to the semblable symmetries [2, 6]. Moreover, all lattice vibrations in 2DMs, especially the interlayer modes, can be deduced from their symmetries, and those for typical 2DMs are listed in Table 10.1. In general, not all S and LB modes can be observed in Raman spectroscopy because of Raman inactivities or zero Raman intensities in special Raman configurations, as well as the weak electron-phonon interactions.

Here, we present an insight into the frequencies of these interlayer vibrations. For the S and LB modes in 2DMs, the relative displacements between the atoms in each rigid layer can be ignored. Therefore, each rigid layer can be treated as a single ball and only the nearest interlayer coupling is considered, as shown in Fig. 10.2a, c, when we calculate the frequencies of the S and LB modes in 2DMs.

Table 10.1 Symmetries, irreducible representations of Raman modes at Γ and interlayer vibrations in 1L, ONL and ENL 2DMs and the corresponding bulk LMs. All E are doubly degenerate

Materials	Thickness	Symmetry	Irreducible representations	Interlayer vibrations
Graphenes	1L	D_{6h}^1	$E_{2g}+E_{1u}+B_{2g}+A_{2u}$	–
	ONL	D_{3h}^1	$(N\text{-}1)A_1'+(N+1)A_2''$ $+(N+1)E'+(N\text{-}1)E''$	$\frac{(N-1)}{2}(A_1'+A_2''+E'+E'')$
	ENL	D_{3d}^3	$NE_g+NE_u+NA_{1g}+NA_{2u}$	$\frac{N}{2}(E_g+A_{1g})+(\frac{N}{2}-1)(E_u+A_{2u})$
	Bulk	D_{6h}^4	$2(E_{2g}+E_{1u}+B_{2g}+A_{2u})$	$E_{2g}+B_{2g}$
2H-MX_2	1L	D_{3h}^1	$A_1'+E''+2(A_2''+E')$	–
(M = Mo,W,	ONL	D_{3h}^1	$\frac{3N-1}{2}(A_1'+E'')+\frac{3N+1}{2}(A_2''+E')$	$\frac{N-1}{2}(A_1'+E''+A_2''+E')$
X = S,Se,Te)	ENL($N\neq 2$)	D_{3d}^3	$\frac{3N}{2}(A_{1g}+A_{2u}+E_g+E_u)$	$\frac{N}{2}(A_{1g}+E_g)+(\frac{N}{2}-1)(A_{2u}+E_u)$
	bulk	D_{6h}^4	$A_{1g}+2A_{2u}+B_{1u}+2B_{2g}$ $+E_{1g}+2E_{1u}+E_{2u}+2E_{2g}$	$E_{2g}+B_{2g}$
$Bi_2Se(Te)_3$	ONL	D_{3d}^3	$\frac{5N-1}{2}(A_{1g}+E_g)+\frac{5N+1}{2}(A_{2u}+E_u)$	$\frac{N-1}{2}(A_{1g}+E_g+A_{2u}+E_u)$
	ENL	D_{3d}^3	$\frac{5N}{2}(A_{1g}+E_g+A_{2u}+E_u)$	$\frac{N}{2}(A_{1g}+E_g)+(\frac{N}{2}-1)(A_{2u}+E_u)$
	Bulk	D_{3d}^5	$2A_{1g}+2E_g+3A_{2u}+3E_u$	–
BP	ONL	D_{2h}^7	$N(2A_g+B_{1g}+2B_{2g}+B_{3g}$ $+A_u+2B_{1u}+B_{2u}+2B_{3u})$	$\frac{N-1}{2}(A_g+B_{2g}+B_{3g}$ $+B_{1u}+B_{2u}+B_{3u})$
	ENL	D_{2h}^{11}	$N(2A_g+B_{1g}+2B_{2g}+B_{3g}$ $+A_u+2B_{1u}+B_{2u}+2B_{3u})$	$\frac{N}{2}(A_g+B_{2g}+B_{3g})$ $+(\frac{N}{2}-1)(B_{1u}+B_{2u}+B_{3u})$
	Bulk	D_{2h}^{18}	$2A_g+B_{1g}+2B_{2g}+B_{3g}$ $+A_u+2B_{1u}+B_{2u}+2B_{3u}$	$A_g+B_{1u}+B_{3u}$
β-GaS(Se)	Bulk	D_{6h}^4	$2(E_{2g}+E_{1u}+B_{2g}+A_{2u}$ $+A_{1g}+B_{1u}+E_{1g}+E_{2u})$	$E_{2g}+B_{2g}$
ϵ-GaS(Se)	Bulk	D_{3h}^1	$4(A_1'+E''+A_2''+E')$	$A_1'+E'$
GeS(Se), SnS(Se)	Bulk	D_{2h}^{16}	$4A_{1g}+4B_{1u}+2B_{1g}+2A_u$ $+4B_{2g}+4B_{3u}+2B_{3g}+2B_{2u}$	$A_g+B_{2g}+B_{3g}$
CuS(>55k)	Bulk	D_{6h}^4	$2A_{1g}+4A_{2u}+4B_{2g}+2B_{1u}$ $+2E_{1g}+4E_{1u}+4E_{2g}+2E_{2u}$	$E_{2g}+B_{2g}$
CuS(<55k)	Bulk	D_{2h}^{17}	$6A_g+4B_{1g}+2B_{2g}+6B_{3g}$ $+2A_u+5B_{1u}+5B_{2u}+3B_{3u}$	$A_g+B_{2g}+B_{3g}$
2H-$SnS(Se)_2$	Bulk	D_{3d}^3	$A_{1g}+2A_{2u}+E_g+2E_u$	$A_{1g}+E_g$
4H-$SnS(Se)_2$	Bulk	C_{6v}^4	$3(A_1+B_1+E_1+E_2)$	A_1+E_2
6Ha-$SnS(Se)_2$	Bulk	C_{3v}	$9(A_1+E)$	A_1+E
6Hb-$SnS(Se)_2$	Bulk	D_{3d}^3	$4A_{1g}+5A_{2u}+4E_g+5E_u$	$A_{1g}+E_g$
GaTe	Bulk	C_{2h}^3	$6(2A_g+B_g+A_u+2B_u)$	$2A_g+B_g$
$ReS(Se)_2$	Bulk	C_i	$18(A''+A')$	$2A''+A'$

This model is known as the linear chain model (LCM) [21]. Assuming the interlayer interactions per unit area between two adjacent rigid layers as $\alpha_0^{\parallel}$ and $\alpha_0^{\perp}$ for the S and LB modes, respectively, the frequencies and atomic displacements of the S and LB modes in NL-2DMs can be calculated by solving linear homogenous equations as follows [13, 21]:

$$\omega_i^2 \mathbf{u}_i = \frac{1}{2\pi^2 c^2 \mu} \mathbf{D} \mathbf{u}_i, \tag{10.1}$$

where $\mathbf{u}_i$ is the phonon eigenvector of the mode i with frequency ω_i, μ is the mass of each rigid layer per unit area, $c = 3.0 \times 10^{10} \mathrm{cm} \cdot \mathrm{s}^{-1}$ is the speed of light and $\mathbf{D}$ is the shear or layer-breathing part of the force constant matrix. By diagonalizing the corresponding N×N (tridiagonal) dynamics matrix, the frequency ω_i of the ith vibrational mode is given by

$$\omega_{S,i} = \frac{1}{\pi c}\sqrt{\alpha_0^{\parallel}/\mu}\sin\left(\frac{i\pi}{2N}\right) \tag{10.2a}$$

$$\omega_{LB,i} = \frac{1}{\pi c}\sqrt{\alpha_0^{\perp}/\mu}\sin\left(\frac{i\pi}{2N}\right), \tag{10.2b}$$

where $i = 1,2,\ldots, N$-1. We usually denote these N-1 S modes and N-1 LB modes as $S_{N,N-i}$ and $LB_{N,N-i}$, respectively, where N is the layer number and i is the number of phonon branches. The $i = N-1$ branch, i.e., S_{N1} and LB_{N1}, corresponds to the modes with the highest frequency, whereas the $i = 1$ branch, i.e., $S_{N,N-1}$ and $LB_{N,N-1}$, correspond the modes with the lowest frequency. The corresponding ith displacement eigenvector $\nu_j^{(i)}$ is given by

$$\nu_j^{(i)} = cos[\frac{i(2j-1)\pi}{2N}], \tag{10.3}$$

where j labels the layers. Figure 10.2a, b shows the frequencies and corresponding normal mode displacements of the S and LB modes in 2-4LG, and those for MoS_2 are shown in Fig. 10.2c, d, in which R and IR represent Raman active and infrared active, respectively.

In the case of bulk 2DMs, $N \to \infty$, $\omega(S_{bulk}) = \frac{1}{\pi c}\sqrt{\alpha_0^{\parallel}/\mu}$ and $\omega(LB_{bulk}) = \frac{1}{\pi c}\sqrt{\alpha_0^{\perp}/\mu}$. Thus, the equations for $\omega(S_{N,N-i})$ and $\omega(LB_{N,N-i})$ can be reduced to

$$\omega(S_{N,N-i}) = \omega(S_{bulk}) sin(i\pi/2N), \tag{10.4}$$

and

$$\omega(LB_{N,N-i}) = \omega(LB_{bulk}) sin(i\pi/2N), \tag{10.5}$$

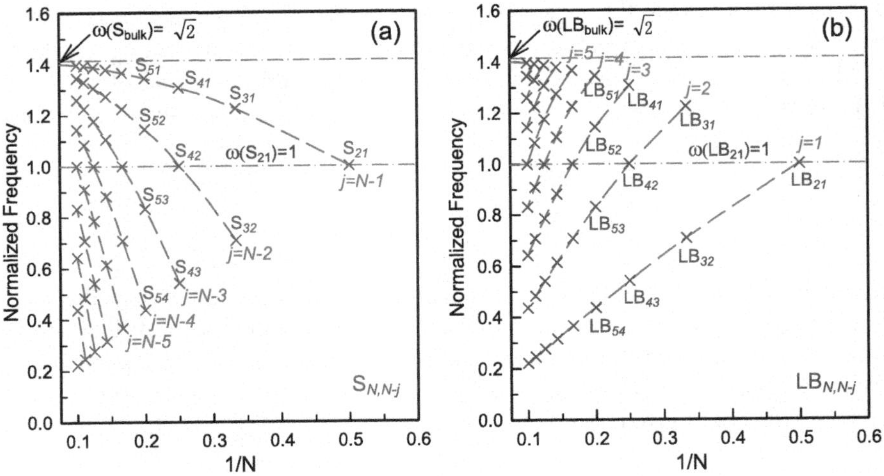

Fig. 10.3 Frequency of the S (**a**) and LB (**b**) modes in NL-2DMs as a function of inverse N, where $\omega(S_{21})$ and $\omega(LB_{21})$ are normalized to 1. (Reproduced with permission from Ref. [2])

respectively. In bilayer 2DMs (2L-2DMs), $\omega(S_{21})=\frac{1}{\sqrt{2}}\omega(S_{bulk})$, $\omega(LB_{21})=\frac{1}{\sqrt{2}}\omega(LB_{bulk})$. That is, once the $\omega(S_{bulk})$ and $\omega(LB_{bulk})$ are normalized to $\sqrt{2}$, $\omega(S_{21})$ and $\omega(LB_{21})$ is equal to 1. Figure 10.3a shows the branches ($i = N$-1,N-2,...) for the S modes, and Fig. 10.3b shows the branches (i = 1,2,...) for the LB modes.

The LCM provides a convenient way to describe the interlayer Raman modes in 2DMs, and once the frequencies of the bulk materials are ascertained, the frequencies of the interlayer modes in the corresponding NL-2DMs can be determined. This method can be extended to the 2D alloys and the heterostructures. Based on $\omega(S_{bulk})$, the shear modulus can be obtained by multiplying $\alpha_0^{\parallel}$ with the interlayer distance. For MoS_2, $\omega(S_{bulk})$ ~32.5 cm^{-1}, $\mu = 3.0\times10^{-7}$ g cm^{-2}, so the interlayer force constant $\alpha_0^{\parallel} = 2.82\times10^{19}$ N/m^3 and the shear modulus is ~18.9 GPa. The shear modulus of other typical 2DMs, as well as the in-plane constants (a), interlayer distance (d), $\omega(S_{bulk})$, and interlayer force constant per unit area for the S modes ($\alpha_0^{\parallel}$) are summarized in Table 10.2 [2].

As discussed above, only the nearest-neighbor interlayer coupling is considered in the LCM. However, in some special 2DMs, such as MLG and tMLG, the LCM with only nearest-neighbor interlayer interactions may be insufficient for reproducing the frequencies of the LB modes [14]. Then, it is necessary to introduce an interlayer force constant between the next-nearest neighbor layers in the LCM, which is referred as 2LCM. This will be discussed in detail in the Chap. 1 on the Raman spectroscopy of graphene materials.

Table 10.2 In-plane lattice constant (a), interlayer distance (d), $\omega(S_{bulk})$, interlayer force constants per unit area ($\alpha_0^{\parallel}$), shear modulus (C_{44}) determined by the C mode frequency and that (C_{44}^*) obtained by other methods

Crystal	a (10^{-8} cm)	d (10^{-8} cm)	$\omega(S_{bulk})$ (cm^{-1})	$\alpha_0^{\parallel}$ (10^{19} N/m^3)	C_{44} (GPa)	C_{44}^*(GPa)
MoS_2	3.15	6.15	32.5	2.82	18.9	18.6[a]
$MoSe_2$	3.29	6.46	25.5	2.60	16.8	
WS_2	3.18	6.25	27.5	3.16	19.7	
WSe_2	3.28	6.48	23.8	3.06	19.8	
GaSe	3.75	7.96	19.3	1.44	11.5	9.0 ± 1[b]
GaS	3.59	7.75	22.4	1.38	10.6	9.96 ± 0.15[b]
CuS	3.79	8.17	19.3	1.73	14.2	
Graphite	2.46	3.35	43.5	1.28	04.3	4.6[c]
BN	2.50	3.33	52.2	1.83	06.1	
$NbSe_2$	3.44	6.27	28.4	2.92	18.3	19.0[a]
GeSe	4.38[d]	5.41	40.0[f]	4.26	23.1	
GeSe	3.83[e]	5.41	39.0[g]	4.05	21.9	
Bi_2Te_3	4.38	2.61	25.46	4.57	11.93	
Bi_2Se_3	4.14	2.58	17.96	5.26	13.57	
$ReSe_2$	6.7[h]/6.6[i]	6.70	17.11/18.67	1.78/1.94	11.96/13.04	
ReS_2[j]	6.38/6.52	6.20	24.75/19.52	2.54/1.55	15.75/9.61	
ReS_2[k]	6.43/6.52	6.20	21.35	1.89	11.72	

[a]Ref. [22]
[b]Ref. [23]
[c]Ref. [24]
[d]c = 4.38
[e]b = 3.83
[f]Along *c* axis
[g]Along *b* axis
[h]Along *a* axis
[i]Along *b* axis
[j]Anisotropic-like structure
[k]Isotropic-like structure

10.3 Techniques for ULF Raman Spectroscopy

Compared with the intralayer vibrations in high-frequency regions, the interlayer modes are usually much lower in frequency($<100\,cm^{-1}$) and closer to the Rayeleigh line, the measurements of which are limited in the Raman system with a monochromator and a notch or edge filters. In recent years, in order to access the ULF Raman signal, several techniques have been applied to the confocal Raman system [25–31]. Here, we only introduce two typical Raman setups for ULF detection: multiple cascaded monochromators [32–35] and single monochromators with notch filters (NFs) based on volume Bragg grating (VBG) techniques [13, 14, 21].

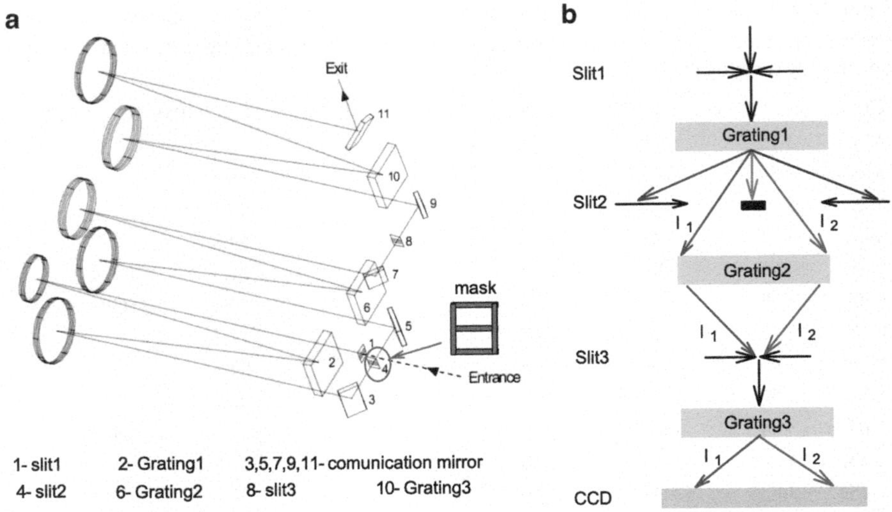

Fig. 10.4 Diagram of products (**a**) and a schematic (**b**) of the tri-grating spectrometer

The standard apparatus to access a Raman signal in the ULF region is to use a subtractive mode of double or triple cascaded high-resolution monochromators, which provides flexible operations down to 5–10 cm^{-1} at different wavelengths [31, 36]. Figure 10.4a shows a diagram of the products in the triple grating spectrometer and Fig. 10.4b is its schematic diagram. The tri-grating spectrometer consists of two parts, in which the first and second grating (1800 grooves/mm) together with a mask (denoted as slit2) and intermediate slit (slit3) are included in the first part to serve as a notch filter and the third grating behaves as a monochromator. The first two gratings serve as a set of aberration-corrected holographic gratings. In general, the scattering light is collected and imaged onto the entrance slit of the spectrometer. Once the scattering light traverses the first grating, it becomes spatially separated according to the various wavelengths. Then, the Rayleigh line is removed by the mask while the Raman scattering light passes through the slit without any obstructions, as shown in Fig. 10.4b. Grating2 is used to converge all the Raman scattering light together, and slit3 removes the stray light caused by diffraction from the mask. Meanwhile, this slit forms the entrance to the second part of the spectrometer. The third grating disperses the scattering light again and then the light is focused onto the detectors. If the mask is removed, the tri-grating spectrometer can serve as an additive mode to achieve superior high resolution. Although the ULF Raman signals are accessible via the multiple cascaded high-resolution monochromators [31, 36–42], the throughput of this setup is really low, at least one order of magnitude lower than that of a single monochromator, resulting in more accumulated time and limitations on potential applications in various 2DMs.

Recently, large advances of VBG-based NFs (BNFs) has enabled the detection of ULF Raman modes even when a single stage Raman spectrometer is used [21]. The

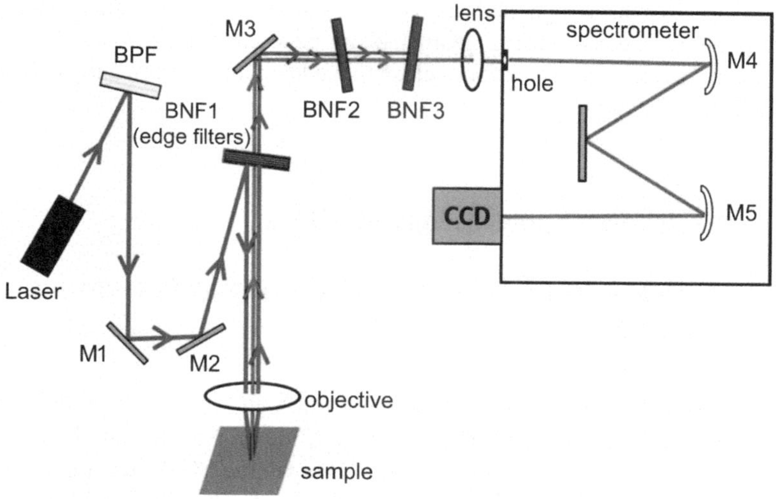

Fig. 10.5 Schematic diagram of a single monochromator with three BNFs [21]

narrow bandwidth (~5–10 cm^{-1}) and high transmittance (up to 80–90% depending on the laser wavelength) for each filter make it possible to measure ULF Raman signals with a high signal-to-noise ratio via easy operation. To effectively reject the Rayleigh line, 3–4 VBG-based NFs with an optical density of 3 or 4 and a spectral bandwidth of ~5–10 cm^{-1} are usually used. The modified configuration of the three BNFs in a single monochromator is shown in Fig. 10.5 and a similar arrangement can be implemented for other spectrometers. The spectral width of VBG-based bandpass filters (BPFs) can be as small as 5–10 cm^{-1} to remove the plasma lines of the lasers. Because the BPF is a reflecting filter, at least two mirrors are necessary to align the laser excitation to the center of the BNFs. The excitation light is then focused on the samples and at the same time the scattered light including the Rayleigh line is collected by the objective. Three BNFs are usually well tuned to reach deep blocking at the laser line up to OD ~9–12. Afterwards, the ULF Raman signals are effectively and selectively guided into the monochromator. The throughput of the Raman setup with this configuration is much higher than that of the tri-grating spectrometer. This setup has been used to probe the S modes of AB-MLGs [21], the ULF interlayer vibrations in various 2DMs [13, 14, 43–47] and the acoustic phonons in nanostructures [48]. Recently, ULF Raman measurements down to 2 cm^{-1} have been accessible at 488 nm excitation, approaching the Brillouin scattering region [49].

In addition, the greater the number of BNFs that are used to obtain fine laser line attenuation, the lower the transmittance for the ULF Raman signals will be. This implies that a monochromator with BNFs may be limited in detecting very weak ULF Raman modes. Thus, a new Raman configuration with nano-edge longpass filters has been introduced [50]. The ULF Raman modes down to 10 cm^{-1} can be detected with high throughput and easy operation via this Raman system. Moreover, a cross-polarized backscattering geometry can be used to suppress the Rayleigh signal and obtain Raman spectra very close to the laser line, significantly enhancing the signal-to-noise ratio.

10.4 ULF Raman Spectroscopy of 2DMs

10.4.1 N-Dependent ULF Raman Spectroscopy in Isotropic NL-2DMs

As elucidated in Sect. 10.2, the frequencies of the S and LB modes in bulk 2DMs and 2L-2DMs are directly related to $\alpha_0^{\parallel}$ and $\alpha_0^{\perp}$, therefore, one can obtain all the frequencies of the S and LB modes in *N*L-2DMs via the LCM once the S and LB modes are detected in 2L-2DMs and bulk counterpart. Thus, the LCM provides a convenient way to describe the interlayer Raman modes in 2DMs and this method can be extended to various 2DMs, including 2D crystals, 2D alloys and vdWHs. The intralayer force constants in each rigid layer are much larger than $\alpha_0^{\parallel}$ and $\alpha_0^{\perp}$. They can be revealed from the frequencies of intralayer optical phonons in 1L-2DMs, [44]. Take in-plane isotropic MoS_2 as an example, based on the diatomic chain model (DCM), [44] the relative displacements between molybdenum and the sulfur layers in the rigid S-Mo-S trilayer are very small for all S and LB modes, ~0.6% in 2L-MoS_2, and the relative displacement decreases with increasing N. Thus, all the atoms in each rigid S-Mo-S trilayer can be considered a single ball with a mass of $M = m_{Mo} + 2m_S$. The balls are coupled with each other by the interlayer coupling of $\alpha_0^{\parallel}$ (~2.82×10^{19} N/m^3) for the S modes and of $\alpha_0^{\perp}$ (~8.9×10^{19} N/m^3) for the LB modes, which can be deduced from the peak positions of the S_{21} and LB_{21} modes in 2L-MoS_2 and the S_{bulk}, LB_{bulk} mode in bulk MoS_2. The eigen-equations related to the N-dependent interlayer vibrations can be analytically solved from Eqs. 10.4 and 10.5. Figure 10.6a shows the calculated frequencies of the branches ($i = N$-1,N-2,...) of the S modes and Fig. 10.6b shows those of the branches (i = 1,2,...) of the LB modes in *N*L-MoS_2 based on the LCM and the DCM. Figure 10.6c depicts the Raman spectra of the S and LB modes in 1–20L MoS_2, whose experimental frequencies are summarized in Fig. 10.6a, b in circles. The sizes of the circles indicate the relative intensities of the corresponding modes. The experimental results are in consistent with those from the LCM and DCM, indicating that the interactions between MoS_2 flakes and substrate are sufficiently weak to be neglected. Furthermore, it is easy to indicate that the S modes in *N*L-MoS_2 primarily come from the branches of $i = N$-1,N-3,N-5, whereas the LB modes primarily come from the branches of i = 1,3,5, with the result that the frequencies of the S and LB modes in the Raman spectra undergo an opposite tendency with increasing N. In bulk materials, the LB mode is Raman inactive and the S mode at 34 cm^{-1} is Raman active. The peak positions and FWHM of the S_{N1}

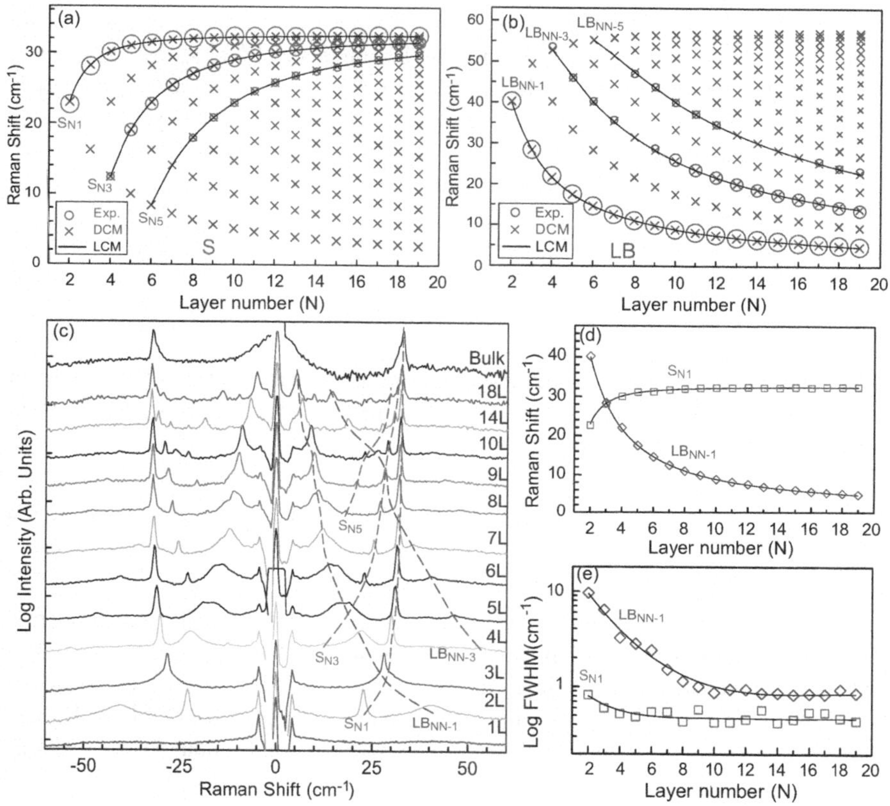

Fig. 10.6 Peak positions of the (**a**) S and (**b**) LB modes of NL-MoS_2 as a function of N. The circles are the experimental peak positions whose diameters indicate the Raman intensity of each mode. The crosses are the results from the DCM, whereas the solid lines are partial results from the LCM. (**c**) Stokes and anti-Stokes Raman spectra of NL-MoS_2 (N = 1–10L, 14L, 18L) and bulk MoS_2. (**d**) Peak positions and (**e**) FWHM of the S_{21} and LB_{21} modes as a function of N. (Reproduced with permission from Ref. [44])

and LB_{NN-1} modes are also represented in Fig. 10.6d, e, respectively. The peak positions of these two branches are in good agreement with those from the LCM. The larger FWHM of the LB modes compared with the S modes can be interpreted as an anharmonic feature of the LB modes with a significant enhancement of phonon-phonon scattering with decreasing N.

The symmetry of MoS_2 flakes determines whether the S and LB modes are Raman active in NL-MoS_2, whereas the intensity of these modes depends on the Raman scattering configurations and the electron-phonon coupling (EPC). Here we only give a detailed explanation of the symmetry and Raman scattering configurations because the calculation of EPC is very complicated. As respect to the irreducible representation from symmetry there are $\frac{N-1}{2}(E' + E'')$ for the S modes and $\frac{N-1}{2}(A_1'+A_2'')$ for the LB modes in ONL-MoS_2 whereas there are $\frac{N}{2}E_g+\frac{N-2}{2}E_u$ for the S modes and $\frac{N}{2}A_{1g}+\frac{N-2}{2}A_{2u}$ for the LB modes in ENL-MoS_2. E′, E″, A_1', A_{1g} and E_g are Raman active, whereas E_u and A_{2u} are infrared active and A_2'' is silent, indicating that E_u, A_{2u} and A_2'' cannot be detected via Raman spectroscopy.

Even the Raman active modes are not always observed in Raman spectra. For example, the Raman tensor of the E″ mode can be written as

$$R(E'') = \begin{pmatrix} 0 & 0 & 0 \\ 0 & 0 & c \\ 0 & c & 0 \end{pmatrix} \tag{10.6}$$

Under the backscattering configuration, assuming the wavevectors of the incident and scattering light are $\overrightarrow{e_i} = (\cos\theta, \sin\theta, 0)$ and $\overrightarrow{e_s} = (\cos\delta, \sin\delta, 0)$, the Raman intensity of the E″ mode is

$$I(E'') \propto \left| \overrightarrow{e_s} \cdot R \cdot \overrightarrow{e_i}^T \right|^2 = 0 \tag{10.7}$$

Thus, I(E″) is always zero in the backscattering configuration.

Other TMDs show structure and symmetry similar to MoS_2 flakes, indicating that the above discussions can be applied to these materials by replacing the $\omega(LB_{21})$ and $\omega(S_{21})$ with the corresponding values of other 2L-TMDs. This is also true for all the in-plane isotropic 2DMs. However, whether they are observable in Raman spectroscopy requires detailed analysis of their symmetries, Raman tensor and EPC.

10.4.2 Stacking-Order Dependent ULF Raman Spectroscopy in Isotropic 2DMs

The stacking order in few-layer 2DMs results in different structural symmetries and layer-to-layer interactions, and thus paves a way to controlling their electronic and optical properties. Indeed, the stacking order governs the symmetry and afterwards the Coulomb interactions, which are crucial for mediating novel physical properties, such as the semiconductor-to-metal transition, valley polarization, superconductivity, magnetism, and the charge density wave [51–55]. For example, AB-3LG is semi-metallic similar to a monolayer, whereas ABC-3LG is predicted to be a tunable band gap semiconductor under an external electric field. Interest in stacking effects has been rejuvenated with these phenomena emerging in few-layer 2DMs, which provides one more way to modify the properties, making 2DMs a promising material for next-generation electronics and optoelectronics. Before taking good control of the 2DMs with various stacking orders, it is crucial to develop a simple and powerful way to distinguish these different stacking sequences in 2DMs. ULF Raman spectroscopy is most often used for stacking-order characterizations.

Although the most recent studies in 2DMs focus on the AB stacking or 2H stacking phase, the stacking effects on the interlayer Raman modes have intrigued many researchers, [53, 56, 57] leading to an in-depth understanding of stacking phases, including that of 2H, 3R and 1T (Fig. 10.1b) and their mixtures. The so-called AB and ABC stackings are also named 2H and 3R in TMDs, respectively. TMDs are usually denoted MX_2. Among these three common phases, 2H and 3R are the most stable and can coexist in 2DM flakes grown via chemical vapour deposition (CVD), whereas the 1T-structure is "metastable" and infrequent in most natural 2DMs. Therefore, here by stating the stacking order, we refer to a traditional way to define 2H- and 3R-stacking sequences.

The first difference between these various stacking orders is in their symmetries. The space group of AB-*N*LG depends strongly on *N*, in which D_{3h} for ONLG and D_{3d} for ENLG (see Sect. 10.2 for details). In contrast to AB-*N*LG, ABC-*N*LG belongs to D_{3d} [42, 58]. Furthermore, the symmetries of *N*L 2H-MX_2 also depends on *N*, which are similar to the *N*LG (as shown in Table 10.1). However, with respect to the *N*L 3R-MX_2, the symmetry reduces to C_{3v}. The differences in symmetry can lead to different assignments of each S and LB modes and thus different Raman activity, indicating that the ULF Raman spectroscopy will be significantly distinct, e.g., S_{N1} are observable in AB-*N*LG but not in ABC-*N*LG due to the Raman-inactivity or weak EPC [58]. Second, various stacking orders may result in different relative displacements between the adjacent layers which have a good effect on the individual bond polarizabilities. According to the bond polarizability model, the Raman intensities are closely related to the individual bond polarizabilities, [56, 59] implying that different intensity trends of the Raman-active ULF modes are seen in different stacking orders. Using the polarizability model, the highest-frequency S mode is predicted to have the largest Raman intensity, whereas the lowest-frequency S mode in 2H-stacked 2DMs has no intensity. And the trend is opposite in the 3R-stacked systems [42, 56, 59]. Hereafter, we will show the stacking-order dependent ULF Raman spectroscopy in detail, taking $MoSe_2$ as an example [56]. The stacking-order dependent ULF Raman spectroscopy in MLGs is referred to in Chap. 1 on Raman spectroscopy of monolayer and multilayer graphenes.

Figure 10.7a shows two schematics of representative stacking sequences of 2L-$MoSe_2$, 2H and 3R. Apart from the layer shift, the second layer in 2H-$MoSe_2$ is rotated by 180° relative to the first layer. The ULF Raman modes of mechanically exfoliated and CVD-grown $MoSe_2$ in the $\bar{z}(xy)z$ configurations are represented in the Fig. 10.7b, c, respectively. All the observed ULF modes can be assigned as S modes because the LB modes cannot be seen under this configuration, which is determined by the symmetry and the Raman tensor [14]. Here, for brevity, the branches of the S modes are labelled S_1, S_2,..., in order of decreasing frequency, in accordance with the labels stated before. The exfoliated *N*L-$MoSe_2$ are in 2H stacking, considering that it is derived from a single cystal that is also of 2H stacking. In contrast to the 2H stacking, in which only the odd S modes (S_1, S_3, S_5,...) are observable, more S modes can be seen in the CVD-grown $MoSe_2$, which implies that more stacking orders co-exit in the CVD-grown flake, as previously reported [60]. From the bond polarizability, the observed S modes in 2H-$MoSe_2$

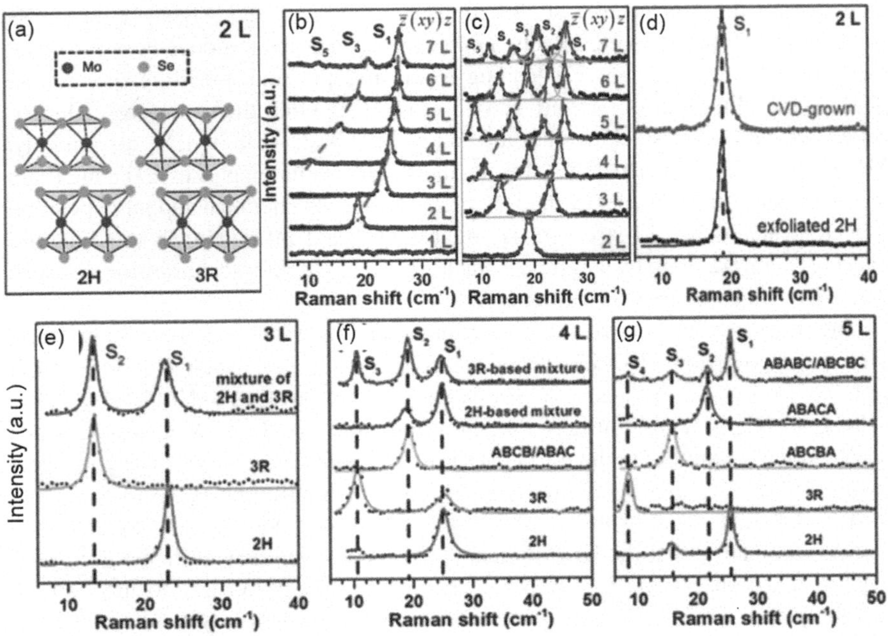

Fig. 10.7 (**a**) Schematics of 2H- and 3R- $MoSe_2$ in 2L. The ULF Raman spectra of (**b**) exfoliated and (**c**) CVD-grown $MoSe_2$ with the layer number ranging from 1L to 7L in $\bar{z}(xy)z$ configurations. Raman modes in the ULF region of CVD-grown (**d**) 2L, (**e**) 3L, (**f**) 4L and (**g**) 5L with different stackings. (Reproduced with permission from Ref. [56])

exhibit a Raman intensity trend that should behave as: $I(S_1)>I(S_3)>I(S_5)$, whereas those in 3R-$MoSe_2$ exhibit exactly the opposite trend [59]. The Raman intensity trend shown in Fig. 10.7c indicates the corresponding samples should have a coexistence of 2H and 3R. From the analysis above, it is easy to identify the stacking sequences from the ULF Raman modes by comparing the frequencies of the observed Raman modes and the Raman intensity trends. As depicted in Fig. 10.7d, because there is only one S mode in 2L-$MoSe_2$, it seems difficult to distinguish the 2H and 3R stackings in 2L-$MoSe_2$. Puretzky et al. found that the absolute Raman intensity of the S_1 modes in 2H-$MoSe_2$ is much larger than that in 3R stacking [61], which paves a new way to clarifying 2L-$MoSe_2$ in 2H and 3R stacking. Figure 10.7e–g shows the ULF Raman spectroscopy of 3–5L $MoSe_2$ with 2H, 3R stacking and their mixtures, which further confirms the Raman intensity trend stated above. Other than 2H and 3R stacking, more stacking orders appear as N increases. And the main stacking order of $MoSe_2$ can be distinguished from the difference in Raman intensity trend of S modes between 2H and 3R stackings. For example, considering a 4L-$MoSe_2$ flake, if the Raman intensity of S_1 mode is larger than that of S_3 mode, the stacking order of this flake can be referred to as 2H-based mixture. This can be extended to all TMDs with various stacking orders. Moreover, the ULF Raman spectroscopy in various stacking orders can be varied from each other based on the bond polarizability model or density functional theory (DFT) calculations [56].

Notably, the characterizations of stacking order are not only applicable in $MoSe_2$ but can also be applied to other 2DMs. The correlations between the stacking order and ULF Raman spectroscopy are increasingly developed in more and more isotropic 2DMs, such as graphene [42, 58] and MoS_2 [53, 57], which sheds light on the potential applications of 2DMs in the optoelectronics devices.

10.4.3 ULF Raman Spectra of In-Plane Anisotropic 2DMs

Various novel physical properties of isotropic 2DMs can be modulated via the layer number and stacking order, as depicted in Sects. 10.4.1 and 10.4.2. In-plane anisotropic 2DMs, such as BP, SnSe, $ReSe_2$ and ReS_2, offer one more degree of freedom to tune the interlayer coupling and thus the optoelectronic properties. The strong in-plane anisotropy in these 2DMs has been proposed for developing new devices with promising properties, which enhances their potential applications in electronics [62] and optoelectronics [63]. Until now, the anisotropic 2DMs can be divided into two categories: one category has strongly buckled honeycomb sheets with ′troughs′ running along the y-axis (***b*** direction), to which BP and SnSe belong [64–66]; the second category can be considered disordered 1T′-structures. In contrast to the commonly seen 1T- or 2H-structures in TMDs with high symmetry, 1T′-structures own extra in-plane metal-metal bonds or charge density wave states, typical examples of which are WTe_2, $ReSe_2$ and ReS_2 [45, 67, 68]. This 1T′ structure, different from the 2H and 3R symmetry in the MoS_2, would lead to random stacking of $ReSe_2$ (ReS_2). Owing to the random stacking in ML-$ReSe_2$ (ReS_2), the symmetry reduces from C_i (monolayer) to C_1 (multilayer), with the result that all the Raman modes including the interlayer modes are Raman active. However, the random stacking also results in weaker interlayer modes in the thicker samples and no observable interlayer modes in bulk materials due to an increasing disorder [45]. In principle, there exist one or more stacking orders for these materials. Qiao et al. revealed two stable stacking orders in ReS_2, namely isotropic-like (IS) and anisotropic-like (AI) NL according to the in-plane isotropic and anisotropic features, respectively, in ULF Raman spectra [47].

In a monolayer 1T′ structure such as ReS_2, a unit cell contains four formula units, including two categories of Re atoms and four categories of S atoms. In addition, the Re atoms are sandwiched by the two S atoms. Owing to the Peierls distortion [69], adjacent Re atoms are bonded in the form of zigzag four-atom clusters, which align along the direction of the lattice vector ***a*** to form Re chains. Figure 10.8a, b represented the schematics of AI-2L-ReS_2 and IS-2L-ReS_2. As shown in the left panel of Fig. 10.8a, b, the Re-Re bond in a unit cell is almost oriented parallel to the direction ***b***, and the top S atoms with these two Re atoms are in the forward direction ***a***, forming the Re-S-Re triangle. The triangle can be chosen to address

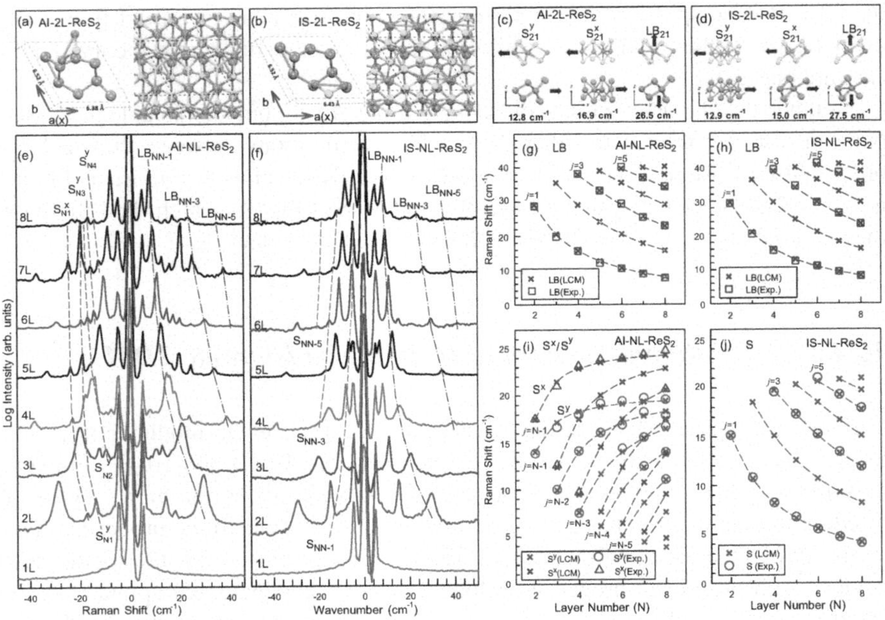

Fig. 10.8 Stacking schematic diagrams and the top view of the crystal structure of AI-2L-ReS_2 (**a**) and IS-2L-ReS_2 (**b**). The atomic displacements schematics and the corresponding frequencies of three ULF interlayer modes in AI- (**c**) and IS-stacked structures (**d**) obtained from DFT calculation are also shown. (**e**) and (**f**) are the Stokes/anti-Stokes ULF Raman spectra for AI-NL-ReS_2 and IS-*N*L-ReS_2 with the layer number ranging from 1L to 8L, respectively. The dashed lines are guides to the eyes for S modes and LB modes. Frequencies of LB modes as a function of *N* in AI- (**g**) and IS-stacked (**h**) ReS_2. The corresponding relation for S modes in AI- (**i**) and IS-stacked structures (**j**) are also shown. The squares, triangles and circles are the experimental data, and the crosses are from the calculations based on the LCM. (Reproduced with permission from Ref. [47])

the structural symmetry of *N*L-ReS_2. For 2L-ReS_2, the Re-Re bond of the top layer has three possible orientations: 0°, 60° and 90° with a given symmetry of the bottom layer. Furthermore, there are four possible relative positions of the two Re-Re bonds of both layers for each orientation, resulting in a total of 12 stacking orders. Among these stacking sequences, AI-2L-ReS_2 is the most stable configuration according to the DFT calculations, whereas IS-2L-ReS_2 is the second most stable configuration. As shown in Fig. 10.8a, b, the Re-Re bond in the top layer of AI-2L-ReS_2 is oriented 60°, whereas it is almost parallel to the ***a*** axes in IS-2L-ReS_2. Although the projection of these two bonds appears symmetric, the bilayer geometry is not exactly identical along the ***a*** and ***b*** axes, which is indicated by the fully relaxed lattice constants of 6.43 (***a***) and 6.52 (***b***).

For anisotropic 2DMs, the two axes in the basal 2D plane are not equal, with the result that the S modes along the two axes are not degenerate. Thus, there are 2(N-1) S modes and N-1 LB modes in *N*L anisotropic 2DMs, in which the S modes can be divided into two categories based on their vibrational directions (x, y). Figure 10.8c, d show the vibrational displacements and the corresponding frequencies of the S modes along ***x***, ***y*** directions and the LB modes. In AI-2L-ReS_2, the theoretical frequencies of the S and LB modes are $\omega(S_{21}^{y}) = 12.8\,\text{cm}^{-1}$ and $\omega(S_{21}^{x}) = 16.9\,\text{cm}^{-1}$, whereas $\omega(LB_{21}) = 26.5\,\text{cm}^{-1}$. The frequency difference between two S modes is $4.1\,\text{cm}^{-1}$, which is larger than the FWHM of the S modes. This means that they can be distinguished in the ULF Raman spectrum. Indeed, not only the frequency difference but also the absolute values of the S and LB modes are in good agreement with those from the experiment, as shown in Fig. 10.8e. In contrast, the theoretical frequency difference between the two S modes is $2.1\,\text{cm}^{-1}$ in IS-2L-ReS_2, which is only half the experimental result and comparable with the FWHM of the S mode, leading to the overlap of these two S modes in Raman spectroscopy, as depicted in Fig. 10.8f. In addition, the LB modes in these two stacked structures are quite close to each other, it is $1.0\,\text{cm}^{-1}$ softer in the AI-2L-ReS_2. Moreover, for ReS_2 with a given N, there are two categories of ULF Raman modes, as shown in Fig. 10.8e, f. Similar to bilayer ReS_2, these two categories correspond to the AI- and IS- stacking structures. Based on the LCM stated in Sect. 10.2, these interlayer modes can be expressed by the S_{21} and LB_{21} modes (see Eqs. 10.4 and 10.5 in Sect. 10.2). Based on LCM, all the ULF Raman modes in isotropic 2DMs can be predicted by Pos(S_{21}) and Pos(LB_{21}). This is also the case for anisotropic 2DMs, by replacing the S_{NN-i} and S_{21} with S_{NN-i}^{x}, S_{NN-i}^{y} and S_{21}^{x}, S_{21}^{y}, respectively. All the LB modes in AI-NL-ReS_2 and IS-NL-ReS_2 as well as those from calculations are summarized in Fig. 10.8g, h. Comparing the experimental and theoretical results, it can be deduced that only the LB_{NN-1}, LB_{NN-3} and LB_{NN-5} (indicated by the dash lines) are observable in both stacking structures. With respect to the S modes in AI-NL-ReS_2 and IS-NL-ReS_2, the frequencies are also summarized in Fig. 10.8i, j. There are two categories of S modes in AI-NL-ReS_2 as previously predicted, in which the interlayer vibrations along the ***x*** (S^x modes) and ***y*** (S^y modes) directions are marked by the dash lines, as demonstrated in Fig. 10.8a. The observed branch for the S^x modes is $i = N$-1, whereas for the S^y mode the observed branches are $i = N$-1, N-2, N-3 and N-4. However, in IS-stacking structures, only the S_{NN-i} branches with $i = 1,3,5$ can be detected and each branch decreases in frequency with increasing N, similar to the corresponding LB modes. This is ascribed to the different symmetries and EPC between the AI and IS-stacking structures.

On the other hand, with the given frequencies of interlayer modes, it is easy to determine the interlayer coupling 2DMs, such as ReS_2. The interlayer force constant can be written as $\alpha_0^{\parallel} = (2\pi^2c^2)\mu(\omega(S_{21}))^2$ and $\alpha_0^{\perp} = (2\pi^2c^2)\mu(\omega(LB_{21}))^2$. From the experimental S_{21}^{x} and S_{21}^{y}, the $\alpha_0^{\parallel}$ along the ***x*** and ***y*** directions are ~90% and ~55% to those of 2L-MoS_2, respectively, whereas in the IS-stacking structure, $\alpha_0^{\parallel}$ is only 67% of that in 2L-MoS_2. Considering $\alpha_0^{\perp}$ in both stacking structures, it is approximately 76% to that in 2L-MoS_2. All these results indicate that the interlayer interactions in AI- and IS-NL-ReS_2 are comparable with those in other 2DMs, which deviates from the previous predictions that ReS_2 lacks interlayer coupling. The revealed strong interlayer coupling and polytypism in NL-ReS_2 stimulate further studies on tuning the novel properties of other anisotropic 2DMs and paves the way to enhancing their potential applications in next-generation electronic or optoelectronic devices with superior characters. In addition, BP is one of the most popular anisotropic 2DMs and has intrigued many researchers. However, it is easily oxidized in air, which significantly suppresses its investigations and applications.

10.4.4 ULF Raman Spectra of Twisted 2DMs

Stacking 2DM flakes can serve the assembly of the artificially structured vdWHs, thus the fundamental properties and potential applications of 2DM flakes have recently attracted the attention of many researchers [5]. Hereafter, we will discuss how the interlayer coupling and ULF Raman modes modulate in two classical types of artificial 2DMs, i.e., twisted 2DMs (this section) and heterostructures (next Sect. 10.4.5). They vary from each other with component types. The twisted 2DMs are structured by stacking the same type of 2DM flakes with different crystal orientations, whereas the heterostructures usually consist of several types of 2DMs.

By assembling an mL-2DM flake on the same type of nL-2DM flake, an (m+n)L-2DM system is formed, which is commonly denoted as t(m+n)L-2DM, such as t(m+n)LG and t(m+n)L-MoS_2. Note that when we are discussing twisted 2DMs, its constituents are always assumed to be in the most stable or natural structure, e.g., Bernal (AB) stacked MLG and 2H-TMDs. The choice for m, n and the twist angle (θ_t) between each mL-, nL-, ... MoS_2 is so immense that it generates a large family of systems with different optical, electronic and vibrational properties. It has already been reported that in gapless graphene, the interlayer coupling depends sensitively on the interlayer twist angle, and plays important roles in modifying the electronic and vibrational properties [13, 14, 50, 70]. The shear coupling at the interface of tMLG is revealed to be ~20% of that in AB-MLG whereas the interlayer breathing force constant approaches ~100% of that in AB-MLG. Furthermore, the Raman intensity of the S, LB and G modes in t(m+n)LG are significantly enhanced for specific excitation energies, which is attributed to the resonance between the θ_t-dependent energies of van Hove singularities (VHSs) in the joint density of states (JDOSs) of all optically allowed transitions [13]. This is also the case in other 2DMs, such as MoS_2 and $MoSe_2$. For MoS_2, the modified layer stacking in twisted samples can lead to a decrease in the interlayer coupling, an enhancement or quenching of the photoluminescence emission yield [71]. Thus, the tunable interlayer coupling and thus the physical properties are accessible by generating artificial twisted MoS_2 with different stacking structures. As stated above, ULF Raman spectroscopy opens the door to directly detecting the interlayer coupling, which is essential to design the 2DMs with promising properties. Here, the ULF Raman modes are used to correlate the relationship between the interfacial coupling and the twisted patterns (so-called Moiré patterns) [17].

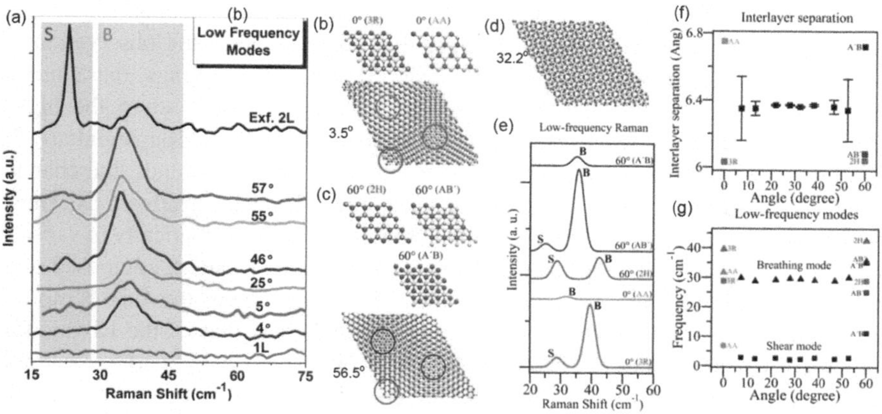

Fig. 10.9 ULF Raman spectra of t(1+1)L-MoS_2 with different twist angles ranging from 0° to 60° and that of 1L, exfoliated bilayer 2H-MoS_2 (Exf. 2L), in which S and B denote the S and LB modes, respectively. Atomic structures of commensurate t(1+1)L-MoS_2 with various twist angles. There are two stacking patches denoted 3R and AA at (**b**) 0° and a mixture of these two stackings when the twisted angle deviates from 0°. Three stacking patches denoted 2H, AB′ and A′B appear when $\theta_t = 60°$ (**c**) and all of these stacking patches can also coexist when θ_t near 60°. When θ_t approaches 30° (**d**), the stacking becomes completely mismatched. (**e**) ULF Raman spectra of different high-symmetry stacking patches from DFT calculations. The interlayer separation (**f**) and ULF Raman modes (**g**) are also shown. (Reproduced with permission from Ref. [17])

Figure 10.9a shows the typical S and LB modes of t(1+1)L-MoS_2 with θ_t between 0° and 60°. The S modes in t(1+1)L-MoS_2 are much weaker than those in 2H-MoS_2, whereas its LB modes show higher intensities. All the modes are redshifted and show an asymmetric profile when the twist angle deviates from 0° and 60°. The interlayer Raman modes show distinct features for different stacking structures. For example, the S mode is absent for certain twist angles (ranging from 20 to 40°), similar to the results of other twisted TMDs [16]. The absence of the S mode is always attributed to the disappearance of local high-symmetry domains. To well elucidate how the twist angle affects the interlayer coupling and thus the ULF Raman modes, it is necessary to clearly depict the stacking structures with various twist angles. Figure 10.9b–d plot the schematics of the possible twisted patterns when the twisted angle is equal to 0°, 3.5°, 60°, 56.5° and 32.2°. When $\theta_t = 0°$, there are two high-symmetry stacking patterns 3R and AA, whereas at 60°, there are three possible stacking patterns denoted 2H, AB′ and A′B. When the twist angles deviate from 0° and 60°, several stacking patterns can coexist in the twisted regions, as indicated by the circles in Fig. 10.9b, c. Furthermore, the patch sizes of high-symmetry configurations are seen to decrease with the twist angle deviating from 0° and 60°. For the twist angle approaching 30°, the stacking does not show the high-symmetry patches, which can usually be assigned to incommensurate patterns. The ULF Raman spectra for the high-symmetry stacking structures from the DFT calculations are also shown in Fig. 10.9e. Interestingly, the S and LB modes show similar Raman intensities in the most stable 2H structures, whereas in other stacking structures, the LB mode is much stronger than the S mode. This observation is in accord with the experimental results. From 2H to other stackings, the S mode becomes weaker and the LB mode becomes stronger. In contrast to the high-frequency vibrations in which the change in the polarizability (i.e., Raman intensity) is primarily contributed by the vibrations of intralayer chemical bonds, the change in the polarizability of interlayer modes is solely due to layer-layer vibrations. Thus, the relative Raman intensities of the S and LB modes change dramatically in various stackings. Turning to θ_t near 0° or 60°, the stacking yields a pattern composed of a mixture of multiple high-symmetry domains. The Raman intensity of the S modes in AB′ and A′B stacking decreased significantly, whereas that of the LB modes increased. Hence, in t(1+1)L-MoS_2 with θ_t near 60° (such as 57° and 55°), the S modes become weaker and the LB modes are greatly enhanced compared with those in 2H-MoS_2. For θ_t near 0°, the Raman intensities of the S and LB modes follow a similar trend. This successfully explains why the S modes are weak and the LB modes are strong in t(1+1)L-MoS_2.

In addition, the average frequencies of all the ULF Raman modes red shift from 2H- to t(1+1)L-MoS_2, which is in line with the experiments. This result can be explained by the weaker interlayer coupling and the increase in the interlayer distance in twisted structures, as depicted by Fig. 10.9f. For θ_t <10° or θ_t >50°, various high-symmetry patches coexist in the overall stacking pattern, leading to distinct frequencies and intensities for the S and LB modes in t(1+1)L-MoS_2. Such changes will in turn modify the relative contributions to the Raman scattering and they are also the reason why the LB modes are always seen as asymmetric. For the twisted angle between 20° and 40°, the stacking patterns are not so commensurate and thus neither rotational nor translation symmetry preserved, with the result that the interlayer coupling varies negligibly. Hence, the interlayer distance is constant and the changes in the S or LB modes are insignificant, as shown in Fig. 10.9g. The absence of local high-symmetry domains due to the mismatched stacking makes the stacking features essentially insensitive to the in-plane shear motion and thus leads to a small overall restoring force. As a result, the frequency of the S mode is so low that it appears in the background of the Rayleigh line and thus beyond the detection limit. In addition, for θ_t ranging from 10° to 20° and 40° to 50°, the stacking arrangements are between a mixture near 0° and 60° and the mismatched stacking near 30°. In this region, the experimental and theoretical results sometimes deviate. To sum up, the ULF Raman spectra vary dramatically in t(1+1)L-MoS_2 with different twist angles, which provides a new way to identify the interlayer coupling in twsited 2DMs with various θ_t. More specifically, the ULF Raman modes share a similar tendency in other t(1+1)L-TMDs, such as in twisted $MoSe_2$ [16].

Moreover, apart from the t(1+1)L-2DM, there is a large family of 2DMs with twisted structures due to the variety of m, n and more constituents. However, the investigations of interlayer modes as well as the novel physical properties are still lacking and require further studies.

10.4.5 ULF Raman Spectra of vdWHs

With the development of material growth and transferring techniques, different types of 2DMs can be stacked onto each other to form vdWHs [72–75]. The vdWHs provides a multitude of applications because it becomes possible to create artificial materials that combine several unique properties for use in novel multi-tasking applications [76, 77]. The strong interlayer coupling in vdWHs can be revealed by the strong photoluminescence of spatially indirect transitions in MoS_2/WSe_2 heterostructures, in which the coupling at the interface can be tuned by inserting dielectric layers such as h-BN [78]. The interlayer coupling can also directly affect the observed S and LB modes, which implies that it is one of the most powerful tools for measuring the interfacial coupling in vdWHs. The S modes are absent in the Raman spectra of $MoSe_2/MoS_2$ and MoS_2/WSe_2, whereas the LB modes are seen in the heterostructure regions with direct interlayer contact and an atomically clean interface [75]. This is the behaviour that is expected by considering the interlayer interactions in vdWHs. The lattice mismatch between the MoS_2 and WSe_2 layers reaches to $\sim$4%, so that the lateral displacements do not produce any overall restoring force. However, the vertical displacements can create a finite restoring force due to the vdW interaction between them. The vdWHs formed by two 2DM flakes from different categories, such as TMDs and graphene, are essential for revealing the intrinsic nature of interfacial coupling in vdWHs.

The MoS_2/NLG vdWHs can be prepared by transferring MoS_2 flakes onto NLG flakes, or by transferring NLG flakes onto MoS_2 flakes [20]. In general, the annealing process is necessary after the transfer to form good interfacial coupling because the two as-transferred heterostructures may not couple with each other. However, an appropriate annealing time is very important to obtain these vdWHs with high quality. For MoS_2/NLG, it is revealed that 30 min in Ar gas at 300° is the best condition for removing the moisture and impurities as well as for maintaining the high Raman intensity of interlayer vibrations. To understand the interfacial coupling in MoS_2/nLG, the MoS_2 with layer number m (mLM) is transferred onto the nLG followed by annealing. Similar to other vdWHs, there are no S modes for 1LM/1LG, indicating weak interfacial coupling between the two constituents. However, in contrast to the ULF Raman spectra in tMLG, $MoSe_2/MoS_2$, MoS_2/WSe_2 and twisted MoS_2, [13, 14, 16, 17, 75] the LB modes cannot be observed in 1LM/1LG. The ULF Raman spectra of 2LM/nLG (n = 1,2,. . . 6,8) are shown in Fig. 10.10a. The S_{21} mode of the 2LM constituent located at the same position in all spectra, which confirms the absence of interfacial shear coupling in mLM/nLG. In addition, because the S modes of the nLG constituent may be overlapped by other Raman modes due to their weak intensity, they are usually not observable. Similar to the LB modes in tMLG, several LB branches can be detected in 2LM/nLGs, whose frequencies red-shift with increasing n, as indicated by the dashed lines. Because the interfacial coupling between 2LM and nLG is comparable to the interlayer interactions in 2LM or nLG, new LB modes appear as n increases. Considering 2LM/nLG as an overall system with $N = n + 2$,

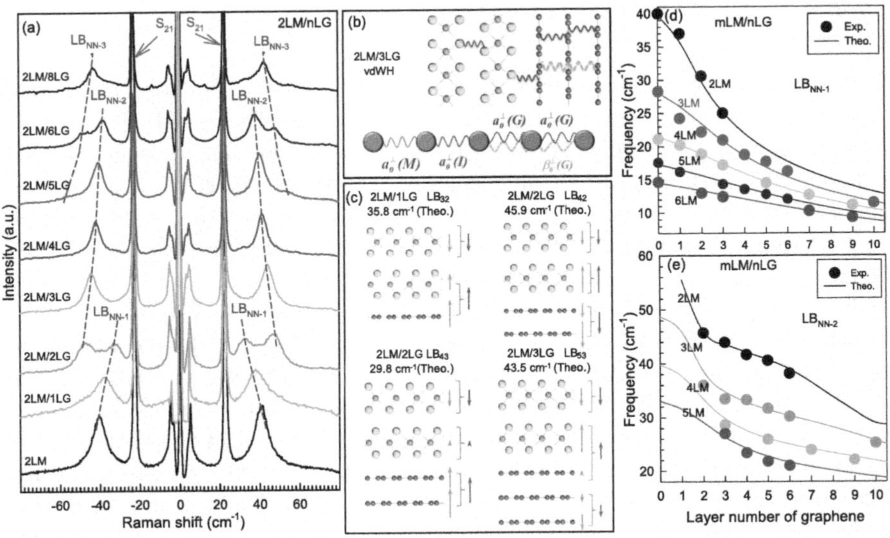

Fig. 10.10 (**a**) Stokes/anti-Stokes Raman spectra of 2LM/nLG in the S and LB peak spectral ranges. (**b**) Schematic diagram of the LCM for LB modes in 2LM/3LG, in which the next nearest LB coupling in the 3LG constituent is considered. (**c**) Normal mode displacements of the observed LB modes in 2LM/1LG, 2LM/2LG and 2LM/3LG. The corresponding calculated (Theo.) frequencies are indicated. (**d**) Pos(LB_{NN-1}) and Pos(LB_{NN-2}) in mLM/nLG as a function of n. The solid lines show the theoretical trends of Pos(LB_{NN-1}) and Pos(LB_{NN-2}) on n based on the LCM. (Reproduced with permission from Ref. [20])

the LB modes can be reproduced by the LCM with the nearest LB force constant in 2LM of $\alpha_0^{\perp}(\mathrm{M}) = 84 \times 10^{18}$ N/m^3 and that in nLG of $\alpha_0^{\perp}(\mathrm{G}) = 106.5 \times 10^{18}$ N/m^3 and $\beta_0^{\perp}(\mathrm{G}) = 9.5 \times 10^{18}$ N/m^3 (Fig. 10.10b). The interfacial LB force constant between 2LM and nLG constituents is referred to as a constant $\alpha_0^{\perp}(\mathrm{I})$ in 2LM/nLG, and the total layer number N is n+2. In this case, three branches of LB_{NN-1}, LB_{NN-2} and LB_{NN-3} can be observed in Fig. 10.10a. Based on the experimental frequencies, $\alpha_0^{\perp}(\mathrm{I})$ is estimated as 60 $\times 10^{18}$ N/m^3, smaller than $\alpha_0^{\perp}(\mathrm{M})$ but larger than $\alpha_0^{\perp}(\mathrm{G})/2$. According to the LCM, the corresponding normal displacements of each layer can also be obtained for each LB mode, and those of the LB_{NN-1} mode in 2LM/1LG and 2LM/2LG as well as of LB_{NN-2} in 2LM/2LG and 2LM/3LG are shown in Fig. 10.10c. Because the unit mass of 1LG is approximately 1/4 of that in 2LM, the 1LG are treated as a perturbation to the 2LM in 2LM/1LG. Thus, 1LG exhibits atomic displacements similar to those of its nearest MoS_2 layer for the LB_{32} mode. In addition, similar behaviour can be found in mLM/1LG [20]. With n of nLG increasing, the nLG cannot be regarded as the perturbations to the 2LM for the normal mode displacements of the LB mode. However, for LB_{42} and LB_{43}, the top two graphene layers can be treated as a unit and thus these two modes exhibit atomic displacements similar to those of LB_{31} and LB_{32} in 3LM. With respect to the LB_{53} in 2LM/3LG, the two top graphene layers can also be seen as a unit and vibrate out-of-phase with the third graphene layer whose atomic displacements are in-phase with the nearest MoS_2 layer. Thus, the normal displacements of LB_{53} are similar to those of LB_{31} in 3LM. Therefore, the analysis of atomic displacements for each LB mode enhance the understanding of the interfacial coupling effects on the generation of new LB modes beyond those of the two constituents. Further investigations of mLM/nLG have been performed and the results are depicted in Fig. 10.10d, e. The LCM with $\alpha_0^{\perp}(\mathrm{I})=60\ \times 10^{18}$ N/m^3 is enough to reproduce the frequencies of the LB_{NN-1} and LB_{NN-2} modes, indicating an almost uniform interlayer LB coupling in mLM/nLG with various twist angles. In addition, the Raman spectroscopy of nLG/mLM was also investigated and the LB modes were revealed to locate at peak positions almost identical to those of mLM/nLG, which implies that the interfacial coupling in mLM/nLG vdWHs is not so sensitive to the stacking order or twist angle. Similar results can also be seen in other heterostructures, such as $MoSe_2/MoS_2$ [75].

10.5 Conclusion

In this chapter, we discussed the interlayer vibrations in 2DMs and vdWHs. The discussions starts after introducing the stacking orders and symmetries of the large family of 2DMs. Interlayer interactions play a direct role in modifying their optical and electronic properties in 2DMs and vdWHs. The ULF interlayer modes are more sensitive to vdW coupling, and their frequencies and Raman activities depend on the layer number, stacking order and symmetry. We addressed the physical origins of interlayer vibrations and then described the direct methods to probe the fundamental S and LB modes. Finally, several examples of the ULF Raman spectroscopy in various 2DMs were reviewed. The frequencies of interlayer vibrations can be defined as a rigorous function of the layer number, whereas the stacking orders and anisotropies play an important role in their Raman activities. In particular, the interlayer vibrations in twisted 2DMs or vdWHs exhibit different behaviours, which can potentially be utilized to distinguish the interfacial coupling and therefore to obtain a good understanding of these systems and their attractive applications.

Acknowledgements We acknowledge support from the National Key Research and Development Program of China (Grant No. 2016YFA0301204), the National Natural Science Foundation of China (Grant No. 11474277, 11874350 and 11434010), and the Beijing Municipal Science and Technology Commission.

References

1. K.S. Novoselov, D. Jiang, F. Schedin, T.J. Booth, V.V. Khotkevich, S.V. Morozov, A.K. Geim, Proc. Natl. Acad. Sci. USA **102**(30), 10451 (2005)
2. X. Zhang, Q.H. Tan, J.B. Wu, W. Shi, P.H. Tan, Nanoscale **8**, 6435 (2016)

3. A.K. Geim, K.S. Novoselov, Nat. Mater. **6**(183), 183 (2007)
4. Q.H. Wang, K. Kalantar-Zadeh, A. Kis, J.N. Coleman, M.S. Strano, Nat. Nanotechnol. **7**(11), 699 (2012)
5. A.K. Geim, I.V. Grigorieva, Nature **499**(7459), 419 (2013)
6. X. Zhang, X.F. Qiao, W. Shi, J.B. Wu, D.S. Jiang, P.H. Tan, Chem. Soc. Rev. **44**, 2757 (2015)
7. K.S. Novoselov, A. Mishchenko, A. Carvalho, A.H. Castro Neto, Science **353**(6298), 9439 (2016)
8. S.N. Shirodkar, U.V. Waghmare, Phys. Rev. Lett. **112**, 157601 (2014)
9. X.L. Li, W.P. Han, J.B. Wu, X.F. Qiao, J. Zhang, P.H. Tan, Adv. Funct. Mater. **27**, 1604468 (2017)
10. L. Liang, J. Zhang, B.G. Sumpter, Q. Tan, P.H. Tan, V. Meunier, ACS Nano **11**(12), 11777 (2017)
11. J.B. Wu, M.L. Lin, X. Cong, H.N. Liu, P.H. Tan, Chem. Soc. Rev. **47**, 1822 (2018)
12. V. Carozo, C.M. Almeida, E.H.M. Ferreira, L.G. Cançado, C.A. Achete, A. Jorio, Nano Lett. **11**(11), 4527 (2011)
13. J.B. Wu, X. Zhang, M. Ijäs, W.P. Han, X.F. Qiao, X.L. Li, D.S. Jiang, A.C. Ferrari, P.H. Tan, Nat. Commun. **5**, 5309 (2014)
14. J.B. Wu, Z.X. Hu, X. Zhang, W.P. Han, Y. Lu, W. Shi, X.F. Qiao, M. Ijiäs, S. Milana, W. Ji, A.C. Ferrari, P.H. Tan, ACS Nano **9**(7), 7440 (2015)
15. K. Liu, L. Zhang, T. Cao, C. Jin, D. Qiu, Q. Zhou, A. Zettl, P. Yang, S.G. Louie, F. Wang, Nat. Commun. **5**, 4966 (2014)
16. A.A. Puretzky, L. Liang, X. Li, K. Xiao, B.G. Sumpter, V. Meunier, D.B. Geohegan, ACS Nano **10**(2), 2736 (2016)
17. S. Huang, L. Liang, X. Ling, A.A. Puretzky, D.B. Geohegan, B.G. Sumpter, J. Kong, V. Meunier, M.S. Dresselhaus, Nano Lett. **16**(2), 1435 (2016)
18. K.S. Novoselov, A.H.C. Neto, Phys. Scripta **2012**(146), 014006 (2012)
19. Y. Gong, J. Lin, X. Wang, G. JShi, S. Lei, Z. Lin, X. Zou, G. Ye, R. Vajtai, B.I. Yakobson, H. Terrones, M. Terrones, B.K. Tay, J. Lou, S.T. Pantelides, Z. Liu, W. Zhou, P.M. Ajayan, Nat. Mater. **13**(12), 1135 (2014)
20. H. Li, J.B. Wu, F. Ran, M.L. Lin, X.L. Liu, Y. Zhao, X. Lu, Q. Xiong, J. Zhang, W. Huang, H. Zhang, P.H. Tan, ACS Nano **11**, 11714 (2017)
21. P.H. Tan, W.P. Han, W.J. Zhao, Z.H. Wu, K. Chang, H. Wang, Y.F. Wang, N. Bonini, N. Marzari, N. Pugno, G. Savini, A. Lombardo, A.C. Ferrari, Nat. Mater. **11**, 294 (2012)
22. J.L. Feldman, J. Phys. Chem. Solids **42**(11), 1029 (1981)
23. M. Gatulle, M. Fischer, A. Chevy, Phys. Status Solidi B **119**(1), 327 (1983)
24. R. Nicklow, N. Wakabayashi, H.G. Smith, Phys. Rev. B **5**, 4951 (1972)
25. B. Yang, M.D. Morris, H. Owen, Appl. Spectrosc. **45**(9), 1533 (1991)
26. P.J. Horoyski, M.L.W. Thewalt, Appl. Spectrosc. **48**(7), 843 (1994)
27. S.G. Belostotskiy, Q. Wang, V.M. Donnelly, D.J. Economou, N. Sadeghi, Appl. Phys. Lett. **89**(25) (2006)
28. H. Okajima, H.O. Hamaguchi, Appl. Spectrosc. **63**(8), 958 (2009)
29. C. Moser, F. Havermeyer, Appl. Phys. B **95**(3), 597 (2009)
30. S. Lebedkin, C. Blum, N. Stürzl, F. Hennrich, M.M. Kappes, Rev. Sci. Instrum. **82**(1), 013705 (2011)
31. A.F.H. van Gessel, E.A.D. Carbone, P.J. Bruggeman, J.J.A.M. van der Mullen, Plasma Sources Sci.and T. **21**(1), 015003 (2012)
32. J. Verble, T. Wietling, P. Reed, Solid State Commun. **11**(8), 941 (1972)
33. R.J. Nemanich, G. Lucovsky, S.A. Solin, in *Proceedings of the International Conference on Lattice Dynamics*, Paris, 1977, pp. 619–621
34. T. Sekine, T. Nakashizu, K. Toyoda, K. Uchinokura, E. Matsuura, Solid State Commun. **35**(4), 371 (1980)
35. S. Sugai, T. Ueda, Phys. Rev. B **26**, 6554 (1982)
36. P.H. Tan, D. Bougeard, G. Abstreiter, K. Brunner, Appl. Phys. Lett. **84**(14), 2632 (2004)

37. Y. Zhao, X. Luo, H. Li, J. Zhang, P.T. Araujo, C.K. Gan, J. Wu, H. Zhang, S.Y. Quek, M.S. Dresselhaus, Q. Xiong, Nano Lett. **13**(3), 1007 (2013)
38. G. Plechinger, S. Heydrich, J. Eroms, D. Weiss, C. Schller, T. Korn, Appl. Phys. Lett. **101**(10), 101906 (2012)
39. H. Zeng, B. Zhu, K. Liu, J. Fan, X. Cui, Q.M. Zhang, Phys. Rev. B **86**, 241301 (2012)
40. M. Boukhicha, M. Calandra, M.A. Measson, O. Lancry, A. Shukla, Phys. Rev. B **87**, 195316 (2013)
41. Y. Zhao, X. Luo, J. Zhang, J. Wu, X. Bai, M. Wang, J. Jia, H. Peng, Z. Liu, S.Y. Quek, Q. Xiong, Phys. Rev. B **90**, 245428 (2014)
42. C.H. Lui, Z. Ye, C. Keiser, E.B. Barros, R. He, Appl. Phys. Lett. **106**(4), 041904 (2015)
43. P.H. Tan, J.B. Wu, W.P. Han, W.J. Zhao, X. Zhang, H. Wang, Y.F. Wang, Phys. Rev. B **89**, 235404 (2014)
44. X. Zhang, W.P. Han, J.B. Wu, S. Milana, Y. Lu, Q.Q. Li, A.C. Ferrari, P.H. Tan, Phys. Rev. B **87**, 115413 (2013)
45. H. Zhao, J. Wu, H. Zhong, Q. Guo, X. Wang, F. Xia, L. Yang, P. Tan, H. Wang, Nano Res. **8**(11), 3651 (2015)
46. J.B. Wu, H. Wang, X.L. Li, H. Peng, P.H. Tan, Carbon **110**, 225 (2016)
47. X.F. Qiao, J.B. Wu, L. Zhou, J. Qiao, W. Shi, T. Chen, X. Zhang, J. Zhang, W. Ji, P.H. Tan, Nanoscale **8**, 8324 (2016)
48. M. Miscuglio, M.L. Lin, F. Di Stasio, P.H. Tan, R. Krahne, Nano Lett. **16**(12), 7664 (2016)
49. X.L. Liu, H.N. Liu, J.B. Wu, H.X. Wu, T. Zhang, W.Q. Zhao, P.H. Tan, Rev. Sci. Instrum. **88**(05), 053110 (2017)
50. M.L. Lin, F.R. Ran, X.F. Qiao, J.B. Wu, W. Shi, Z.H. Zhang, X.Z. Xu, K.H. Liu, H. Li, P.H. Tan, Rev. Sci. Instrum. **87**(5), 053122 (2016)
51. Z. Li, K.F. Mak, E. Cappelluti, T.F. Heinz, Nat. Phys. **7**(12), 944 (2011)
52. D. Xiao, G.B. Liu, W. Feng, X. Xu, W. Yao, Phys. Rev. Lett. **108**, 196802 (2012)
53. J. Yan, J. Xia, X. Wang, L. Liu, J.L. Kuo, B.K. Tay, S. Chen, W. Zhou, Z. Liu, Z.X. Shen, Nano Lett. **15**(12), 8155 (2015)
54. T. Ritschel, J. Trinckauf, K. Koepernik, B. Buchner, M.v. Zimmermann, H. Berger, Y.I. Joe, P. Abbamonte, J. Geck, Nat. Phys. **11**(4), 328 (2015)
55. X. Xu, W. Yao, D. Xiao, T.F. Heinz, Nat. Phys. **10**(5), 343 (2014)
56. X. Luo, Y. Zhao, J. Zhang, S.T. Pantelides, W. Zhou, S. Ying Quek, Q. Xiong, Adv. Mater. **27**, 4502 (2015)
57. J.U. Lee, K. Kim, S. Han, G.H. Ryu, Z. Lee, H. Cheong, ACS Nano **10**(2), 1948 (2016)
58. X. Zhang, W.P. Han, X.F. Qiao, Q.H. Tan, Y.F. Wang, J. Zhang, P.H. Tan, Carbon **99**, 118 (2016)
59. X. Luo, X. Lu, C. Cong, T. Yu, Q. Xiong, S. Ying Quek, Sci. Rep. **5**, 14565 (2015)
60. X. Lu, M.I.B. Utama, J. Lin, X. Gong, J. Zhang, Y. Zhao, S.T. Pantelides, J. Wang, Z. Dong, Z. Liu, W. Zhou, Q. Xiong, Nano Lett. **14**(5), 2419 (2014)
61. A.A. Puretzky, L. Liang, X. Li, K. Xiao, K. Wang, M. Mahjouri-Samani, L. Basile, J.C. Idrobo, B.G. Sumpter, V. Meunier, D.B. Geohegan, ACS Nano **9**(6), 6333 (2015)
62. L. Li, Y. Yu, G.J. Ye, Q. Ge, X. Ou, H. Wu, D. Feng, X.H. Chen, Y. Zhang, Nat. Nanotechnol. **9**(5), 372 (2014)
63. F. Xia, H. Wang, D. Xiao, M. Dubey, A. Ramasubramaniam, Nat. Photon. **8**(12), 899 (2014)
64. J. Qiao, X. Kong, Z.X. Hu, F. Yang, W. Ji, Nat. Commun. **5**(4475), 4475 (2014)
65. L.D. Zhao, S.H. Lo, Y. Zhang, H. Sun, G. Tan, C. Uher, C. Wolverton, V.P. Dravid, M.G. Kanatzidis, Nature **508**(7496), 373 (2014)
66. X. Ling, L. Liang, S. Huang, A.A. Puretzky, D.B. Geohegan, B.G. Sumpter, J. Kong, V. Meunier, M.S. Dresselhaus, Nano Lett. **15**(6), 4080 (2015)
67. I. Pletikosić, M.N. Ali, A.V. Fedorov, R.J. Cava, T. Valla, Phys. Rev. Lett. **113**, 216601 (2014)
68. M.N. Ali, J. Xiong, S. Flynn, J. Tao, Q.D. Gibson, L.M. Schoop, T. Liang, N. Haldolaarachchige, M. Hirschberger, N.P. Ong, R.J. Cava, Nature **514**(7521), 205 (2014)
69. M. Kertesz, R. Hoffmann, J. Am. Chem. Soc. **106**(12), 3453 (1984)
70. C. Cong, T. Yu, Nat. Commun. **5**, 4709 (2014)

71. T. Jiang, H. Liu, D. Huang, S. Zhang, Y. Li, X. Gong, Y.R. Shen, W.T. Liu, S. Wu, Nat. Nanotechnol. **9**, 825 (2014)
72. M.P. Levendorf, C.J. Kim, L. Brown, P.Y. Huang, R.W. Havener, D.A. Muller, J. Park, Nature **488**, 627 (2012)
73. Y. Gong, S. Lei, G. Ye, B. Li, Y. He, K. Keyshar, X. Zhang, Q. Wang, J. Lou, Z. Liu, R. Vajtai, W. Zhou, P.M. Ajayan, Nano Lett. **15**(9), 6135 (2015)
74. Z. Liu, L. Ma, G. Shi, W. Zhou, Y. Gong, S. Lei, X. Yang, J. Zhang, J. Yu, K.P. Hackenberg, A. Babakhani, J.C. Idrobo, R. Vajtai, R. Vajtai, P.M. Ajayan, Nat. Nanotechnol. **8**, 119 (2013)
75. C.H. Lui, Z. Ye, C. Ji, K.C. Chiu, C.T. Chou, T.I. Andersen, C. Means-Shively, H. Anderson, J.M. Wu, T. Kidd, Y.H. Lee, R. He, Phys. Rev. B **91**, 165403 (2015)
76. W.J. Yu, Y. Liu, H. Zhou, A. Yin, Z. Li, Y. Huang, X. Duan, Nat. Nanotechnol. **8**(12), 952 (2013)
77. M. Massicotte, P. Schmidt, F. Vialla, K.G. Schädler, A. Reserbat-Plantey, K. Watanabe, T. Taniguchi, K.J. Tielrooij, F.H.L. Koppens, Nat. Nanotechnol. **11**(1), 42 (2015)
78. H. Fang, C. Battaglia, C. Carraro, S. Nemsak, B. Ozdol, J.S. Kang, H.A. Bechtel, S.B. Desai, F. Kronast, A.A. Unal, G. Conti, C. Conlon, G.K. Palsson, M.C. Martin, A.M. Minor, C.S. Fadley, E. Yablonovitch, R. Maboudian, A. Javey, Proc. Natl. Acad. Sci. USA **111**(17), 6198 (2014)

Chapter 11
Raman Imaging of Two Dimensional Materials

Xuhong An, Zhenhua Ni, and Zexiang Shen

Abstract Raman imaging is a powerful technique that can provide the spatial distribution of the properties of the micro-/nano- material. Different parameters of the Raman peaks, e.g. height/area, position, full width at half maximum (FWHM), and also ratios/differences between peaks, can be used to construct the Raman imaging and provide valuable information for the study of 2D materials and heterostructure. In this chapter, we will introduce the basic principle of Raman imaging, and also its application in the study of 2D materials, including the effects of thickness and stacking configurations, heterostructure and interlayer coupling, defects, strain. We will also show that Raman imaging is an ideal tool to study the growth mechanism of CVD graphene.

11.1 Introduction to Raman Imaging

Raman imaging is a powerful technique for the study of the properties of micro- or nano- materials. The basic principle of Raman imaging is shown in Fig. 11.1: Raman spectrum is acquired at each pixel of the desired region through the control of a motorized or piezo stage (a–b), and a series of images could be generated based on different parameters of the Raman peaks, e.g. height/area, position, full width at half maximum (FWHM), and also ratios/differences between peaks (c). For example, the Raman images obtained from the intensity of the G peak, the FWHM of 2D peak, and the intensity ratio of 2D and G peaks of graphene are shown in Fig. 11.1d–f. It can be clearly seen that different information of the graphene sheets with different thicknesses can be achieved from the Raman images, e.g. the G peak intensity as

X. An · Z. Ni (✉)
School of Physics, Southeast University, Nanjing, China
e-mail: zhni@seu.edu.cn

Z. Shen (✉)
Division of Physics and Applied Physics, School of Physical and Mathematical Sciences, Nanyang Technological University, Singapore, Singapore
e-mail: zexiang@ntu.edu.sg

P.-H. Tan (ed.), *Raman Spectroscopy of Two-Dimensional Materials*, Springer Series in Materials Science 276, https://doi.org/10.1007/978-981-13-1828-3_11

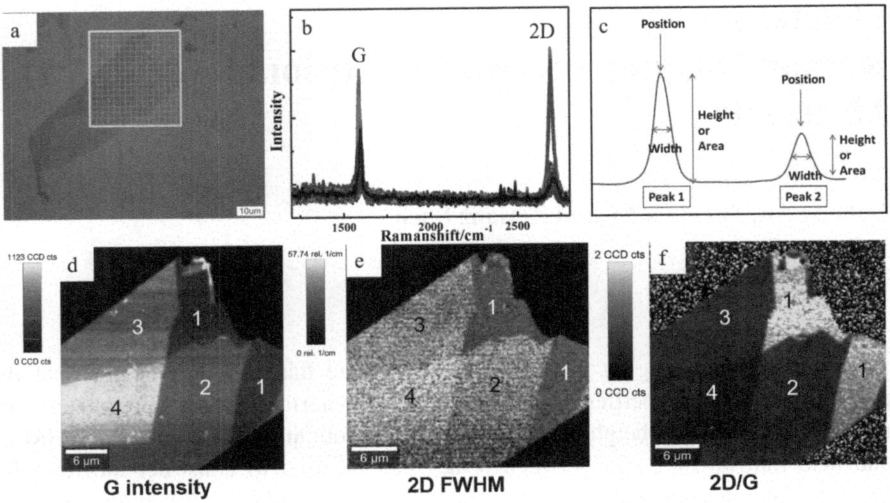

Fig. 11.1 Schematic for obtaining the Raman images. The laser spot is scanning across the designed area point by point (**a**). The Raman spectrum is collected at each position (**b**). (**c**) Different information could be achieved from the Raman peaks, including the position, FWHM, height/area intensity, and are used to construct the Raman images. The Raman images of the graphene sheets obtaining from the intensity of G peak (**d**), FWHM of 2D peak (**e**) and the intensity ratio between 2D and G peaks (**f**). The thickness of graphene sheet is labeled in the Raman images

well as the FWHM of 2D peaks increase, while the intensity ratio of 2D and G peaks drops with the increase of graphene thickness. It can also be revealed that the graphene sheet with same thickness is very uniform, without any winkles or adlayers.

The most important advantages of Raman imaging technique are: (1) It can provide the spatial distribution of the properties of the samples across a large area. Since the parameters for acquiring the Raman spectrum are exactly the same at each point, the peak intensity as well as other information can be directly compared. (2) The spatial information obtained by Raman imaging can be correlated with other imaging techniques, e.g. optical microscope, scanning electron microscope (SEM), atomic force microscope (AFM), and other spectroscopic imaging techniques such as photoluminescence (PL) imaging. (3) Real time evolution of the properties could be investigated through Raman imaging, e.g. the phase transition, strain, growth. You will know exactly where and when the above phenomena start to happen, and how it changes with time.

The spatial resolution of Raman imaging is normally between hundreds of nanometers and several microns, which is diffraction limited and determined by the excitation wavelength, the numerical aperture (NA) of the objectives, the confocal system, and the step sizes for scanning. A good Raman imaging system could achieve a spatial resolution of ~300 nm, without the use of near field techniques, e.g. tip enhanced Raman scattering (TERS), or micro spheres/lens, which is enough for the study of materials in micron-sizes. With the development of high performance CCD and the optimization of the optics, the integration time required for obtaining a smooth Raman spectrum could be as short as milliseconds, and therefore greatly minimizes the time for acquiring Raman image. Line-scanning technique is also developed to obtain the Raman image line by line, instead of point by point. In a word, the Raman imaging technique has been well developed recent years and is widely adapted in the study of nano- and micro- materials.

The properties of two dimensional (2D) materials are strongly influenced by different factors, such as interlayer interactions (stacking orders or heterostructures), defects, and strain. The above factors are usually localized or restricted to certain areas of the samples. For example, different defects could appear in 2D materials, including vacancies, edges, grain boundaries (GBs), and adatoms. These defects are unlikely to be located by randomly acquiring the single Raman spectrum point by point, which are also very time-consuming. In addition, the strain in 2D materials could be unevenly distributed and affect the properties of the materials locally. Therefore, it would be ideal to have an imaging technique to study the spatial distribution of the properties of the 2D materials, i.e. Raman imaging.

This chapter will introduce the study of 2D materials by using Raman imaging technique. It will cover the thickness, interlayer interaction, heterostructure, defects, strain, and chemical vapor deposition (CVD) growth mechanisms of 2D materials. The organization of the chapter is as follows: In Sect. 11.2, we focus on the effects of thickness and stacking orders on 2D materials by Raman imaging. We will show how to identify the thickness and various stacking orders (including the twist angle between different layers) in graphene and transition-metal dichalcogenides (TMDs), as well as heterostructure (graphene/MoS_2, vertical and lateral heterostructures from TMDs monolayer); In Sect. 11.3, we will discuss the application of Raman imaging in the characterization of defects such as point defects, edges, and GBs in graphene and TMDs; In Sect. 11.4, the strain in 2D materials are investigated by Raman imaging. In Sect. 11.5, the CVD growth mechanism of graphene is probed by Raman imaging through the use of isotopic labels. We hope this chapter would contribute to the better understanding of 2D materials and also demonstrate that Raman imaging is a powerful technique for the study of nano−/micro- materials.

11.2 Raman Imaging of 2D Materials with Different Thicknesses and Stacking Configurations

The electronic, optical, mechanical, and thermal properties of 2D materials are strongly dependent on their thicknesses. The electronic properties of 2D materials manifest stepwise variations and finally approach that of the three-dimensional counterpart with a further increase in layer number [1, 2]. For example, the carriers in single layer graphene (SLG) are massless Dirac fermions, while those in multi-layer graphene are massive [3]. The band gap of Bernal stacked bilayer graphene

(BLG) can be tuned by applied perpendicular electric field [4]. Additionally, there is a transition from indirect band gap to direct band gap for many TMDs materials when its number of layers change from multilayer to single layer, and results in the great improvement of luminescence efficiency [5–7]. There is also a variation of band gap from ~0.3 to ~1.8 eV from bulk to single layer of black phosphorus (BP) [8]. In addition, the distinct properties of multilayer 2D materials are closely associated to its particular stacking configurations. For instance, phenomena such as tunable energy difference between Van Hove Singularities (VHSs) [9] and chirality tunneling [10] have been discovered in twisted bilayer graphene (TBG). Experimentally, the electro- and magneto- transport properties for folded graphene sheets (non-AB stacking) are observed which are not similar to those of AB stacked bilayers [11]. For MoS_2, Raman vibrations, photoluminescence and valley polarization are also mediated by stacking configurations [12, 13]. Therefore, a simple, efficient and nondestructive mean to determine the thickness and stacking configurations of 2D materials are highly anticipated for materials growth, fundamental research, as well as applications. Raman imaging technique can achieve such purpose through the variations of intensity, position and FWHM of major features/peaks of 2D materials with different thickness and stacking configurations.

11.2.1 Raman Imaging of 2D Materials with Different Thicknesses

There are two most intense Raman features in graphene, i.e. G and 2D peaks. The G peak is due to the doubly degenerate zone center E_{2g} mode and the 2D peak is the second order of zone-boundary phonons [14]. Figure 11.2 [15] shows the typical Raman spectra and image of graphene with thickness from 1 to 4 layers on 285 nm SiO_2/Si substrate. The optical image and Raman spectra are shown in Fig. 11.2a, b. Figure 11.2c shows the Raman image plotted by the intensity of G peak. The differences between graphene samples with different thicknesses are very obvious. The intensities of the G peak along the dash and dot-dash lines are shown in Fig. 11.2d. The intensity of G peak increases almost linearly with the increase of graphene thickness. The Raman G peak originates from the in-plane vibration of sp^2 carbon atoms, and hence its intensity increases when more carbon atoms are probed in multilayer graphene. Generally, the intensity of G peak on SiO_2/Si multilayer substrate as a function of thickness can be well described by the multilayer interference of incident light and the multi-reflection of Raman signals in graphene/SiO_2/Si system based on Fresnel equations [16]. It increases almost linearly up to 9 layers, and then decreases with thickness in the range of ~5–50 nm. Additionally, the intensity of G peak of bulk graphite is apparently weaker than that of BLG on this specific substrate. Overall, the variation of G peak intensity can be used to determine the number of layers of graphene sheets, especially in

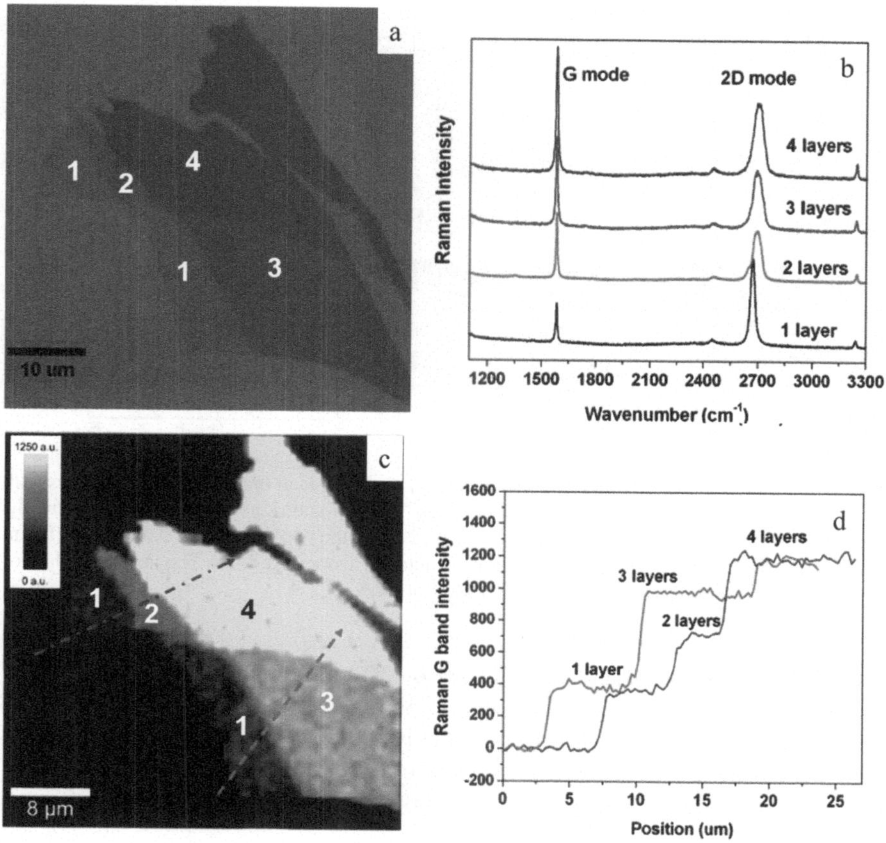

Fig. 11.2 (**a**) Optical image of graphene sheets with 1–4 layers. (**b**) Raman spectra of graphene sheets with 1–4 layers. (**c**) Raman image plotted by the intensity of G peak. (**d**) The cross sections of the dot-dash and dash lines shown in (**c**)

the Raman imaging configuration, where the spectra are collected from different positions almost simultaneously.

For AB-stacked few layer graphene (FLG), the width of 2D peak also provides a quantitative guide to distinguish the layer number (1 to 5 layers). According to the theoretical model of double resonance, the splitting of the electronic band structure in FLG is responsible for the stepwise broadening of 2D peaks. Figure 11.3a is the optical image of a typical FLG with 1–4 layers [17]. The Raman images of the FWHM of 2D peaks corresponding to the optical image are shown in Fig. 11.3b, c, where the brighter color represents wider 2D peak, and hence thicker graphene flakes. With the increase of layer number, the FWHM of 2D peak is larger, as indicated in Fig. 11.3d. For SLG, FWHM of 2D peak is $\sim 27.5 \pm 3.8$ cm^{-1}. It is a sharp and symmetric peak and can be well fitted by one Lorentzian peak. On the other hand, the 2D peaks of thicker graphene flakes are broader and can be

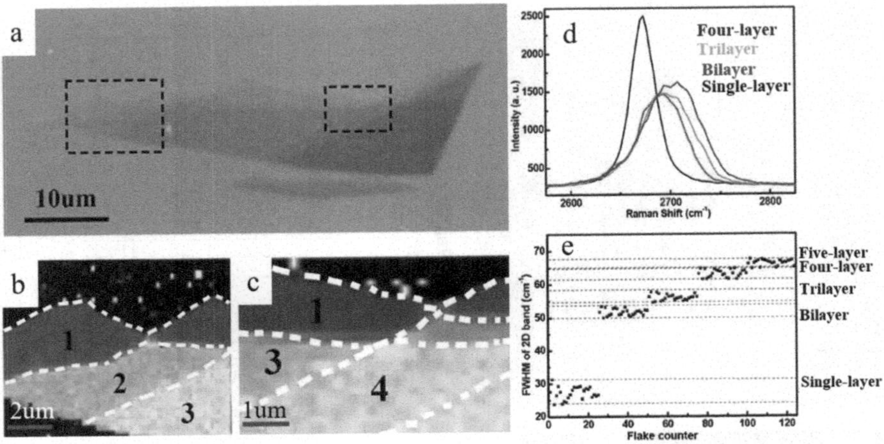

Fig. 11.3 Raman characterizations of AB-stacked FLG. (**a**) Optical image of graphene flakes with layer number from 1 to 4. (**b** and **c**) Raman images from left and right boxes in (**a**), respectively, according to the FWHM of 2D peak. Note that, for clarity, the two images are scaled by different brightness. (**d**) The 2D peaks of FLG with layer number from 1 to 4. (**e**) The statistical data of FWHM of 2D peaks with respect to different layer number

fitted by multiple subpeaks. The FWHM of 2D peak for 5 to 2 layers graphene is 66.1 ± 1.4 cm^{-1}, 63.1 ± 1.6 cm^{-1}, 56.2 ± 1.6 cm^{-1}, 51.7 ± 1.7 cm^{-1} (Fig. 11.3e), respectively. More importantly, there is non-overlapping between various thicknesses of FLG so that the value of FWHM of 2D peaks can be regarded as a direct standard to identify the layer numbers of AB stacked multilayer graphene flakes.

Similarly, for other 2D materials, Raman imaging can also provide reliable features to identify the layer number. Taking MoS_2 as an example, four first-order Raman active modes at ~32 cm^{-1} (E^2_{2g}), ~286 cm^{-1} (E_{1g}), ~383 cm^{-1} (E^1_{2g}) and ~408 cm^{-1} (A_{1g}) in bulk MoS_2 can be observed. The E^2_{2g} mode is not easily obtained owing to the limit of Rayleigh line rejection filter, and the E_{1g} mode is forbidden in back-scattering experiment on a basal plane [18]. The in-plane E^1_{2g} mode and the out-of-plane A_{1g} mode are commonly used to identify the layer number of MoS_2. The optical image and Raman spectra of MoS_2 with 1–4 layers are shown in Fig. 11.4a, b [18]. As can be seen, the frequency of A_{1g} peak increases (blue shift) while that of the E^1_{2g} peak decreases (red shift) with the increase of layer number [19]. The Raman images with the intensity and width of E^1_{2g} peak are depicted in Fig. 11.4c, d [18]. Unfortunately, the contrast of these two images are poor, especially for multilayers, which means that it is difficult to differentiate layer numbers of MoS_2 with peak intensity and width. However, the frequencies of E^1_{2g} and A_{1g} peaks are strongly dependent on the variations of layer number, as shown in Fig. 11.4e, f [18], where the brighter color represents the blue shift of peak. Uniform color contrast is presented in each region which precisely corresponds to different

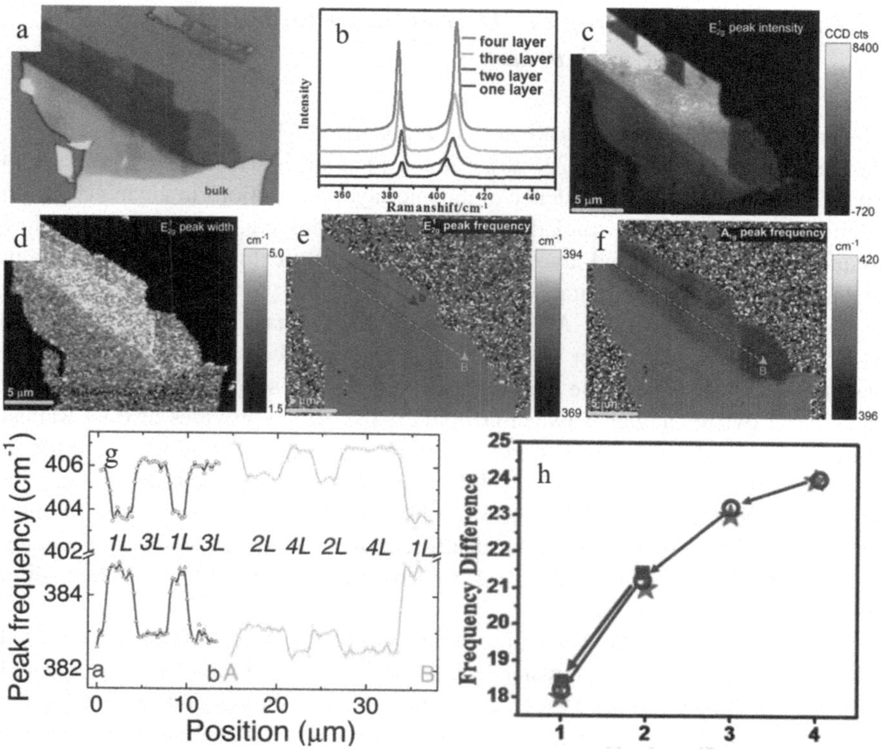

Fig. 11.4 (**a**) Optical image of MoS_2 with 1–4 layers. (**b**) Raman spectra of MoS_2 with 1–4 layers. Raman images of the E^1_{2g} peak intensity (**c**) and width (**d**). Raman images of E^1_{2g} (**e**) and A_{1g} peak (**f**) frequencies. (**g**) The E^1_{2g} and A_{1g} peak frequency profiles along the lines a-b and A-B drawn in (**e**) and (**f**). (**h**) The frequency differences between E^1_{2g} and A_{1g} peaks of 1–4 layer MoS_2

thicknesses. It is worth noting that for E^1_{2g} peak, the brighter color corresponds to thinner MoS_2 layer and for A_{1g} peak, it is opposite. Figure 11.4g [18] shows the frequency profiles of E^1_{2g} and A_{1g} peaks along the a-b and A-B dashed lines shown in Fig. 11.4e, f. As shown in Fig. 11.4h [19], the frequency differences of E^1_{2g} and A_{1g} peaks are ~18, ~21, ~23, and ~24 cm^{-1} for 1–4 layer MoS_2, respectively. Hence, the variation of E^1_{2g} and A_{1g} peak frequencies can be used to accurately determine the layer number of ultrathin MoS_2 flake, instead of the intensities or widths of the peaks. Such variation can also be found in other TMDs materials such as $MoSe_2$, and WS_2 [20, 21], which fully demonstrates the capability of Raman spectra and imaging in the study of TMDs.

In a short summary, Raman spectrum is an intrinsic characteristic of 2D materials, and the advantage of Raman imaging in determining the number and the distribution of layers of 2D materials is its independence on the substrate used, especially for flexible substrate or suspend case.

11.2.2 Raman Imaging of 2D Materials with Different Stacking Configurations

The properties of 2D materials are also strongly affected by stacking configurations through interlayer coupling [22, 23]. Raman imaging is an efficient approach to investigate the stacking dependence of the interlayer interactions in 2D materials. In addition to the commonly studied high frequency modes, e.g. E_{2g}, A_{1g}, the ultralow frequency shear modes and layer breathing modes also exhibit strikingly different features for different configurations [24].

As mentioned above, for AB-stacked FLG, the G peak becomes more intense and 2D peak become broader with increasing layer number. However, for arbitrary stacking graphene layers, the electronic structure has a strong dependence on the stacking or twisted angles between layers [25–28], which provides distinct Raman features. The FLG with twisted angles can be fabricated by folding one layer of graphene on above another (1 + 1 folded graphene) [68]. They also commonly appear in epitaxial grown FLG on SiC [29], or CVD grown multilayer samples [30]. Figure 11.5 gives the stacking disorder induced Raman spectra and imaging of graphene layers, including 1 + 1 folded graphene and CVD graphene film containing TBG domains.

Figure 11.5a, c [31] show the optical images of SLG before and after folding, and the schematic drawings are shown in (b, d). The 1 + 1 folded samples were prepared by gently flushing de-ionized water across the surface of the substrate containing the target graphene sheet [31]. The Raman images of G and 2D peak position and area intensity are shown in Fig. 11.5e, h [31], where brighter color represents higher frequency and stronger intensity. Compared with SLG, the frequency of 2D peak for 1 + 1 folded graphene has a blue shift of $\sim$12 cm^{-1} and its intensity becomes stronger, which is due to the change of electronic band structure of 1 + 1 folded graphene or TBG [32]. While there is almost no change for G peak position, the intensity of G peak for folded graphene is about twice as those of SLG and similar to that of BLG because they both contain two graphene layers. These features can be further confirmed in the Raman spectra in Fig. 11.5i [31], where the Raman spectra of SLG, 1 + 1 folded graphene, and Bernal stacked BLG samples are presented. The shape of 2D peak for 1 + 1 folded graphene is sharp and symmetric like that of SLG, which can be fitted by only one Lorentzian curve. This shows that the electronic structure of 1 + 1 folded graphene is similar to that of SLG, and without the splitting of energy bands. On the contrary, broader 2D peak is obtained for BLG, which can be fitted by four peaks deriving from the splitting of valence and conduction bands [33]. Additionally, the intensity ratio of 2D to G peak for 1 + 1 folded graphene is higher comparing with SLG, partially caused by the different resonance conditions between folded graphene and SLG.

Figure 11.5j–q is the optical and Raman images of a CVD grown monolayer graphene film with irregular TBG grains [14]. The optical image (shown in Fig. 11.5j) as well as Raman image of the G peak intensity (shown in Fig. 11.5k) could clearly reveal the single layer and bilayer domains. For most of the bilayer domains,

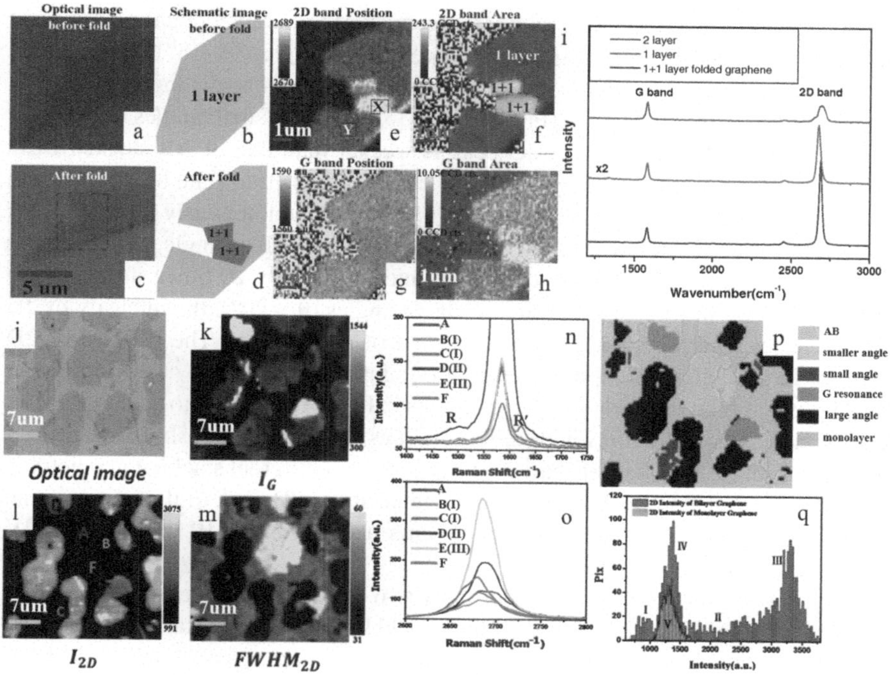

Fig. 11.5 Optical images of SLG before (**a**) and after (**c**) folding. Schematics of SLG before (**b**) and after folding (**d**). Raman images obtaining from the 2D (**e**) and G (**g**) peak positions, 2D (**f**) and G (**h**) peak area intensity. "X" and "1 + 1" denote the folded parts. (**i**) Raman spectra of BLG, SLG and 1 + 1 folded graphene. (**j**–**q**) The Raman characterizations of CVD graphene film containing irregular bilayer grains. Optical image (**j**) and Raman images obtaining from the G peak intensity (**k**), 2D peak intensity (**l**), and 2D peak FWHM (**m**) of the same sample; all the scale bars are 7 μm. The G (**n**) and 2D peak (**o**) Raman spectra correspond to points A-F in (**l**). The distribution of various twist angles (**p**) and the histogram distribution of the 2D peak integrated intensity (**l**) corresponding to the area in figure (**j**)

the Raman intensity is about twice as that of single layer. However, some of the bilayer domains (point D) has G peak intensity tens of times as that of single layer (point F), which corresponds to TBG with the G peak resonance at a particular twist angle [34]. Such an obvious enhancement in the G peak can also be found in Fig. 11.5n. This is due to the match of excitation energy with the energy difference between valance and conduction VHSs in TBG. The equation to calculate the critical G resonance angle θ is as follows:

$$\theta = 3aE_{laser}/\hbar v_F 4\pi \tag{11.1}$$

Where a demonstrates the lattice parameter (2.46 Å), $\hbar$ describes the decreased Planck constant and v_F shows the Fermi velocity in SLG (approximately 10^6 ms^{-1}). Hence, the G resonance for TBG with $E_{laser} = 2.33$ eV (532 nm) happens when the twist angle is ~12^0. According to the strong and weak interlayer coupling, the twist angle in TBG can be divided into three types: small angles ($\theta < 8^0$), intermediate angles ($8^0 < \theta < 20^0$) and large angles ($\theta > 20^0$), which can be distinguished by the shape and intensity of 2D peak (Fig. 11.5l, o) [32, 35]. For small angles ($\theta < 8^0$, the point B and C), the intensity of 2D peak is weaker than that of AB stacked BLG (Point A). For intermediate angles ($8^0 < \theta < 20^0$), a transition state between strong and weak interlayer coupling, the intensity of 2D peak monotonously increases with the increase of twist angle. Point D belongs to the intermediate angle, and this point also has a striking enhancement of G peak intensity, nearly 20 times stronger than that of SLG as shown in Fig. 11.5n. For large angles ($\theta > 20^0$), the intensity of 2D peak reaches saturation and is 2–3 times that of single layer. The FWHM of 2D peak of large angles also approaches that of single layer (30 cm^{-1}) on account of decoupling of bilayer graphene. Large regions in Fig. 11.5l belong to the large angle, e.g. point E. Hence, regions with different types are unambiguously extracted and shown in Fig. 11.5p. The distribution of samples with different stacking geometry is also clearly labeled in the histogram of the intensity of 2D peak of the same sample (Fig. 11.5q), including I (small angle), II (intermediate angle), III (large angle), IV (AB stacked) and V (single layer).

For trilayer graphene, there are normally two stacking configurations, ABA and ABC as shown in Fig. 11.6a, b [36], which could be identified by Raman spectra and imaging. Optical image, contrast image, and G peak intensity image in Fig. 11.6c–e [36] reveal that mechanically exfoliated graphene sheet is very uniform and has the same layer number of 3. The FWHM of 2D peak varies from 40 to 68 cm^{-1} and also its Raman image presents apparently two parts, suggesting different stacking orders (Fig. 11.6f) [36]. Lui et al. demonstrated that the FWHM of 2D peak of ABC stacked graphene is ~10 cm^{-1} broader than that of ABA trilayers [37], and therefore the brighter part is ABC stacked graphene while the darker part is ABA stacked graphene. The Raman spectra of 2D peak also shows great difference between ABA and ABC stacked orders in Fig. 11.6g, h [36]. Although both of them can be fitted by six subpeaks, the subpeaks of ABC stacked graphene have an evident blue shift comparing with those of ABA stacked graphene [36]. Furthermore, the stacking order transformation between ABC and ABA stacked graphene by triazine decoration is shown in Fig. 11.6i–n since the Bernal (ABA)-stacked graphene trilayer is considered to be thermodynamically more stable than the rhombohedral (ABC) counterpart. It can be clearly seen from the variations of 2D width image and AFM topography that there is no change for ABA stacked domain, but there is a stacking transformation from ABC to ABA for ABC stacked domain after triazine decoration. Such transformation begins from the small ABC domain such as the triangle area in Fig. 11.6i. These results can be further confirmed by the Raman spectra of 2D and M peaks in Fig. 11.6m, n. Cong et al. have shown that the M peak at ~1740 cm^{-1} exhibits a single peak in the ABA-stacked trilayer, but it splits into two peaks in the ABC-stacked trilayer [38]. After triazine decoration, the FWHM of 2D peak for ABC stacked domain is apparently narrower and similar to that of ABA

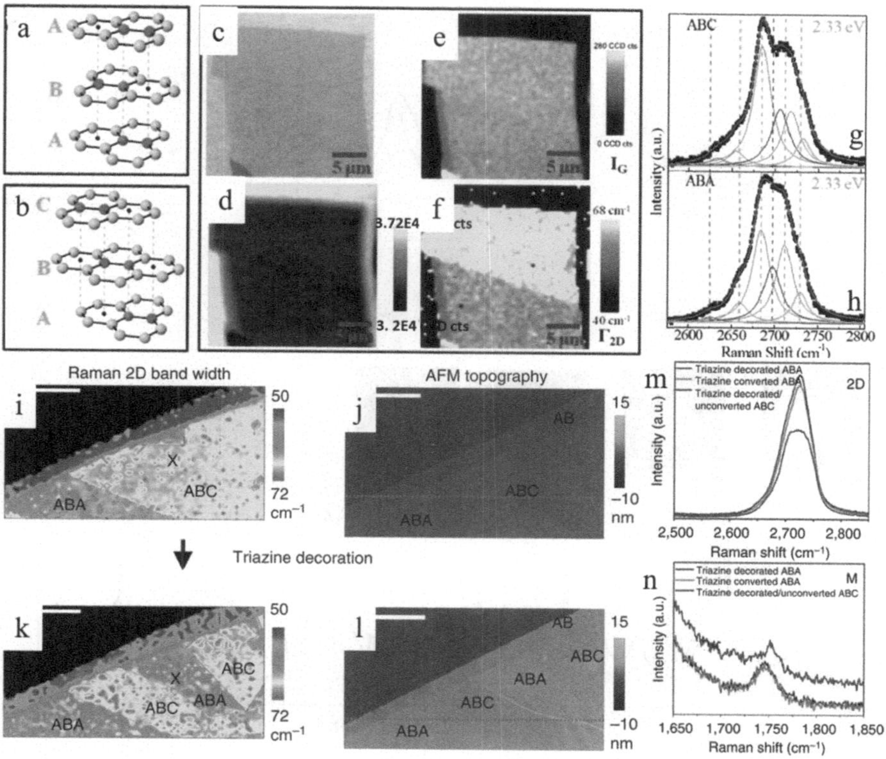

Fig. 11.6 (**a** and **b**) Stacking structures of trilayer graphene with (**a**) ABA and (**b**) ABC stacking sequences. (**c**)Optical and (**d**) contrast images of a trilayer graphene sheet. Raman images of the G peak intensity (**e**) and the 2D peak width (**f**) of the trilayer graphene sheet obtained at $E_{laser} = 2.33$ eV. (**g** and **h**) Raman spectra of the 2D peak of ABC- and ABA-stacked trilayer graphene. (**i**–**l**) Raman images of the FWHM of 2D peak (**i** and **k**) and AFM images (**j** and **l**) before and after triazine decoration, respectively. (**m** and **n**) Raman spectra of 2D and M peaks for the triazine decorated ABA, triazine converted ABA and triazine decorated/unconverted ABC-stacking domains

stacked domain, and unconverted ABC stacked domain still has a broader FWHM. On the other hand, M peak with two peaks for ABC stacked domain becomes a single peak and is identical with that ABA stacked domain after triazine decoration.

Next, we continue to discuss the stacking configurations of TMDs. For bilayer TMDs system, there are two typical stacking polytypes (AA and AB). An optical image containing both AA- and AB-stacked bilayer MoS_2 grown by CVD is shown in Fig. 11.7a [39]. The corresponding Raman spectra in the high-frequency range (350–440 cm^{-1}) are shown in Fig. 11.7b [39], where the intensity of both E_{2g}^1 and A_{1g} peaks for AB stacking is higher than that of AA stacking and no significant frequency changes are observed in the two high frequency modes. However, the ultralow frequency (ULF) Raman modes show distinct behaviors (Fig. 11.7c [39]). The in-plane shear mode (SM) locates at ~22.6 (~22.8) cm^{-1} and the out-of-plane

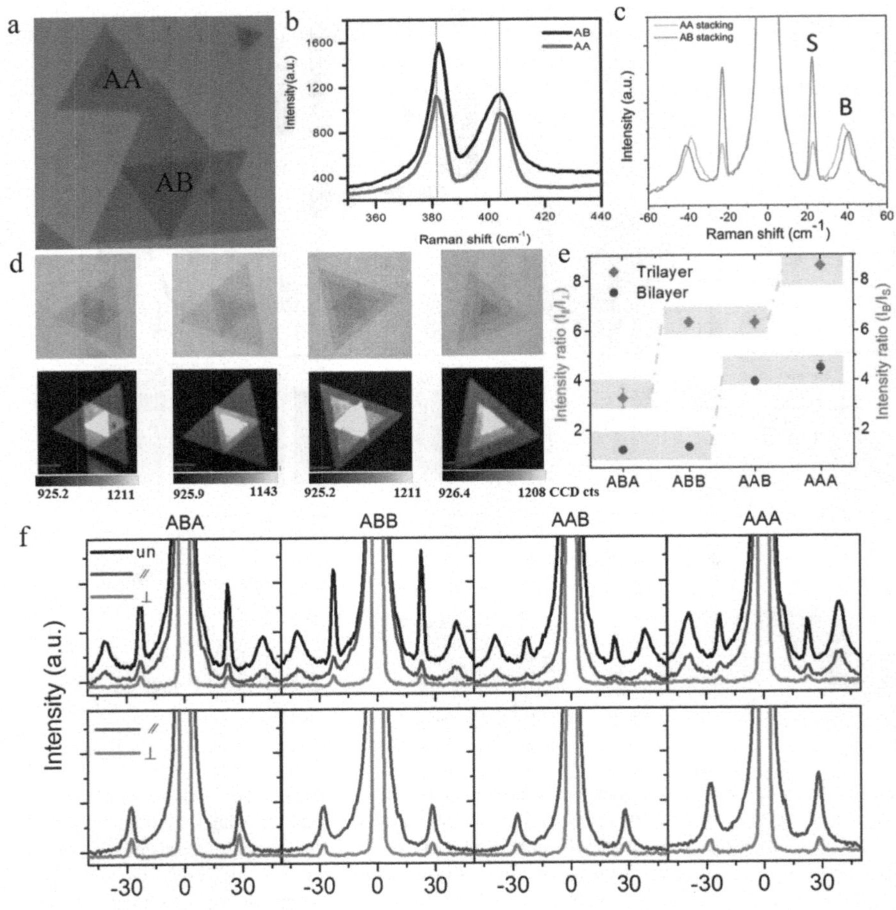

Fig. 11.7 Stacking-dependent Raman behaviors of bilayer and trilayer MoS_2. (**a**) Optical image of CVD grown bilayer MoS_2 with AA and AB stacking. High-frequency (**b**) and ULF (**c**) Raman spectra of AA- and AB-stacked bilayer MoS_2. Structural characterization of ABA, ABB, AAB, and AAA trilayer MoS_2 (**d–f**). (**d**) Optical micrograph (Upper panel) and the corresponding Raman intensity images from d (Lower panel). (**e**) The intensity ratio I_b/I_s of BM and SM (dots) for bilayer MoS_2 and $I_{//}/I_{\perp}$ (diamonds) for trilayer MoS_2. (**f**, Upper panel) The ULF Raman spectra of the second layer of trilayer MoS_2. (**f**, Lower panel) The ULF Raman spectra of the third layer of trilayer MoS_2 in (**d**)

breathing mode (BM) locates at ~41.6 (~38.7) cm^{-1} for AB (AA), respectively. These two modes can also be used to determine the stacking configurations of trilayer MoS_2. Figure 11.7d [24] shows the optical (upper) and Raman images (lower) of trilayer MoS_2 with four different stackings, ABA, ABB, AAB, and AAA. According to the relative intensity ratio (I_b/I_s) of BM and SM of bilayer MoS_2 in Fig. 11.7e and also the upper panel of Fig. 11.7f [24], the first two layers of trilayer MoS_2 system can be divided into two types: the ABA and ABB stackings with

I_b/I_s = ~1 could be regarded as AB stacking for the bilayer, while the AAB and AAA stackings with I_b/I_s = ~4 should be regarded as AA stacking, respectively. However, it can be seen from the lower panel of Fig. 11.7f [24] that for the third layer of trilayer MoS_2, the BM and SM are nearly combined into one peak in the ULF range, which demonstrates that the stacking orders of the third layer cannot be straightforward determined by I_b/I_s. Therefore, the polarization measurements are further used to determine the stacking configuration of trilayer MoS_2. Notably, the SM appears in both $\overline{Z}(xx)$ $Z(\perp)$ and $\overline{Z}(xy)$ $Z(\|)$ configurations. This is very different from the BM, which can only be found in the $\overline{Z}(xx)$ $Z(\|)$ configuration. Here, $I_{//}$ is the summation of the parallel components of both SM and BM, while $I_\perp$ contains only the perpendicular component of SM. The intensity ratio $I_{//}/I_\perp$ sharply decreases for AAA stacking (~8), followed by AAB (ABB) stacking (~6), and ABA stacking (~3) with the variations of the stacking patterns in Fig. 11.7e [24]. Thus, Raman imaging technique can be used as a fast and convenient nondestructive method to identify the stacking sequence of 2D materials, which is otherwise very difficult and time consuming by using other techniques [39].

11.2.3 Raman Imaging of 2D Heterostructures

Heterostructures derived from 2D materials such as graphene, hexagonal boron nitride and TMDs provide a novel platform for exploring new physical phenomena and devices [40–43]. For instance, TMDs heterostructures are especially exciting for novel optoelectronic and photovoltaic applications because they exhibit optical bandgap in the range from near-infrared to visible and extremely strong light-matter interactions [44]. A photodetector based on the graphene/MoS_2 heterostructure is able to offer a high photogain larger than 10^8 [45]. The subtle chemistry and interactions within its layer structure are critical for the application of 2D heterostructures in functional devices. Therefore, it is necessary to differentiate heterostructure from single layer materials and confirm the formation of heterostructure by nondestructive and free contamination tool. Here, as examples, heterostructures such as graphene/MoS_2, vertical and lateral WS_2/MoS_2 are probed by Raman imaging.

Figure 11.8 [45] presents Raman characterizations of graphene/MoS_2 heterostructures. The optical images are shown in Fig. 11.8a, e, and the Raman spectra of G peak of graphene are shown in Fig. 11.8b. It is noted that the G peak of SLG on SiO_2/Si is at ~1585.4 cm^{-1} and it is broadened with a red shift to ~1581.7 cm^{-1} on MoS_2. Thus, the variations in G peak demonstrate that electron concentration increases, namely, the Fermi level of graphene is increased in heterostructure with light exposure, as compared to the known *p*-doped graphene by PMMA [46]. On the contrary, there is no significant variations in D peak (the Raman image and spectra of D peak are not present here), which means that the photoelectrons produced by the Raman laser are infused into graphene. This conclusion is consolidated by the Raman images of the G peak position (Fig. 11.8c) and the FWHM (Fig. 11.8d).

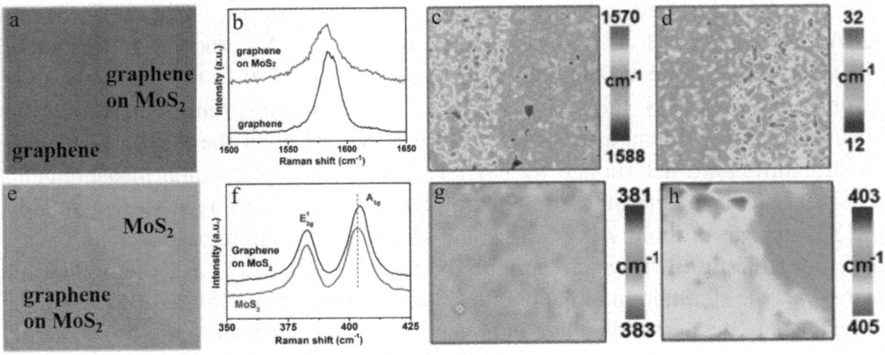

Fig. 11.8 Raman images and representative Raman spectra of graphene/MoS_2 heterostructure. (**a**) and (**e**), The optical images of different heterostructure samples. Raman spectra (**b**) and Raman images (G peak position and FWHM) representative for graphene and graphene on MoS_2 (**c** and **d**). (**f**) to (**h**) Raman spectra and Raman images for MoS_2 (position of $E^1{}_{2g}$ and A_{1g}) and that covered with graphene. The excitation source is 473 nm laser

On the other hand, A_{1g} peak of MoS_2 in contact with graphene has a blue shift of ~1 cm^{-1} and becomes narrower and stronger comparing with those of pristine MoS_2 (Fig. 11.8f), which demonstrate that electron concentration reduces in MoS_2, namely, photo-excited electrons are transferred from MoS_2 to graphene while photo-excited holes are kept in the MoS_2 layer. This can also be verified by the images of the positions of $E^1{}_{2g}$ (Fig. 11.8g) and A_{1g} (Fig. 11.8h).

Figure 11.9 [47] gives the schematic, optical and Raman characterizations of vertically stacked and in-plane WS_2/MoS_2 heterostructures (a). Firstly, for vertically stacked bilayer WS_2/MoS_2 heterostructure, Raman spectra (11.9c) taken from the bright region (such as points 1 and 2 in Fig. 11.9b) only show the $E^1{}_{2g}$ (~383.9 cm^{-1}) and A_{1g} (~405.3 cm^{-1}) peaks of monolayer MoS_2 [48, 49], which suggests that the lower layer is single layer MoS_2 [50]. Nevertheless, for bilayer area, namely, the dark region such as points 3 and 4, there are two extra peaks, the overlapping 2LA (M) and $E^1{}_{2g}$ (~356.8 cm^{-1}) and A_{1g} (~418.5 cm^{-1}) peaks of upper layer WS_2. The formation of vertically stacked bilayer WS_2/MoS_2 heterostructure can be further verified by Raman intensity images of the $E^1{}_{2g}$ peaks of both MoS_2 (~384 cm^{-1}) and WS_2 (~357 cm^{-1}) as shown in Fig. 11.9d, e, respectively. The intensity of $E^1{}_{2g}$ peak of MoS_2 is relatively weaker in the center of the image (Fig. 11.9d) because it is at the bottom of the WS_2 layer. As for in-plane WS_2/MoS_2 heterostructure, an optical image is shown in Fig. 11.9f. The formation of in-plane WS_2/MoS_2 heterostructure, where triangular monolayer MoS_2 domain serves as the core and WS_2 serves as the shell layer, is demonstrated by Raman spectra (g) and image (h) of the intensity of $E^1{}_{2g}$ peaks at ~351 cm^{-1} (WS_2) and ~381 cm^{-1} (MoS_2). Meanwhile, WS_2 and MoS_2 share the same crystal orientation at WS_2/MoS_2 in-plane heterostructure, which is exhibited in high-resolution transmission electron microscope (TEM) in Fig. 11.9i.

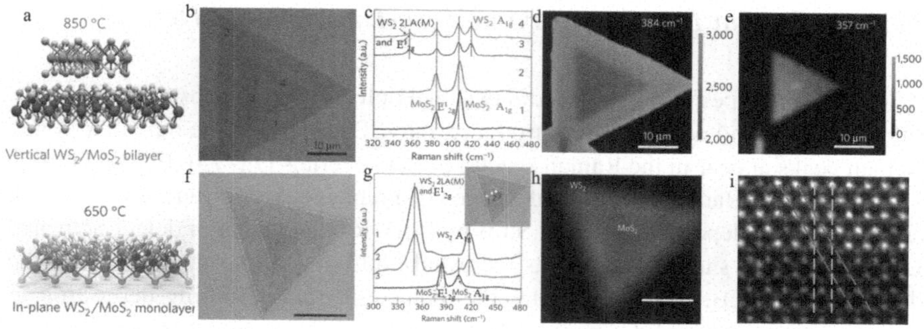

Fig. 11.9 (**a**) Schematic and the morphologies of the vertically stacked and in-plane WS_2/MoS_2 heterostructures. Optical and Raman characterizations of vertically stacked (**b**–**e**) and in-plane (**f**–**h**) WS_2/MoS_2 heterostructure. (**b** and **f**) the optical images of heterostructure. (**c**) Raman spectra of vertically stacked heterostructure taken from the four points marked in (**b**). (**d** and **e**) Raman intensity images obtained from the ~384 cm^{-1} and ~357 cm^{-1} peaks, respectively. (**g**) Raman spectra of in-plane heterostructure. (**h**) Combined Raman intensity image at ~351 cm^{-1} (WS_2) and ~381 cm^{-1} (MoS_2), showing the core-shell structure with WS_2 as the shell and MoS_2 as the core. (**i**) High resolution TEM image of the in-plane heterostructure. The vertical and slant dash lines depict the atomic planes along the armchair and zigzag directions, respectively, indicating the WS_2 and MoS_2 domains share the same crystal orientation in-plane heterostructure

Raman imaging can not only differentiate the materials in heterostructure, but also be used to confirm the formation of heterostructure with vertical or in-plane stacking. These would contribute to the exploration of new physical phenomena and applications of two dimensional materials in novel devices.

11.3 Raman Imaging of Defects in 2D Materials

There are various intrinsic and extrinsic defects in 2D materials. These defects are mainly divided into two types according to dimensionality: point/zero-dimensional (Stone-Wales defects [51–54], adatoms [55–58], vacancy [59, 60], substitutional impurities [61–64]) and one-dimensional defects (GBs [65–67] and edges [68–71]), which can significantly affect the electronic, optical, mechanical, and thermal properties of 2D materials. For example, the carrier mobility of 2D materials strongly depends on the numbers of defects and also its crystalline grain size, due to the scattering from localized defects and GBs. Besides, if the location of defects can be controlled, novel 2D materials can be prepared for p-n junction diodes. Thus, the statistics and analysis of the amount, type and location of defects are very important for the fundamental study and device applications of 2D materials. Raman imaging provides a very effective method to study the spatial distribution and also the characteristics of the defects.

11.3.1 Raman Imaging of Point/Zero-Dimensional Defects

After different types of point/zero-dimensional defects such as substitutional impurities, vacancy and adatoms are introduced into 2D materials, the defect-activated Raman peaks appear in the Raman spectra of 2D materials [72–75]. Figure 11.10a–f [72] show the Raman spectra and images of graphene with different layers with and without the deposition of 5 nm SiO_2 by PLD. Raman image of defect-activated D peak intensity shows that defects are evenly introduced into graphene after the deposition of SiO_2, which illustrates the damage to sample during deposition or the interaction between SiO_2 and graphene, e.g. vacancies, dislocation and dangling bonds (Fig. 11.10a). In addition, while the G peak intensity image shows no obvious change before and after deposition (Fig. 11.10d, f), the intensity of the D peak decreases with the increase of graphene thickness and is invisible for bulk graphite, demonstrating that defects are more easily introduced into thinner graphene sheets (Fig. 11.10c, e). Figure 11.10g, h [74] show the Raman features of hydrogenated graphene. There are five regions determined by the intensity of G peak and noted as 1 L (single layer), 2 L (bilayer), 3 L (trilayer layers), 4 L (four layers), and more than four layers. The observation of D peak shows that defects are introduced into graphene by hydrogen plasma treatment which leads to the formation of C-H sp^3 bonds. According to the Raman image of D peak, the intensity decreases with the increase of graphene thickness except for 1 L. Such variations in D peak intensity of the hydrogenated graphene layers with the thickness is apparently different from the monotonous variations of D peak intensity induced by oxidation and particle bombardment [72, 75]. It illustrates the highest amount of sp^3 defect in 2LG, follows by 3LG, then 4LG, and more than four layers, while the least is in 1LG, which were further demonstrated by XPS characterizations. The hydrogen coverage is $\sim$16.67% for SLG, and $\sim$45.5% and $\sim$35.3% for 3 L and 5 L, respectively, which further suggests the hydrogenation barriers are strongly dependent on the layer number of graphene. Figure 11.10i [73] shows Raman spectra of nitrogen doped graphene (NG), defective graphene (DG) by Ar^+ bombardment as well as pristine graphene. Defect-activated D ($\sim$1348 cm^{-1}) and D′ ($\sim$1620 cm^{-1}) peaks appear in NG and DG, and D′ peak is more evident in NG. This shows the different nature of defect could be identified by $I_D/I_{D'}$. As shown in Fig. 11.10j, the distributions of the intensity of G peak demonstrate the uniformity of NG film, while those of D peak (Fig. 11.10k) demonstrate the non-uniform distribution of nitrogen related defects. Figure 11.10l–n [76] shows graphene functionalized by diazonium salts. Large amount of sp^3 defects exist in graphene on SiO_2 substrate while sparse covalent functionalization could happen for graphene on octadecyltrichlorosilane (OTS) substrate (Fig. 11.10l), which can be further confirmed by spatial Raman image of the intensity ratio of D and G peaks (I_D/I_G) (Fig. 11.10n). Graphene on OTS-covered areas presents a low functionalization indicated by narrower stripes while that on SiO_2 substrate presents a high functionalization indicated by the wider stripes. Therefore, the different functionalization on different substrates can be clearly monitored by Raman imaging through the defect-activated D peak.

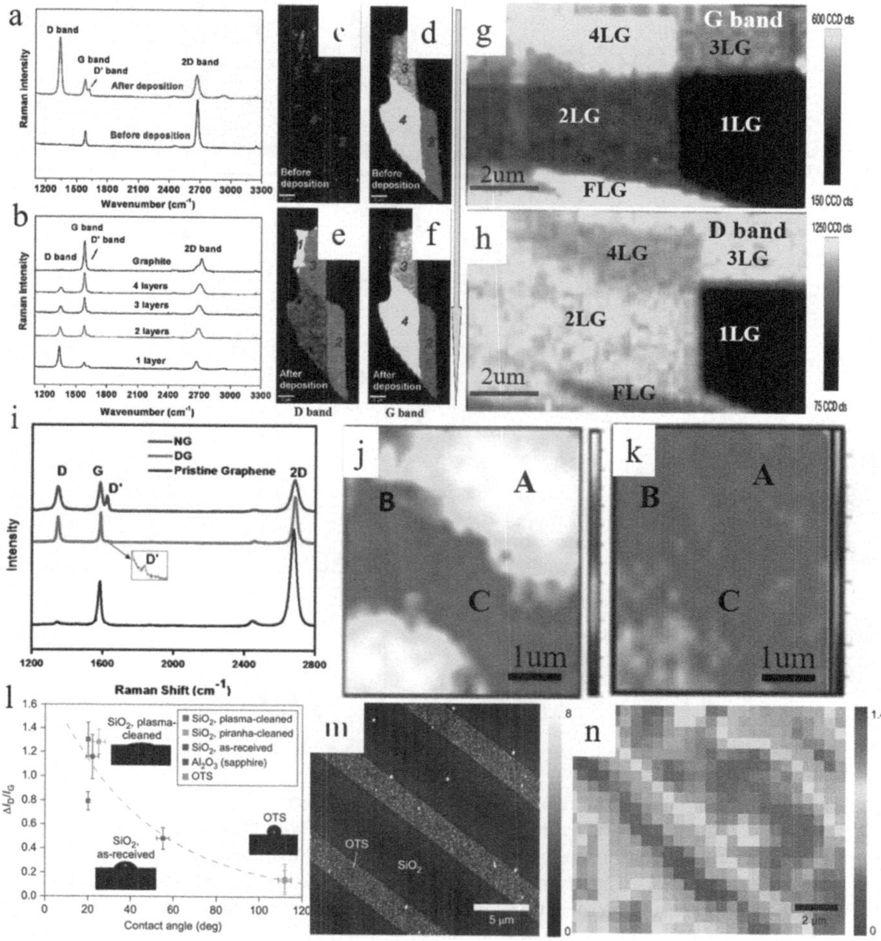

Fig. 11.10 (**a**) Raman spectra of SLG before and after the deposition of 5 nm SiO_2. (**b**) Raman spectra of graphene with different thicknesses after SiO_2 deposition. Raman images of D (**c**, **e**) and G peak (**d** and **f**) intensity of graphene sheets before and after 5 nm SiO_2 deposition. Raman images of G (**g**) and D peak (**h**) intensity of the pristine and hydrogenated graphene sheet. (**i**) Typical Raman spectra of pristine graphene, DG and NG. Raman images of intensities of G (**j**) and D peaks (**k**) of a NG film. (**l**) Intensity ratio of Raman D and G peaks (I_D/I_G) after diazonium functionalization plotted as a function of water contact-angle of the substrate before graphene deposition. AFM topographic image (**m**) and Raman spatial image of I_D/I_G intensity ratio (**n**) after diazonium functionalization

11.3.2 Raman Imaging of One-Dimensional Defects

One-dimensional defects, such as edges and GBs, commonly exist in 2D materials, and could be easily identified by Raman imaging. There are two types of edges in graphene, i.e. zigzag (Z-edge) and armchair (A-edge). In addition, CVD-grown

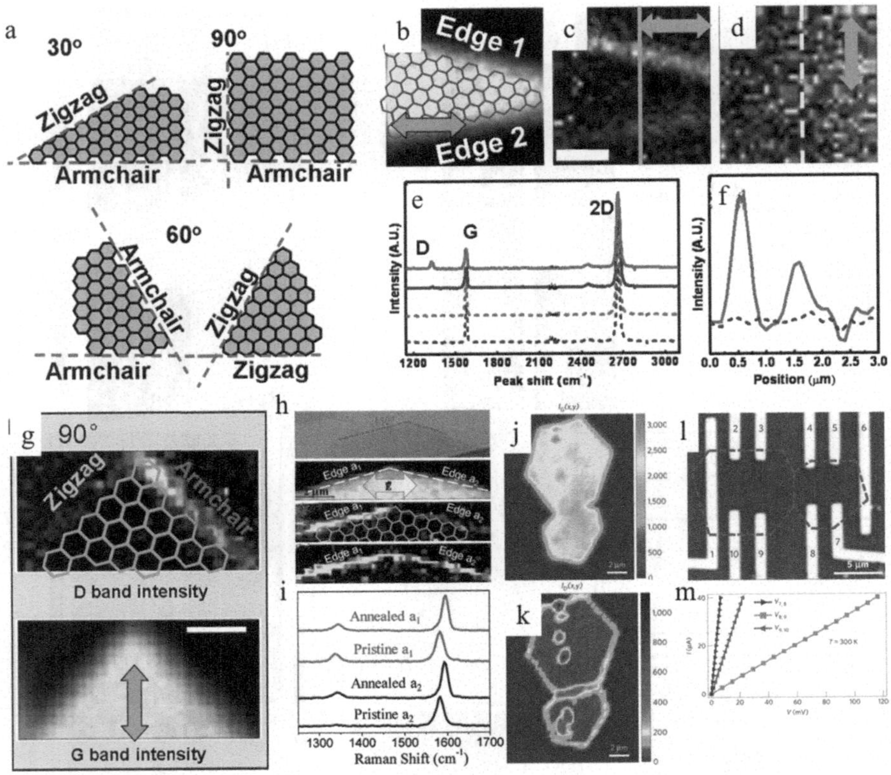

Fig. 11.11 (**a**) Illustration of the relationship between angles and the chiralities of the adjacent edges. (**b**) Raman image constructed by the intensity of G peak with the expected arrangement in gray. (**c**) and (**d**) are Raman images constructed by the D peak intensity with excitation laser in horizontal and vertical polarizations, respectively. (**e**) Raman spectra taken from edge 1 and edge 2 with horizontal polarization (solid lines) and vertical polarization (dashed lines), respectively. (**f**) The D peak intensity profile obtained from the solid and dotted lines in (**c**) and (**d**). (**g**) Raman images of graphene with 90° edges. (**h**) Optical image and Raman images of SLG with 150° edge. The three Raman images from up to down are obtained from the G peak intensity, D peak intensity (before annealing), D peak intensity (after annealing at 300°), respectively. (**i**) Raman spectra of two edges in (**h**) before and after annealing. (**j** and **k**) Raman image of D and G peak intensities for two coalesced CVD-graphene grains. (**l**) Optical image of a device with multiple electrodes (numbered 1–10) contacting two coalesced graphene grains. (**m**) Representative room-temperature I–V curves measured within each graphene grain and across the GB

graphene is polycrystalline and GB has great influence on its properties. Through Raman imaging, the defect-activated D peak is detected at the edges and GBs, and the intensity of the D peak can be used to distinguish the types of edges.

Figure 11.11 shows the schematic diagram, Raman images and spectra of graphene with different edge configurations [71]. A majority of the angles of grpahene sheet are distributed around $n*30°$ (n, an integer 0–6), which illustrates two edge structures, zigzag and armchair for the arrangement of carbon atoms.

Uniform distribution of the G peak intensity indicates a high quality of graphene (~30°) as shown in Fig. 11.11b [71]. The D peak intensity is only presented at edges 1 and 2 in horizontal polarization, and vanishes in the vertical polarization as shown in Fig. 11.11c, d, f [71]. Since D peak is proportional to $[\vec{e} \times \vec{p}]$, where $\vec{e}$ is the polarization of incident laser and $\vec{p}$ is the momentum on photoexcited electrons or holes being relative to Dirac point, the intensity of D peak is the strongest when $\vec{e}$ is parallel to the edge. According to Raman spectra and image of edge 1 and 2 in horizontal polarization in Fig. 11.11c, e [71], the intensity of D peak at edge 1 is ~1.7 times higher than that of edge 2. Therefore, edge 1 and edge 2 is identified as A-edge and Z-edge, respectively, because the double resonance process can only be fulfilled at the A-edge. The case is similar for 90° edge (Fig. 11.11g), where the right edge with stronger D peak intensity is corresponding to the armchair geometry. More interestingly, the edges can also undergo structure rearrangement at high temperature. Figure 11.11h [77] shows the optical and Raman images of SLG with angle of 150^0. Through Raman imaging, edge 1 and edge 2 correspond to A-edge and Z-edge, respectively. In addition, Z-edge can undergo rearrangement to meet double resonance conditions and form A-edge segments after annealing treatment at 300°, which is demonstrated by Raman image and Raman spectra shown in Fig. 11.11h, i. As for FLG, the rearrangement of edge structure is even more complicate and the formation of inactive sp^2 hybridized smooth loop configuration could be observed at Z-edge of BLG [78]. Therefore, Raman imaging can be employed to investigate the chiralities, rearrangement of edges of graphene by monitoring the variation of D peak intensity. Raman imaging can also clearly identify the locations of GBs by the variations of the D peak intensity as shown in Fig. 11.11j–m [79]. GBs is actually a disorder edge, and it also has an evident intervalley double resonance scattering. Therefore, the D peak intensity is very strong at GBs, in addition to other point defects within the grain. The presence of GBs inevitably affects the optical, electronic and thermal properties of 2D materials. On one hand, the electronic transport properties are significant affected at GBs as shown in Fig. 11.11l, m. *I*–*V* curves measurements suggest that the slope at GBs (squares) is smaller than that within grains (triangles), which corresponds to larger resistance at GBs. On the other hand, the GBs are also active centers, which is beneficial for designing of nanometer-scale highly sensitive chemical detectors. Ultrahigh sensitivity is demonstrated due to a synergetic combination of gas molecules accumulation and the presence of a sharp onset energy [66].

Different from graphene, other 2D materials such as TMDs do not exhibit obvious defect-activated Raman peaks at edge and GB. In such case, variations in the intensity and frequency of Raman peak can be detected by Raman imaging at the edge and GB of TMDs. Raman images in Fig. 11.12b, d clearly reveal the cracks/edges formed by annealing because of the reaction between O_2 and MoS_2 [80], which is further demonstrated by PL image (a) and AFM (e) [80]. Interestingly, the PL intensity dramatically increases at the crack position and A_{1g} peak has a blue shift caused by *p*-type doping [81, 82]. There are a lot of dangling bonds on Mo and S atoms at the cracks of MoS_2, and these dangling bonds can be regarded as defects,

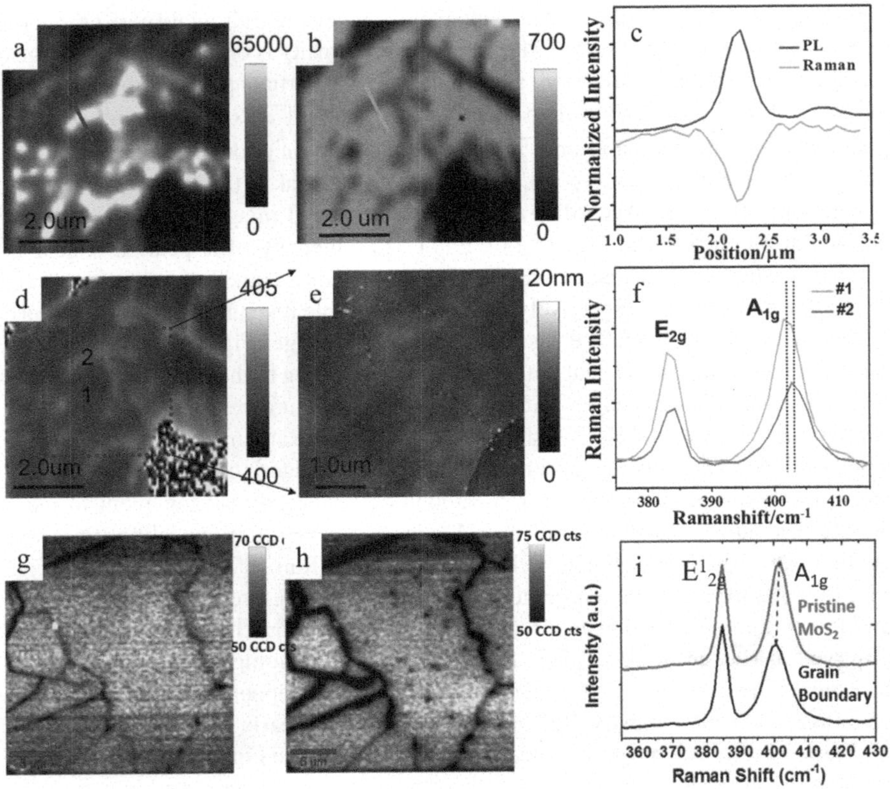

Fig. 11.12 (**a**) PL and (**b**) Raman A_{1g} peak intensity images of monolayer MoS_2 after annealing. (**c**) Normalized intensity profiles of a crack in (**a**) and (**b**). (**d**) Raman image of the A_{1g} peak frequency. (**e**) AFM image corresponding to the dash square in (**d**). (**f**) Raman spectra taken from locations #1 and #2 in (**d**). Raman images of $E^1{}_{2g}$ (**g**) and A_{1g} (**h**) peaks intensity. (**i**) Averaged Raman spectra at GB and non-GB regions

which are very active centers for molecular adsorption. As a consequence, O_2 can adsorb on the cracks/defects with much stronger binding energy and also introduce *p*-type doping comparing with that on an ideal MoS_2 surface. The efficient charge transfer as well as the suppression of nonradiative recombination of excitons at defect sites, together result in the ultrahigh PL at the edges of MoS_2. Raman imaging can also be employed to clearly identify the GBs of TMDs. The intensities of $E^1{}_{2g}$ and A_{1g} peaks of MoS_2 decrease at GB as shown in Fig. 11.12g, h [83]. According to the average Raman spectrum shown in Fig. 11.12i [83], both the intensity of A_{1g} mode and the peak separation of $E^1{}_{2g}$ and A_{1g} decrease for grain boundaries. Such variations suggest that *n*-doped are introduced in GBs of MoS_2.

11.4 Strain in 2D Materials Monitored by Raman Imaging

The strain, especially uniaxial strain, can dramatically modify the electronic and optical properties of 2D materials. A band-gap opening could be achieved in strained graphene due to the breaking of sublattice symmetry [84, 85]. The PL of TMDs can be modulated by strain engineering due to change of band structure [86]. However, the distribution of strain in 2D material is usually non-uniform. Raman imaging in 2D materials could provide information to help the better understanding of crystal and band structure of strained sample [87, 88].

Controlled uniaxial strain could be introduced into 2D materials by stretching the deformable substrates they sit on, fabricating wrinkle and so on. Figure 11.13 gives Raman images and spectra of 2D materials as a function of strain. In graphene, the frequency of 2D peak has a red shift both in 1 and 3 layer graphene with the increase of strain as shown in Fig. 11.13a [89], which is caused by the elongation of the carbon-carbon bonds and lowing of vibrational frequency. Frequency of the 2D peak shows linearly decrease with the increase of strain, with a slope of $-27.8\ cm^{-1}/\%$ for SLG and $-21.9\ cm^{-1}/\%$ for three-layer graphene as shown in Fig. 11.13b [89]. After strain releasing, the 2D peak almost backs to the original position. Such invertible and quick recovery property suggests the excellent flexibility of graphene, which is beneficial to its application as an ultrasensitive strain senor. Raman image of 2D peak position can even monitor the whole process for applying strain, as shown in Fig. 11.13c [89]. Compared with three-layer graphene, SLG has a relative larger shift of 2D peak. Because the strain on graphene sheets was applied by pulling the flexible substrate, which interacts with graphene through van der Waals force. Hence, it would be much easier for strain to be transferred to thinner samples. Figure 11.13d–f [90] gives the schematic diagram of the fabrication of wrinkled MoS_2 with local uniaxial strain. The maximum uniaxial tensile strain ε is collected on top of the winkles, which can be calculated as $\varepsilon \sim \pi^2 h\delta/(1-\sigma^2)\lambda^2$, where σ is 0.125, the Possion's ratio of MoS_2, h is the number of layer, δ and λ are height and width of wrinkle. It is evident that there are red shifts of $E^1{}_{2g}$ and the A_{1g} modes in strained MoS_2 (wrinkled MoS_2) as shown in Fig. 11.13f. Moreover, the red shift of A_{1g} peak is smaller than that of the $E^1{}_{2g}$ mode since A_{1g} mode corresponds to the vibration of the sulfur atoms in antiphase out-of-plane, while $E^1{}_{2g}$ mode corresponds to the vibration of sulfur and molybdenum atoms in antiphase parallel to the crystal plane. The $E^1{}_{2g}$ Raman peak shifts by $-1.7\ cm^{-1}$ per % strain as shown in Fig. 11.13e. Raman image of $E^1{}_{2g}$ further illustrates that strain could be efficiently introduced to MoS_2 in wrinkles as shown in Fig. 11.13g [90].

In addition, non-uniform strain is sometimes intrinsic in CVD-grown 2D materials, e.g. TMDs on SiO_2, graphene on Cu/SiC. Raman image of 2D peak position could be used to monitor the distribution of strain in graphene. Figure 11.14a–d [91] gives the variations of 2D peak for graphene on SiC substrate. The frequency

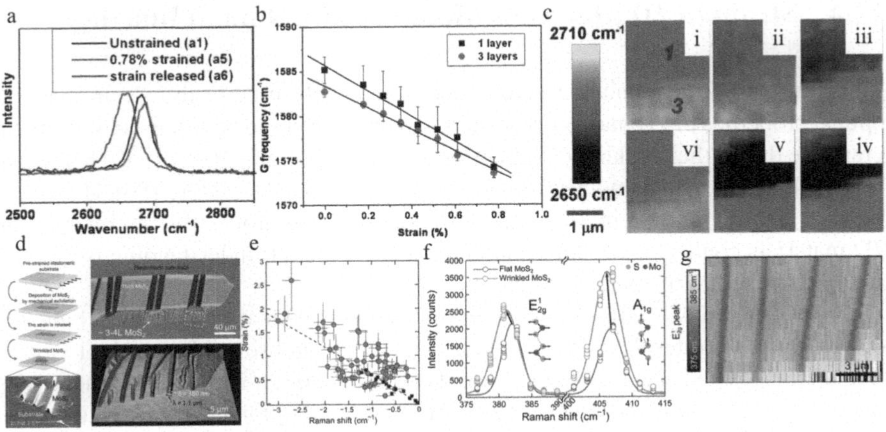

Fig. 11.13 (**a**) The Raman spectra of SLG (unstrained, strained and strain released). (**b**) The frequency of 2D peak of single-and three-layer graphene as a function of uniaxial strain. (**c**) 2D frequency Raman images of unstrained (i), 0.18% (ii), 0.35% (iii), 0.61% (iv), and 0.78% (v) strained, and relaxed (vi) graphene. (**d**) Schematic diagram of the fabrication process of wrinkled MoS_2 monolayers. (**e**) Raman shift of the $E^1{}_{2g}$ vibrational mode as a function of strain. The black squares show the results obtained by applying uniform tensile strain to few-layer MoS_2. (**f**) Raman spectra measured on a flat and on a wrinkled region of a 4-layer-thick MoS_2 flake. (**g**) Raman image of wrinkled MoS_2

Fig. 11.14 (**a**) Raman 2D peak position as a function of lateral position obtained in a Raman line scan over a monolayer epitaxial graphene on SiC. (**b**) Individual spectra obtained in positions as indicated by dots shown in (**a**). (**c** and **d**) A comparison of Raman image of the 2D peak position and AFM image of graphene grown on SiC substrate. Raman images of the intensity (**e**) and position (**g**) of $E^1{}_{2g}$ mode. PL images of the intensity (**f**) and position (**h**). Scalar bar, 3um

of 2D peak is in the range of 2689–2754 cm^{-1} as shown in Fig. 11.14a, which are explained by local compressive stain caused by the lattice mismatch between graphene and SiC surface as well as a large difference in the coefficient of thermal expansion [92, 93]. Therefore, higher and lower local strain corresponding to higher and lower positions of 2D peak are observed. Raman image of 2D peak frequency reflects the distribution of strain in graphene on SiC substrate as shown in Fig. 11.14c. Combined with the surface topography by AFM (Fig. 11.14d), it is clearly revealed that strain in epitaxial graphene is strongly affected by the morphology of the SiC substrate, e.g. the terrace step-edges. Strain can also be introduced into CVD-grown TMDs on SiO_2/Si substrate. Figure 11.14e–h [86] shows the Raman images of monolayer MoS_2 triangles from edge to center with a tensile strain produced by fast cooling. A uniform intensity (with 5% variation) over the triangle is described both in $E^1{}_{2g}$ and A_{1g} mode, while PL intensity is non-uniform and up to 200% higher at edge and some area. In the case of peaks position, blue shift of ~1.0 cm^{-1} in $E^1{}_{2g}$ and ~0.4 cm^{-1} in A_{1g} are achieved from edge to center and PL peak has a blue shift of ~8 meV. These apparent contrast suggest that tensile strain exists in MoS_2 and local strain are extended from edge to center, which may be caused by mismatch between MoS_2 and substrate. The strain induced by the process of fast cooling in MoS_2 can be removed by transferring to a new substrate.

11.5 Probing the Grown Mechanism of CVD Graphene by Raman Imaging

Graphene can be prepared by various methods including mechanical exfoliation [94], CVD [95–99], epitaxial growth on SiC [100–102], and chemical exfoliation [103–105] and so on. Among all these approaches, mechanical exfoliation produces graphene with very high quality, but it is limited both in size and quantity and hard to control the number of layers. Epitaxially grown graphene on SiC is difficult to be transferred to other substrates and SiC substrate is very expensive [100, 101]. Large quantities graphene can be obtained by chemical exfoliation but their electrical transport properties are relatively poor. CVD with simple operation and low cost is regarded as a significant mean for the production of graphene film. It provides large area and high-quality graphene that can be easily transferred to other substrates. This method is widely used to prepare transparent conductive thin film, transistors and so on [106]. However, many critical factors, such as the growth mechanism, growth rate and stacking orders, could strongly influence the quality and physical properties of CVD graphene. Raman imaging plays a vital role in monitoring these factors by isotopically labelled mean, namely, introducing $^{12}CH_4$ and $^{13}CH_4$ sequentially as gas source during CVD growth. The different atomic mass for isotope results in different vibration frequency as shown in Fig. 11.15a [107], where the G and 2D peaks of ^{13}C-graphene are located at much lower position

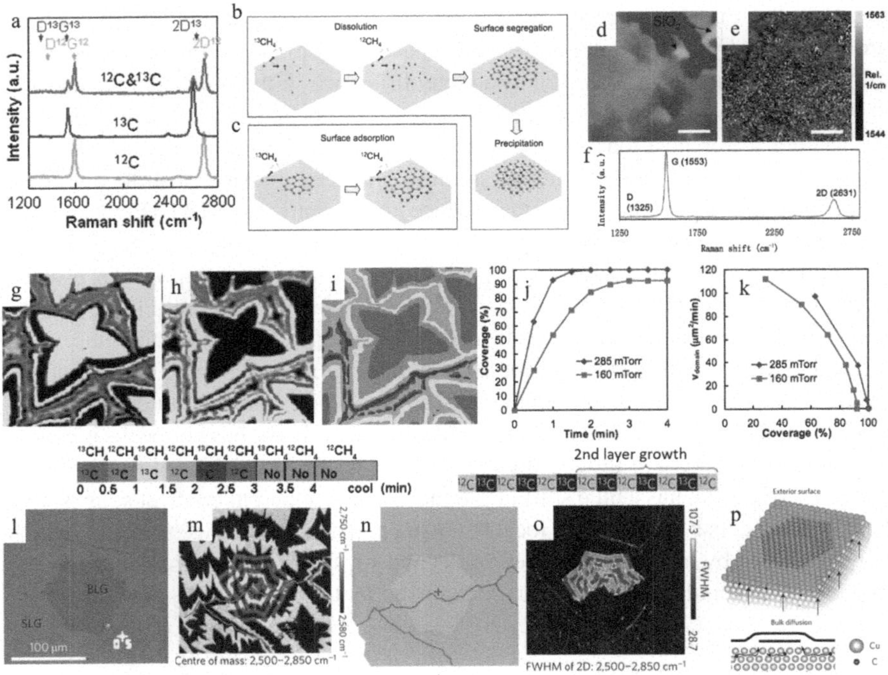

Fig. 11.15 (**a**) Raman spectra from ^{12}C-graphene (bottom), ^{13}Cgraphene (middle), and the junction of ^{12}C- and ^{13}C-graphene (upper), respectively. (**b** and **c**) Schematic diagrams of two typical growth mechanisms for sequential input of C isotopes. (**d–f**) Optical image of graphene on a SiO_2/Si wafer (**d**) and the corresponding Raman image of the frequency of the G peak (**e**). (**f**) Typical Raman spectrum of a FLG film grown on Ni. Scale bars are 5 μm. (**g–i**) Micro-Raman characterization of the isotope-labeled graphene grown on Cu substrate. Raman images of the intensities of G^{13}-band (g), and G^{12}-band (**h**). (**i**) The isotope distributions with methane dosing sequences and times. (**j**) Graphene coverage as a function of methane exposure time. (**k**) Average graphene domain area growth rate as a function of coverage. Optical images (**l**) and the corresponding Raman images (**m** and **o**) of the graphene bilayer domains. (**n**) Schematics of 2nd layer nucleation sites (crosses). The thick lines in (**n**) are 1st graphene layer domain boundaries. (**p**) Schematics of the subsurface diffusion for the growth of 2nd layer graphene

than those of ^{12}C-graphene. Raman imaging can then clearly monitor the formation of graphene with mix or separated isotope sources by the positions of G or 2D peaks.

The grown mechanisms of CVD graphene mainly include the (1) carburization and precipitation mechanism and (2) surface growth mechanism according to the different carbon solubility in metals as shown in Fig. 11.15b, c [107]. For metals with large carbon solubility such as Co and Ni, carbon can diffuse into the metal substrate at high temperature and then precipitate, nucleate and grow during cooling. The thickness of graphene obtained by this growth mechanism is non-uniform and normally contains more than one-layer due to the variation of temperature, as shown in the optical image in Fig. 11.15d [107], where $^{13}CH_4$ and $^{12}CH_4$ are introduced sequentially. It can be seen from the G peak frequency image in

Fig. 11.15e [107] that no distinguishable separation of isotopes was observed, namely, the distribution of ^{12}C and ^{13}C is uniform. It corresponds to the position of ~1553 cm^{-1} for G peak, which is composed by 45% ^{13}C and 55% ^{12}C. These results support the carburization and precipitation mechanism for CVD growth graphene on Ni substrate [108–111]. On the other hand, for Cu substrate with low carbon solubility, $^{13}CH_4$ and $^{12}CH_4$ are alternately introduced and decomposed at high temperature. Carbon atoms are then absorbed at the surface of Cu substrate to form graphene islands, and further form continuous and uniform graphene domain. Figure 11.15g–i show Raman images of a CVD grown graphene constructed by the G peak intensities corresponding to ^{13}C (~1520 cm^{-1}) and ^{12}C (~1580 cm^{-1}) frequencies [112]. Each ring represents the growth of graphene in half minute with different isotope source. It can be seen that graphene growth terminated after the sixth dose at methane partial pressure of 160 mTorr (Fig. 11.15i), and Cu surface is only ~90% covered by graphene even though CH_4 is continually supplied. On the other hand, if the methane partial pressure is increased to 285 mTorr, uniform, continue and large area graphene with single layer can be obtained by the surface self-limited growth mechanism after about 1.5 min, as shown in Fig. 11.15j. In addition to above growth mechanism, graphene can also grow on insulator substrates such as SiO_2, sapphire and boron nitride [113–118], carbon source decomposes at high temperature, absorbs at the surface of substrate at low temperature and then nucleates, forms graphene film. However, graphene obtained by these growth mechanisms normally has a smaller size, more defects and poor electronic properties. Raman imaging can also be used to monitor the growth rate of CVD graphene by the variations of isotopes ring density. It can be seen from Raman images (g-i) that isotopes ring density is higher in the beginning and then decreases generally with the increase of growth time. These variations in isotopes rings density correspond to high growth rate in the beginning and then slow growth rate with the increase of growth time, as also shown in Fig. 11.15g–k [112]. Raman image also demonstrates that higher growth rate is obtained at the tips of the lobes (1.2 μm/min), instead of lower growth rate between the lobes and near the end of the growth process (0.1 μm/min). The growth rate at the graphene joined points is much lower than at the tip of the growing front [119].

In the case of the growth order of FLG, there is always controversy that the growth of 2nd layer is above or below 1st layer. Figure 11.15l–n [120] give the information of bilayer graphene growth. It can be seen from the Raman image that isotopic rings are presented at the 1st layer which is controlled by the surface self-limited mechanism. Isotopic rings are also observed in the 2nd layer, where C from bulk diffuses into the bottom of the 1st layer to support the 2nd growth. For 2nd layer domains, it nucleates under originally nucleated position of 1st layer and grows as a single crystal. Meanwhile, the stacking order between 1st and 2nd is always Bernal stacking as further indicated by Fig. 11.15p [120]. However, when the growth of 2nd layer across the GBs of the 1st layer domain, it straightforward leads to the missing of Bernal stacking, namely, stacking disorder. This is further described by the decrease of the 2D FWHM (25–40 cm^{-1}) in the Raman image (Fig. 11.15o). These conclusions are beneficial to scalable synthesis of large-area

Bernal-stacked BLG. In a word, Raman imaging can be used to monitor the growth mechanism, growth rate and growth order of CVD graphene by isotopic labelled technique, which is very important for its application in transparent conductive thin film, transistors and so on.

11.6 Conclusions

We have discussed the Raman imaging technique and also its application in the study of 2D materials, including the effects of thickness and stacking configurations, heterostructure and interlayer coupling, defects, strain. We have also demonstrated that Raman imaging is an ideal tool to study the growth mechanism of CVD graphene. The further development of Raman imaging technique could be the improvement of spatial resolution with the help of near field optics, the correlation of Raman imaging with other high-resolution imaging techniques, e.g. AFM, SEM. Some of the above tasks have already been carried out and promising results are achieved. The further development of Raman imaging technique could certainly help on the better understanding of 2D materials and other nano−/micro- structures.

Acknowledgments This work was supported by the National Key Research and Development Program of China (No. 2017YFA0205700), NSFC (61774034), and the Strategic Priority Research Program of Chinese Academy of Sciences, Grant No. XDB30000000.

References

1. A. Grüneis, C. Attaccalite, L. Wirtz, et al., Tight-binding description of the quasiparticle dispersion of graphite and few-layer graphene. Phys. Rev. B **78**, 205425 (2008)
2. B. Partoens, F.M. Peeters, From graphene to graphite: electronic structure around theKpoint. Phys. Rev. B **74**, 075404 (2006)
3. A.K. Geim, K.S. Novoselov, The rise of graphene. Nat. Mater. **6**, 183–191 (2007)
4. Y. Zhang, T.T. Tang, C. Girit, et al., Direct observation of a widely tunable bandgap in bilayer graphene. Nature **459**, 820–823 (2009)
5. L. Liu, S.B. Kumar, Y. Ouyang, et al., Performance limits of monolayer transition metal dichalcogenide transistors. IEEE Trans. Electron Devices **58**, 3042–3047 (2011)
6. R. Coehoorn, C. Haas, J. Dijkstra, et al., Electronic structure of $MoSe_2$, MoS_2, and WSe_2. I. Band-structure calculations and photoelectron spectroscopy. Phys. Rev. B **35**, 6195–6202 (1987)
7. A. Kuc, N. Zibouche, T. Heine, Influence of quantum confinement on the electronic structure of the transition metal sulfide TS_2. Phys. Rev. B **83**, 245213 (2011)
8. J. Yang, R. Xu, J. Pei, et al., Optical tuning of exciton and trion emissions in monolayer phosphorene. Light Sci. Appl. **4**, e312 (2015)
9. I. Brihuega, P. Mallet, H. Gonzalez-Herrero, et al., Unraveling the intrinsic and robust nature of van Hove singularities in twisted bilayer graphene by scanning tunneling microscopy and theoretical analysis. Phys. Rev. Lett. **109**, 196802 (2012)
10. W.Y. He, Z.D. Chu, L. He, Chiral tunneling in a twisted graphene bilayer. Phys. Rev. Lett. **111**, 066803 (2013)

11. H. Schmidt, T. Lüdtke, P. Barthold, et al., Tunable graphene system with two decoupled monolayers. Appl. Phys. Lett. **93**, 172108 (2008)
12. D. Xiao, G.-B. Liu, W. Feng, et al., Coupled spin and valley physics in monolayers of MoS 2 and other group-VI dichalcogenides. Phys. Rev. Lett. **108**, 196802 (2012)
13. K.F. Mak, K. He, J. Shan, et al., Control of valley polarization in monolayer MoS2 by optical helicity. Nat. Nanotechnol. **7**, 494–498 (2012)
14. Y. Chen, L. Meng, W. Zhao, et al., Raman mapping investigation of chemical vapor deposition-fabricated twisted bilayer graphene with irregular grains. Phys. Chem. Chem. Phys. **16**, 21682–21687 (2014)
15. Z. Ni, H. Wang, J. Kasim, et al., Graphene thickness determination using reflection and contrast spectroscopy. Nano Lett. **7**, 2758–2763 (2007)
16. Y. Wang, Z. Ni, Z. Shen, et al., Interference enhancement of Raman signal of graphene. Appl. Phys. Lett. **92**, 043121 (2008)
17. Y. Hao, Y. Wang, L. Wang, et al., Probing layer number and stacking order of few-layer graphene by Raman spectroscopy. Small **6**, 195–200 (2010)
18. H. Li, Q. Zhang, C.C.R. Yap, et al., From Bulk to Monolayer MoS2: evolution of Raman scattering. Adv. Funct. Mater. **22**, 1385–1390 (2012)
19. Y. Liu, H. Nan, X. Wu, et al., Layer-by-layer thinning of MoS2 by plasma. ACS Nano **7**, 4202–4209 (2013)
20. J.C. Shaw, H. Zhou, Y. Chen, et al., Chemical vapor deposition growth of monolayer MoSe2 nanosheets. Nano Res. **7**, 511–517 (2015)
21. A. Berkdemir, H.R. Gutiérrez, A.R. Botello-Méndez, et al., Identification of individual and few layers of WS2 using Raman spectroscopy. Sci. Rep. **3**, 1755 (2013)
22. Y. Lee, D. Tran, K. Myhro, et al., Competition between spontaneous symmetry breaking and single-particle gaps in trilayer graphene. Nat. Commun. **5**, 5656 (2014)
23. W. Zhang, J. Yan, C.H. Chen, et al., Molecular adsorption induces the transformation of rhombohedral- to Bernal-stacking order in trilayer graphene. Nat. Commun. **4**, 2074 (2013)
24. J. Yan, J. Xia, X. Wang, et al., Stacking-dependent interlayer coupling in trilayer MoS(2) with broken inversion symmetry. Nano Lett. **15**, 8155–8161 (2015)
25. M.-Y. Choi, Y.-H. Hyun, Y. Kim, Angle dependence of the Landau level spectrum in twisted bilayer graphene. Phys. Rev. B **84**, 195437 (2011)
26. A. Luican, G. Li, A. Reina, et al., Single-layer behavior and its breakdown in twisted graphene layers. Phys. Rev. Lett. **106**, 126802 (2011)
27. Y. Wang, Z. Ni, L. Liu, et al., Stacking-dependent optical conductivity of bilayer graphene. ACS Nano **4**, 4074–4080 (2010)
28. W. Yan, M. Liu, R.F. Dou, et al., Angle-dependent van Hove singularities in a slightly twisted graphene bilayer. Phys. Rev. Lett. **109**, 126801 (2012)
29. J. Hass, F. Varchon, J.-E. Millan-Otoya, et al., Why multilayer graphene on 4 H− SiC (000 1‾) behaves like a single sheet of graphene. Phys. Rev. Lett. **100**, 125504 (2008)
30. A. Reina, X. Jia, J. Ho, et al., Large area, few-layer graphene films on arbitrary substrates by chemical vapor deposition. Nano Lett. **9**, 30–35 (2008)
31. Z. Ni, Y. Wang, T. Yu, et al., Reduction of Fermi velocity in folded graphene observed by resonance Raman spectroscopy. Phys. Rev. B **77**, 235403 (2008)
32. K. Kim, S. Coh, L.Z. Tan, et al., Raman spectroscopy study of rotated double-layer graphene: misorientation-angle dependence of electronic structure. Phys. Rev. Lett. **108**, 246103 (2012)
33. A.C. Ferrari, J. Meyer, V. Scardaci, et al., Raman spectrum of graphene and graphene layers. Phys. Rev. Lett. **97**, 187401 (2006)
34. R.W. Havener, H. Zhuang, L. Brown, et al., Angle-resolved Raman imaging of interlayer rotations and interactions in twisted bilayer graphene. Nano Lett. **12**, 3162–3167 (2012)
35. S. Coh, L.Z. Tan, S.G. Louie, et al., Theory of the Raman spectrum of rotated double-layer graphene. Phys. Rev. B **88**, 165431 (2013)
36. C. Cong, T. Yu, K. Sato, et al., Raman characterization of ABA-and ABC-stacked trilayer graphene. ACS Nano **5**, 8760–8768 (2011)

37. C.H. Lui, Z. Li, Z. Chen, et al., Imaging stacking order in few-layer graphene. Nano Lett. **11**, 164–169 (2011)
38. C. Cong, T. Yu, R. Saito, et al., Second-order overtone and combination Raman modes of graphene layers in the range of 1690–2150 cm^{-1}. ACS Nano **5**, 1600–1605 (2011)
39. J. Xia, J. Yan, Z.X. Shen, Transition metal dichalcogenides: structural, optical and electronic property tuning via thickness and stacking. FlatChem **4**, 1–19 (2017)
40. C. Dean, L. Wang, P. Maher, et al., Hofstadter's butterfly in moire superlattices: A fractal quantum Hall effect. Nature **497**, 598–602 (2013)
41. B. Hunt, J. Sanchez-Yamagishi, A. Young, et al., Massive Dirac fermions and Hofstadter butterfly in a van der Waals heterostructure. Science **340**, 1427–1430 (2013)
42. M. Yankowitz, J. Xue, D. Cormode, et al., Emergence of superlattice Dirac points in graphene on hexagonal boron nitride. Nat. Phys. **8**, 382–386 (2012)
43. W.J. Yu, Y. Liu, H. Zhou, et al., Highly efficient gate-tunable photocurrent generation in vertical heterostructures of layered materials. Nat. Nanotechnol. **8**, 952–958 (2013)
44. O. Lopez-Sanchez, D. Lembke, M. Kayci, et al., Ultrasensitive photodetectors based on monolayer MoS2. Nat. Nanotechnol. **8**, 497–501 (2013)
45. W. Zhang, C.P. Chuu, J.K. Huang, et al., Ultrahigh-gain photodetectors based on atomically thin graphene-MoS2 heterostructures. Sci. Rep. **4**, 3826 (2014)
46. A. Das, S. Pisana, B. Chakraborty, et al., Monitoring dopants by Raman scattering in an electrochemically top-gated graphene transistor. Nat. Nanotechnol. **3**, 210–215 (2008)
47. Y. Gong, J. Lin, X. Wang, et al., Vertical and in-plane heterostructures from WS2/MoS2 monolayers. Nat. Mater. **13**, 1135–1142 (2014)
48. A.M. Van Der Zande, P.Y. Huang, D.A. Chenet, et al., Grains and grain boundaries in highly crystalline monolayer molybdenum disulphide. Nat. Mater. **12**, 554–561 (2013)
49. S. Najmaei, Z. Liu, W. Zhou, et al., Vapor phase growth and grain boundary structure of molybdenum disulfide atomic layers. Nat. Mater. **12**, 754–759 (2013)
50. H. Terrones, E. Del Corro, S. Feng, et al., New first order Raman-active modes in few layered transition metal dichalcogenides. Sci. Rep. **4**, 4215 (2014)
51. F. Banhart, J. Kotakoski, A.V. Krasheninnikov, Structural defects in graphene. ACS Nano **5**, 26–41 (2010)
52. J. Kotakoski, A.V. Krasheninnikov, U. Kaiser, et al., From point defects in graphene to two-dimensional amorphous carbon. Phys. Rev. Lett. **106**, 105505 (2011)
53. J. Ma, D. Alfe, A. Michaelides, et al., Stone-Wales defects in graphene and other planar s p 2-bonded materials. Phys. Rev. B **80**, 033407 (2009)
54. A.J. Stone, D.J. Wales, Theoretical studies of icosahedral C 60 and some related species. Chem. Phys. Lett. **128**, 501–503 (1986)
55. J. Choi, H. Zhang, J.H. Choi, Modulating optoelectronic properties of two-dimensional transition metal dichalcogenide semiconductors by photoinduced charge transfer. ACS Nano **10**, 1671–1680 (2016)
56. O. Cretu, A.V. Krasheninnikov, J.A. Rodriguez-Manzo, et al., Migration and localization of metal atoms on strained graphene. Phys. Rev. Lett. **105**, 196102 (2010)
57. S. Tongay, J. Zhou, C. Ataca, et al., Broad-range modulation of light emission in two-dimensional semiconductors by molecular physisorption gating. Nano Lett. **13**, 2831–2836 (2013)
58. Z.T. Wu, W.W. Zhao, W.Y. Chen, et al., The influence of chemical solvents on the properties of CVD graphene. J. Raman Spectrosc. **46**, 21–24 (2015)
59. H. Qiu, T. Xu, Z. Wang, et al., Hopping transport through defect-induced localized states in molybdenum disulphide. Nat. Commun. **4**, 2642 (2013)
60. W. Zhou, X. Zou, S. Najmaei, et al., Intrinsic structural defects in monolayer molybdenum disulfide. Nano Lett. **13**, 2615–2622 (2013)
61. L. Ci, L. Song, C. Jin, et al., Atomic layers of hybridized boron nitride and graphene domains. Nat. Mater. **9**, 430–435 (2010)

62. H.P. Komsa, J. Kotakoski, S. Kurasch, et al., Two-dimensional transition metal dichalcogenides under electron irradiation: defect production and doping. Phys. Rev. Lett. **109**, 035503 (2012)
63. J. Lu, A. Carvalho, X.K. Chan, et al., Atomic healing of defects in transition metal dichalcogenides. Nano Lett. **15**, 3524–3532 (2015)
64. N. Nemec, D. Tomanek, G. Cuniberti, Contact dependence of carrier injection in carbon nanotubes: an ab initio study. Phys. Rev. Lett. **96**, 076802 (2006)
65. W. Bao, N.J. Borys, C. Ko, et al., Visualizing nanoscale excitonic relaxation properties of disordered edges and grain boundaries in monolayer molybdenum disulfide. Nat. Commun. **6**, 7993 (2015)
66. P. Yasaei, B. Kumar, R. Hantehzadeh, et al., Chemical sensing with switchable transport channels in graphene grain boundaries. Nat. Commun. **5**, 4911 (2014)
67. X. Zou, Y. Liu, B.I. Yakobson, Predicting dislocations and grain boundaries in two-dimensional metal-disulfides from the first principles. Nano Lett. **13**, 253–258 (2013)
68. D. Cao, T. Shen, P. Liang, et al., Role of chemical potential in flake shape and edge properties of monolayer MoS2. J. Phys. Chem. C **119**, 4294–4301 (2015)
69. C. Cong, T. Yu, H. Wang, Raman study on the G mode of graphene for determination of edge orientation. ACS Nano **4**, 3175–3180 (2010)
70. B. Krauss, P. Nemes-Incze, V. Skakalova, et al., Raman scattering at pure graphene zigzag edges. Nano Lett. **10**, 4544–4548 (2010)
71. Y. You, Z. Ni, T. Yu, et al., Edge chirality determination of graphene by Raman spectroscopy. Appl. Phys. Lett. **93**, 163112 (2008)
72. Z.H. Ni, H.M. Wang, Y. Ma, et al., Tunable stress and controlled thickness modification in graphene by annealing. ACS Nano **2**, 1033–1039 (2008)
73. Z. Zafar, Z.H. Ni, X. Wu, et al., Evolution of Raman spectra in nitrogen doped graphene. Carbon **61**, 57–62 (2013)
74. Z. Luo, T. Yu, K.-J. Kim, et al., Thickness-dependent reversible hydrogenation of graphene layers. ACS Nano **3**, 1781–1788 (2009)
75. L. Liu, S. Ryu, M.R. Tomasik, et al., Graphene oxidation: thickness-dependent etching and strong chemical doping. Nano Lett. **8**, 1965–1970 (2008)
76. Q.H. Wang, Z. Jin, K.K. Kim, et al., Understanding and controlling the substrate effect on graphene electron-transfer chemistry via reactivity imprint lithography. Nat. Chem. **4**, 724–732 (2012)
77. Y.N. Xu, D. Zhan, L. Liu, et al., Thermal dynamics of graphene edges investigated by polarized Raman spectroscopy. ACS Nano **5**, 147–152 (2010)
78. D. Zhan, L. Liu, Y.N. Xu, et al., Low temperature edge dynamics of AB-stacked bilayer graphene: naturally favored closed zigzag edges. Sci. Rep. **1**, 12 (2011)
79. Q. Yu, L.A. Jauregui, W. Wu, et al., Control and characterization of individual grains and grain boundaries in graphene grown by chemical vapour deposition. Nat. Mater. **10**, 443–449 (2011)
80. H. Nan, Z. Wang, W. Wang, et al., Strong photoluminescence enhancement of MoS2 through defect engineering and oxygen bonding. ACS Nano **8**, 5738–5745 (2014)
81. B. Chakraborty, A. Bera, D.V.S. Muthu, et al., Symmetry-dependent phonon renormalization in monolayer MoS2transistor. Phys. Rev. B **85**, 161403 (2012)
82. N. Mao, Y. Chen, D. Liu, et al., Solvatochromic effect on the photoluminescence of MoS_2 monolayers. Small **9**, 1312–1315 (2013)
83. M. O'brien, N. Mcevoy, T. Hallam, et al., Transition metal dichalcogenide growth via close proximity precursor supply. Sci. Rep. **4**, 7374 (2014)
84. G. Giovannetti, P.A. Khomyakov, G. Brocks, et al., Substrate-induced band gap in graphene on hexagonal boron nitride: Ab initiodensity functional calculations. Phys. Rev. B **76**, 073103 (2007)
85. R.M. Ribeiro, N.M.R. Peres, J. Coutinho, et al., Inducing energy gaps in monolayer and bilayer graphene: local density approximation calculations. Phys. Rev. B **78**, 075442 (2008)

86. Z. Liu, M. Amani, S. Najmaei, et al., Strain and structure heterogeneity in MoS2 atomic layers grown by chemical vapour deposition. Nat. Commun. **5**, 5246 (2014)
87. M. Huang, H. Yan, T.F. Heinz, et al., Probing strain-induced electronic structure change in graphene by Raman spectroscopy. Nano Lett. **10**, 4074–4079 (2010)
88. M. Mohr, J. Maultzsch, C. Thomsen, Splitting of the Raman2Dband of graphene subjected to strain. Phys. Rev. B **82**, 201409 (2010)
89. Z.H. Ni, T. Yu, Y.H. Lu, et al., Uniaxial strain on graphene: Raman spectroscopy study and band-gap opening. ACS Nano **2**, 2301–2305 (2008)
90. A. Castellanos-Gomez, R. Roldan, E. Cappelluti, et al., Local strain engineering in atomically thin MoS2. Nano Lett. **13**, 5361–5366 (2013)
91. J.A. Robinson, C.P. Puls, N.E. Staley, et al., Raman topography and strain uniformity of large-area epitaxial graphene. Nano Lett. **9**, 964–968 (2009)
92. Z.H. Ni, W. Chen, X.F. Fan, et al., Raman spectroscopy of epitaxial graphene on a SiC substrate. Phys. Rev. B **77**, 200803 (2008)
93. J. Röhrl, M. Hundhausen, K.V. Emtsev, et al., Raman spectra of epitaxial graphene on SiC(0001). Appl. Phys. Lett. **92**, 201918 (2008)
94. K.S. Novoselov, A.K. Geim, S.V. Morozov, et al., Electric field effect in atomically thin carbon films. Science **306**, 666–669 (2004)
95. L.G. De Arco, Y. Zhang, A. Kumar, et al., Synthesis, transfer, and devices of single-and few-layer graphene by chemical vapor deposition. IEEE Trans. Nanotechnol. **8**, 135–138 (2009)
96. Y. Yu, Z. Li, W. Wang, et al., Investigation of multilayer domains in large-scale CVD monolayer graphene by optical imaging. J. Semicond. **38**, 033003 (2017)
97. K.S. Kim, Y. Zhao, H. Jang, et al., Large-scale pattern growth of graphene films for stretchable transparent electrodes. Nature **457**, 706 (2009)
98. X. Li, W. Cai, J. An, et al., Large-area synthesis of high-quality and uniform graphene films on copper foils. Science **324**, 1312–1314 (2009)
99. Q. Yu, J. Lian, S. Siriponglert, et al., Graphene segregated on Ni surfaces and transferred to insulators. Appl. Phys. Lett. **93**, 113103 (2008)
100. C. Berger, Z. Song, T. Li, et al., Ultrathin epitaxial graphite: 2D electron gas properties and a route toward graphene-based nanoelectronics. J. Phys. Chem. B **108**, 19912–19916 (2004)
101. C. Berger, Z. Song, X. Li, et al., Electronic confinement and coherence in patterned epitaxial graphene. Science **312**, 1191–1196 (2006)
102. T. Ohta, A. Bostwick, T. Seyller, et al., Controlling the electronic structure of bilayer graphene. Science **313**, 951–954 (2006)
103. S. Stankovich, D.A. Dikin, R.D. Piner, et al., Synthesis of graphene-based nanosheets via chemical reduction of exfoliated graphite oxide. Carbon **45**, 1558–1565 (2007)
104. S. Park, R.S. Ruoff, Chemical methods for the production of graphenes. Nat. Nanotechnol. **4**, 217–224 (2009)
105. H. Wang, J.T. Robinson, X. Li, et al., Solvothermal reduction of chemically exfoliated graphene sheets. J. Am. Chem. Soc. **131**, 9910–9911 (2009)
106. Y. Zhang, L. Zhang, C. Zhou, Review of chemical vapor deposition of graphene and related applications. Acc. Chem. Res. **46**, 2329–2339 (2013)
107. X. Li, W. Cai, L. Colombo, et al., Evolution of graphene growth on Ni and Cu by carbon isotope labeling. Nano Lett. **9**, 4268–4272 (2009)
108. J. Shelton, H. Patil, J. Blakely, Equilibrium segregation of carbon to a nickel (111) surface: a surface phase transition. Surf. Sci. **43**, 493–520 (1974)
109. L. Isett, J. Blakely, Segregation isosteres for carbon at the (100) surface of nickel. Surf. Sci. **58**, 397–414 (1976)
110. M. Eizenberg, J. Blakely, Carbon monolayer phase condensation on Ni (111). Surf. Sci. **82**, 228–236 (1979)
111. M. Eizenberg, J. Blakely, Carbon interaction with nickel surfaces: monolayer formation and structural stability. J. Chem. Phys. **71**, 3467–3477 (1979)
112. X. Li, C.W. Magnuson, A. Venugopal, et al., Graphene films with large domain size by a two-step chemical vapor deposition process. Nano Lett. **10**, 4328–4334 (2010)

113. H. Bi, S. Sun, F. Huang, et al., Direct growth of few-layer graphene films on SiO2substrates and their photovoltaic applications. J. Mater. Chem. **22**, 411–416 (2012)
114. J. Chen, Y. Wen, Y. Guo, et al., Oxygen-aided synthesis of polycrystalline graphene on silicon dioxide substrates. J. Am. Chem. Soc. **133**, 17548–17551 (2011)
115. M.T. Cole, N. Lindvall, A. Yurgens, Noncatalytic chemical vapor deposition of graphene on high-temperature substrates for transparent electrodes. Appl. Phys. Lett. **100**, 022102 (2012)
116. G. Hong, Q.-H. Wu, J. Ren, et al., Mechanism of non-metal catalytic growth of graphene on silicon. Appl. Phys. Lett. **100**, 231604 (2012)
117. K.-B. Kim, C.-M. Lee, J. Choi, Catalyst-free direct growth of triangular nano-graphene on all substrates. J. Phys. Chem. C **115**, 14488–14493 (2011)
118. N. Lindvall, M.T. Cole, T.J. Booth, et al., Controllable chemical vapor deposition of large area uniform nanocrystalline graphene directly on silicon dioxide. J. Appl. Phys. **111**, 044103 (2012)
119. X. Li, C.W. Magnuson, A. Venugopal, et al., Large-area graphene single crystals grown by low-pressure chemical vapor deposition of methane on copper. J. Am. Chem. Soc. **133**, 2816–2819 (2011)
120. Y. Hao, L. Wang, Y. Liu, et al., Oxygen-activated growth and bandgap tunability of large single-crystal bilayer graphene. Nat. Nanotechnol. **11**, 426–431 (2016)

Correction to: Raman Spectroscopy of Two-Dimensional Materials

Ping-Heng Tan

Correction to:
P.-H. Tan (ed.), *Raman Spectroscopy of Two-Dimensional Materials*, Springer Series in Materials Science 276,
https://doi.org/10.1007/978-981-13-1828-3

Professor Tan's affiliation was incorrect in the original book. The correct affiliation details are updated in the current version of the book as follows:

State Key Laboratory of Superlattices and Microstructures, Institute of Semiconductors, Chinese Academy of Sciences, Beijing, China

Center of Materials Science and Opto-Electronics Engineering & CAS Center of Excellence in Topological Quantum Computation, University of Chinese Academy of Sciences, Beijing, China

The updated version of the chapters can be found at
https://doi.org/10.1007/978-981-13-1828-3
https://doi.org/10.1007/978-981-13-1828-3_1
https://doi.org/10.1007/978-981-13-1828-3_10

P.-H. Tan (ed.), *Raman Spectroscopy of Two-Dimensional Materials*, Springer Series in Materials Science 276, https://doi.org/10.1007/978-981-13-1828-3_12

The manufacturer's authorised representative in the EU is Springer Nature Customer Service Centre GmbH, Europaplatz 3, 69115 Heidelberg, Germany. If you have any concerns regarding our products, please contact ProductSafety@springernature.com

Printed and bound by CPI Group (UK) Ltd, Croydon, CR0 4YY
25/09/2026
02222897-0020